The Role of Fluids in Crustal Processes

NATIONAL ACADEMY PRESS

The National Academy Press was created by the National Academy of Sciences to publish the reports issued by the Academy and the National Academy of Engineering, the Institute of Medicine, and the National Research Council, all operating under the charter granted to the National Academy of Sciences by the Congress of the United States.

Studies in Geophysics

The Role of Fluids in Crustal Processes

Geophysics Study Committee
Commission on Geosciences, Environment, and Resources
National Research Council

NATIONAL ACADEMY PRESS
Washington, D.C. 1990

NATIONAL ACADEMY PRESS 2101 Constitution Avenue, N.W. Washington, DC 20418

NOTICE: The project that is the subject of this report was approved by the Governing Board of the National Research Council, whose members are drawn from the councils of the National Academy of Sciences, the National Academy of Engineering, and the Institute of Medicine. The members of the committee responsible for the report were chosen for their special competences and with regard for appropriate balance.

This report has been reviewed by a group other than the authors according to procedures approved by a Report Review Committee consisting of members of the National Academy of Sciences, the National Academy of Engineering, and the Institute of Medicine.

The National Academy of Sciences is a private, nonprofit, self-perpetuating society of distinguished scholars engaged in scientific and engineering research, dedicated to the furtherance of science and technology and to their use for the general welfare. Upon the authority of the charter granted to it by the Congress in 1863, the Academy has a mandate that requires it to advise the federal government on scientific and technical matters. Dr. Frank Press is president of the National Academy of Sciences.

The National Academy of Engineering was established in 1964, under the charter of the National Academy of Sciences, as a parallel organization of outstanding engineers. It is autonomous in its administration and in the selection of its members, sharing with the National Academy of Sciences the responsibility for advising the federal government. The National Academy of Engineering also sponsors engineering programs aimed at meeting national needs, encourages education and research, and recognizes the superior achievements of engineers. Dr. Robert M. White is president of the National Academy of Engineering.

The Institute of Medicine was established in 1970 by the National Academy of Sciences to secure the services of eminent members of appropriate professions in the examination of policy matters pertaining to the health of the public. The Institute acts under the responsibility given to the National Academy of Sciences by its congressional charter to be an adviser to the federal government and, upon its own initiative, to identify issues of medical care, research, and education. Dr. Samuel O. Thier is president of the Institute of Medicine.

The National Research Council was organized by the National Academy of Sciences in 1916 to associate the broad community of science and technology with the Academy's purposes of furthering knowledge and of advising the federal government. Functioning in accordance with general policies determined by the Academy, the Council has become the principal operating agency of both the National Academy of Sciences and the National Academy of Engineering in providing services to the government, the public, and the scientific and engineering communities. The Council is administered jointly by both Academies and the Institute of Medicine. Dr. Frank Press and Dr. Robert M. White are chairman and vice chairman, respectively, of the National Research Council.

Support for the Geophysics Study Committee was provided by the Department of Energy, the National Science Foundation, and the National Oceanic and Atmospheric Administration.

Library of Congress Cataloging-in-Publication Data

The Role of fluids in crustal processes / Geophysics Study Committee, Commission on Geosciences, Environment, and Resources, National Research Council.
    p. cm. — (Studies in geophysics)
    Based on papers presented at an American Geophysical Union symposium.
    Includes bibliographical references and index.
    ISBN 0-309-04037-X
    1. Earth—Crust. 2. Fluid dynamics. I. National Research Council. (U.S.). Geophysics Study Committee. II. Series.
QE511.R64 1990
551.1'3—dc20   90-35845
    CIP

Copyright © 1990 by the National Academy of Sciences

No part of this book may be reproduced by any mechanical, photographic, or electronic process, or in the form of a phonographic recording, nor may it be stored in a retrieval system, transmitted, or otherwise copied for public or private use, without written permission from the publisher, except for the purposes of official use by the U.S. government.
Printed in the United States of America

Panel on the Role of Fluids in Crustal Processes

JOHN D. BREDEHOEFT, U.S. Geological Survey, *Cochairman*
DENIS L. NORTON, University of Arizona, *Cochairman*
TERRY ENGELDER, The Pennsylvania State University
AMOS M. NUR, Stanford University
JACK E. OLIVER, Cornell University
HUGH P. TAYLOR, JR., California Institute of Technology
SPENCER R. TITLEY, University of Arizona
PETER J. VROLIJK, University of Michigan
JOHN V. WALTHER, Northwestern University
STEPHEN M. WICKHAM, University of Chicago

Geophysics Study Committee

BYRON D. TAPLEY, University of Texas, *Chairman*
RICHARD T. BARBER, Monterey Bay Aquarium Research Institute
ROBIN BRETT, U.S. Geological Survey
RALPH J. CICERONE, University of California, Irvine
* RANA A. FINE, University of Miami
LYNN W. GELHAR, Massachusetts Institute of Technology
* ARNOLD L. GORDON, Lamont-Doherty Geological Observatory
* MARK F. MEIER, University of Colorado
* NORMAN F. NESS, University of Delaware
* THOMAS A. POTEMRA, Applied Physics Laboratory, The Johns Hopkins University
* GEORGE C. REID, National Oceanic and Atmospheric Administration
JOANNE SIMPSON, NASA Goddard Space Flight Center
* ROBERT S. YEATS, Oregon State University

Staff

THOMAS M. USSELMAN

Agency Liaison Representatives

BILAL U. HAQ, National Science Foundation
GEORGE A. KOLSTAD, Department of Energy
NED A. OSTENSO, National Oceanic and Atmospheric Administration

---

* Terms ended June 30, 1989

* Terms began July 1, 1989

Commission on Geosciences, Environment, and Resources

M. GORDON WOLMAN, The Johns Hopkins University, *Chairman*
ROBERT C. BEARDSLEY, Woods Hole Oceanographic Institution
B. CLARK BURCHFIEL, Massachusetts Institute of Technology
RALPH J. CICERONE, University of California, Irvine
PETER S. EAGLESON, Massachusetts Institute of Technology
GENE E. LIKENS, New York Botanical Gardens
JERRY D. MAHLMAN, NOAA Geophysical Fluid Dynamics Laboratory
SCOTT M. MATHESON, Parsons, Behle & Latimer
JACK E. OLIVER, Cornell University
PHILIP A. PALMER, E.I. du Pont de Nemours & Co.
FRANK L. PARKER, Vanderbilt University
DUNCAN T. PATTEN, Arizona State University
MAXINE L. SAVITZ, Garrett Corporation
LARRY L. SMARR, University of Illinois, Urbana-Champaign
STEVEN M. STANLEY, Case Western Reserve University
Sir CRISPIN TICKELL, United Kingdom Representative to the United Nations
KARL K. TUREKIAN, Yale University
IRVIN L. WHITE, New York State Energy and Development Authority
JAMES H. ZUMBERGE, University of Southern California
STEPHEN RATTIEN, *Executive Director*
STEPHEN D. PARKER, *Associate Executive Director*

Studies in Geophysics*

ENERGY AND CLIMATE
Roger R. Revelle, *panel chairman*, 1977, 158 pp.

ESTUARIES, GEOPHYSICS, AND THE ENVIRONMENT
Charles B. Officer, *panel chairman*, 1977, 127 pp.

CLIMATE, CLIMATIC CHANGE, AND WATER SUPPLY
James R. Wallis, *panel chairman*, 1977, 132 pp.

THE UPPER ATMOSPHERE AND MAGNETOSPHERE
Francis S. Johnson, *panel chairman*, 1977, 168 pp.

GEOPHYSICAL PREDICTIONS
Helmut E. Landsberg, *panel chairman*, 1978, 215 pp.

IMPACT OF TECHNOLOGY ON GEOPHYSICS
Homer E. Newell, *panel chairman*, 1979, 136 pp.

CONTINENTAL TECTONICS
B. Clark Burchfiel, Jack E. Oliver, and Leon T. Silver, *panel co-chairmen*, 1980, 197 pp.

MINERAL RESOURCES: GENETIC UNDERSTANDING FOR PRACTICAL APPLICATIONS
Paul B. Barton, Jr., *panel chairman*, 1981, 119 pp.

SCIENTIFIC BASIS OF WATER-RESOURCE MANAGEMENT
Myron B. Fiering, *panel chairman*, 1982, 127 pp.

SOLAR VARIABILITY, WEATHER, AND CLIMATE
John A. Eddy, *panel chairman*, 1982, 104 pp.

---

* Published to date.

CLIMATE IN EARTH HISTORY
Wolfgang H. Berger and John C. Crowell, *panel co-chairmen*, 1982, 198 pp.
FUNDAMENTAL RESEARCH ON ESTUARIES: THE IMPORTANCE OF AN INTERDISCIPLINARY APPROACH
L. Eugene Cronin and Charles B. Officer, *panel co-chairmen*, 1983, 79 pp.
EXPLOSIVE VOLCANISM: INCEPTION, EVOLUTION, AND HAZARDS
Francis R. Boyd, *panel chairman*, 1984, 176 pp.
GROUNDWATER CONTAMINATION
John D. Bredehoeft, *panel chairman*, 1984, 179 pp.
ACTIVE TECTONICS
Robert E. Wallace, *panel chairman*, 1986, 266 pp.
THE EARTH'S ELECTRICAL ENVIRONMENT
E. Philip Krider and Raymond G. Roble, *panel co-chairmen*, 1986, 263 pp.
SEA-LEVEL CHANGE
Roger Revelle, *panel chairman*, 1990, 246 pp.
THE ROLE OF FLUIDS IN CRUSTAL PROCESSES
John D. Bredehoeft and Denis L. Norton, *panel co-chairmen*, 1990, 170 pp.

PREFACE

This report is part of a series, *Studies in Geophysics*, that has been carried out over the past 13 years to provide (1) a source of information from the scientific community to aid policymakers in decisions on societal problems that involve geophysics and (2) assessments of emerging research topics within the broad scope of geophysics. An important part of such reports is an evaluation of the adequacy of current geophysical knowledge and the appropriateness of current research programs in addressing needed information.

The study resulting in this report on the role of fluids in crustal processes is primarily in the latter category—an emerging area of research—but it is not without its applications in understanding (1) the tectonics of the crust, (2) the occurrences and characteristics of mineral and energy resources, and (3) waste disposal. It was initiated by the Geophysics Study Committee in consultation with the liaison representatives of the federal agencies that support the committee, relevant boards and committees within the National Research Council, and members of the scientific community.

The study examines the premise that pore fluids are important in our understanding of geological processes. There is mounting evidence that pore pressure at mid-crustal depths in active tectonic areas is at or near the lithostatic load. This topic presents an exciting proposition that might help unify much of what we know about crustal processes—geological, geophysical, and geochemical.

The preliminary scientific findings of the authored background chapters were presented at an American Geophysical Union symposium. In completing their chapters, the authors had the benefit of discussions at this symposium as well as the comments of several scientific referees. Ultimate responsibility for the individual chapters, however, rests with the authors.

The Overview and Recommendations of the study summarizes the highlights of the chapters and formulates conclusions and recommendations. In preparing the Overview and Recommendations, the panel co-chairmen and the Geophysics Study Committee had the benefit of meetings that took place at the symposium, comments of the panel, several other meetings of the committee, and the comments of scientists, who reviewed the report according to procedures established by the National Research Council's Report Review Committee. Responsibility of the Overview and Recommendations rests with the Geophysics Study Committee and the co-chairmen of the panel.

CONTENTS

Overview and Recommendations   3

Background

1. Mass and Energy Transport in a Deforming Earth's Crust   27
   John D. Bredehoeft and Denis L. Norton
2. Pore Fluid Pressure Near Magma Chambers   42
   Denis L. Norton
3. Evolution and Style of Fracture Permeability in Intrusion-Centered Hydrothermal Systems   50
   Spencer R. Titley
4. Fluid Dynamics During Progressive Regional Metamorphism   64
   John V. Walther
5. Oxygen and Hydrogen Isotope Constraints on the Deep Circulation of Surface Waters into Zones of Hydrothermal Metamorphism and Melting   72
   Hugh P. Taylor, Jr.

6. Hydrothermal Systems Associated with Regional Metamorphism and Crustal Anatexis: Example from the Pyrenees, France — 96
   Stephen M. Wickham and Hugh P. Taylor, Jr.
7. Time-Dependent Hydraulics of the Earth's Crust — 113
   Amos M. Nur and Joseph Walder
8. COCORP and Fluids in the Crust — 128
   Jack E. Oliver
9. Smoluchowski's Dilemma Revisited: A Note on the Fluid-Pressure History of the Central Appalachian Fold-Thrust Belt — 140
   Terry Engelder
10. Fluid-Pressure History in Subduction Zones: Evidence from Fluid Inclusions in the Kodiak Accretionary Complex, Alaska — 148
    Peter Vrolijk and Georgianna Myers
11. Degassing of Carbon Dioxide as a Possible Source of High Pore Pressures in the Crust — 158
    John D. Bredehoeft and Steven E. Ingebritsen

Index — 165

OVERVIEW AND RECOMMENDATIONS

SUMMARY

The central role of H_2O-rich fluids in determining the dynamic conditions in the Earth's crest is apparent in the repeated occurrence of fluid properties in all of the transport equations. This symbolic depiction of the processes shows not only the influence of processes on one another but also that this coupling condition is a consequence of the presence of an often sparse, but essential, occurrence of water in the systems. Numerous examples exist that demonstrate water as an active agent of the mechanical, chemical, and thermal processes that control many geologic processes that operate within the crust.

This study assesses the current scientific understanding of the role of fluids in crustal processes. An important part of such an assessment is an evaluation of the adequacy of the geophysical knowledge base and the opportunities to improve it. Recognition of the role of water as the material that controls the extent of coupling among processes shows promise of reducing the magnitude of the analytical problem to one that focuses on the controlling links in the system.

INTRODUCTION

Fluids play a vital role in virtually all crustal processes. Circulation of fluids has important effects on the transport of chemical constituents and heat, and is the principal control on the formation of hydrothermal ore deposits. The mechanisms by which crustal rocks deform are strongly influenced by the presence of water, as well as by the pore fluid pressure. It has also been suggested that fluids play a very significant role in earthquakes. On a broader scale, pore fluid pressure influences the mechanical processes that control rock deformation in and below the accretionary wedge in subduction zones. The volume of fluid carried to deeper levels in subduction zone complexes appears to influence the rate and depth of melting, which will determine the site of volcanism in the overriding plate.

(See Vrolijk and Myers, Chapter 10, for additional discussion.) Development of rigorous models of crustal dynamics, therefore, require a substantial body of well constrained data as well as in-depth understanding regarding fluids in the crust.

Geologists have traditionally focused on the solids, minerals, and lithologic units in their study of earth processes; the importance and distribution of the fluid phase often has been overlooked. This is in part due to the difficulty of obtaining samples of the fluids. Investigations that have not taken into account the presence of water in the Earth may inadequately describe geophysical mechanisms in the Earth or describe them incorrectly.

Hydrologists and reservoir engineers approach problems associated with the Earth differently than geologists. With hydrologists, the focus is on the fluid. The questions are: **What is the current state of fluid; how is the fluid moving; how is it transporting mass and energy?** Implicit in this approach is the concept that **if one can understand how the fluids move and transport mass and energy, then one can better understand the dynamics of geologic processes within the crust**. Once rock has formed in the Earth's crust, moving fluids and material transport are most likely to produce change in the rocks.

When geologists treat problems of fluid-rock interaction they have generally done so from the perspective of geochemistry (see for example, Fyfe *et al.*, 1978). The geochemical approach lacks information concerning the motion of the fluid. In assembling this report we emphasized formulating the problem in terms of mass and energy transport, an approach rather different from classical geochemistry. Our approach resembles nonequilibrium thermodynamics. To many trained in geochemistry, this approach appears to neglect the chemistry.

Problems of subsurface fluid flow appear to be complex. However, the physics and chemistry are described by a set of conservation statements for mass, energy, and momentum, which leads to a set of coupled partial differential equations. The dependent variables of interest are fluid pressure, chemical composition of the fluid, and fluid temperature or enthalpy. The coupled equations must be solved simultaneously. The equations are difficult to solve analytically; however, **with modern digital computers, solutions to realistic problems of great scientific interest are now readily possible**. While the results of such a dynamic system may be complex, in principal, the physics and chemistry are conceptually simple.

The characteristics of flowing fluid systems are determined by the bulk permeability, pressure gradient, fluid composition, temperature, and available fluid volume. At shallow crustal levels (depths less than several kilometers) these variables are measured directly in wells. However, for rocks at deeper crustal levels, the bulk of available information regarding the behavior of fluids comes from observation of exposed rocks that once resided at deeper crustal levels. These outcrops provide a series of "snapshots" (often multiple exposures) that, when integrated, allow inferences to be drawn concerning the behavior of fluids at different levels of the crust; the magnitude of the relevant variables noted above can only be established within very broad limits. Since fluids present when the strata were at deeper crustal levels have long since escaped (or at least changed) from the rock system, these "snapshots" have the disadvantage of providing only a record of the effect fluids had on the solid crust, rather than providing a sample of the fluid and its role in crustal development. A problem of using fluid inclusions is not so much in interpreting a "snapshot," but in deciding which one, of numerous multiple exposures, is appropriate for the fluid sample preserved.

Present day surface exposures indicate that, at all crustal levels, fluids have been present. Field, isotopic, fluid inclusion, and phase equilibria studies of rocks and mineralized fractures from shallow crustal levels, vein and pegmatitic masses from intermediate levels, and gneissic units from deeper crustal levels, document that fluids were present in significant volumes. These investigations (see e.g., Wickham and Taylor, Chapter 6) have demonstrated that rocks may be chemically modified as a result of fluid migration. Consistent with this fact is the repeated observation that fluids have played a dominant role in the transport and concentration of metals in most ore deposits.

Geophysical studies have also shed light on the nature of deep crustal fluids. Electromagnetic and conductivity soundings have revealed zones of relatively low electrical resistivity in the deep crust, suggesting the presence of a continuous aqueous fluid phase. Seismic reflections and inferred low velocity zones have been interpreted to be regions in which the fluid pressure is anomalously high, in some cases approaching the lithostatic pressure (see Oliver, Chapter 8, this volume). Table 1 summarizes much of the observational geochemical and geophysical indicators that bear on the question of the extent of free water in the Earth's crust. In addition to these indicators, fluids have been directly sampled at about 11 km by the Soviets at the Kola Peninsula drillhole. These data collectively support the existence of fluid circulation and therefore fractures to crustal depths of at least 10 to 15 km and perhaps significantly deeper. Some evidence exists that nonaqueous fluids (e.g., CO_2 and gaseous and liquid hydrocarbons) may also be present.

Despite these results, there is very little quantitative information from depths greater than 5 km regarding the actual processes that lead to fluid migration, heat and mass transport, the development of mineral deposits, and the role of fluids in earthquake events. We remain at the stage in which each new field area provides a relatively unique picture of the processes associated with fluids in the crust. There is an obvious need to sample the deeper crust through drilling.

On the basis of the mechanical properties of crust, we can infer that, locally, large relative volumes of fluid may be expected at shallow crustal depths, and vanishingly small volumes may be present at deep crustal levels. This conclusion reflects the effects of brittle versus ductile behavior of rocks, their pore size and permeability reduction at higher pressures, and the local presence or absence of fluid sources. However, we remain ignorant of how the transition from high fluid volume regimes to low fluid volume regimes occurs. Since the crust must be considered a dynamic system in which rocks at a given crustal depth may be forced to shallower or deeper depths through time, it is also important to understand how the transition from one regime to another can influence the chemical and physical properties of any given rock through time. In addition, we do not know the actual range of volumes to be expected at any particular crustal depth; some studies indicate that some chemical modification of deep crust (which requires significant fluid volumes) has occurred, contrary to inferences based solely on inferred mechanical properties.

One example of how H_2O-rich fluid strongly affects the evolution of conditions in the crust is the central role it plays in hydrothermal systems. At constant volume, a rise in temperature of 1°C can cause a pressure increase of several bars in the local fluid pressure.

TABLE 1 Possible Indicators for the Presence of Free Water in the Earth's Crust as Suggested by Various Investigators (See Table 7.1, Chapter 7, for complete references.)

Indicator	Depth Range
Water table	0 to 2 km
Deep wells	to 12 km
Reservoir induced seismicity	to 12 km
Crustal low velocity zones	7 to 12 km
Crustal electrical conductivity zones	10 to 12 km
Oxygen isotopes	to 12 km
Metamorphism	to >20 km
Crack healing and sealing	???
Formation of hydrothermal ore deposits	to >5 km
Crustal seismic attenuation zones	7 to 12 km
Low stress on faults	0 to 10 km
Silicic volcanism	near surface
Fluid inclusions	???

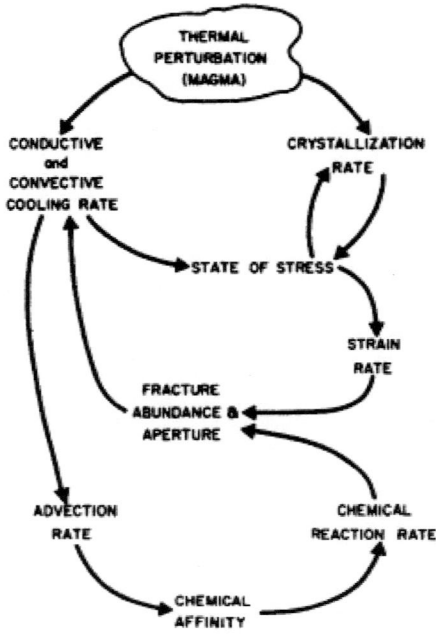

Figure 1
Systematic relationships between principal transport rates and their products. Arrows depict directions of energy, mass flow, and feedback effects of state conditions on rates (from Norton, 1984).

The introduction of a large body of magma into the crust can result in a convecting groundwater system around the magma body, which transports heat away from the magma at supercritical conditions. Because a supercritical fluid is efficient as a heat-transporting medium, the rates of groundwater flow will control the cooling rates of the igneous body. The same convective process will also redistribute chemical components from the pluton to the host environment, including ore-forming constituents. Groundwater flow controls the distribution and grade of ore deposits.

The magnitude of groundwater convection and consequently the extent of mineral alteration around a magma body depends on the permeability of rocks. However, even if the rocks were initially impermeable, the pore fluid pressure generated by heat dispersed into the host rocks is high enough to create hydraulic fractures. Thus, the system can generate its own permeability by making fractures. This fracturing process tends to be episodic. Repeated sequences of fracturing followed by mineral deposition in the fractures are typical of magma environments. As the fractures gradually fill with minerals, permeability decreases, and the fluid flow is retarded. This cycle of processes forms an intricately interconnected feedback system (Figure 1) in which the properties of the fluid phase exert primary control on the evolution of the system.

CONSERVATION EQUATIONS

Physical Parameters

The concept of conservation provides a basis for writing a set of equations that symbolically represent the transport processes involving water in the Earth's crust. For **mass, energy, and momentum**, equations are formulated that express the conservation of that quantity with respect to the local system. These equations (see Chapter 1, for details) describe the rate of change in these quantities with respect to time in a representative rock volume of the system.

During the nineteenth century some carefully conducted experiments revealed a set of empirical laws that express the flux of mass and energy in terms of a driving force and a proportional constant that incorporates a medium's properties. These flux laws form the basis for discussing transport; they include Fick's Law of Diffusion, Fourier's Law of Heat Conduction, DeDonder's Law of Affinity, and Darcy's Law of Fluid Flow. Each has the general form of the flux of a particular quantity, where the flux is proportional to the gradient in a field parameter. The conservation equations for mass, energy, and momentum all derive from one or more of these flux laws. The set of appropriate differential equations to understand the physical processes is summarized in Table 2. Table 2 is set up (column 1) assuming **no coupling**; in other words, each state of fluid-pressure, composition, temperature, and so on, can be derived independently. This is done purely for simplicity; in general, **everything is coupled**. Table 2 illustrates how much information is required to solve any one problem of interest.

In each case the movement of mass and energy involves an empirical relationship and an empirically derived parameter that describes a property of the medium. For example, in groundwater the flow of fluid through a porous medium is described by Darcy's Law, which is an empirical relationship requiring an empirically derived parameter—the "permeability" of the porous medium. For each problem there is a set of such parameters that · describes the physical chemical setting of the problem being considered.

Table 3 lists in more detail the parameters that are required to solve a realistic fluid transport problem. Many of the parameters are both scale and time dependent. The

TABLE 2 Information Required to Address the Major Variables Involved with Understanding Fluid Properties (These individual properties need to be coupled in order to assess the role of fluids in crustal processes.)

System State	Information Required	Output
Pressure (isothermal and no change in fluid chemistry)	Geometry Boundary Conditions Parameters Hydraulic conductivity Porosity Elastic constants Sources and sinks of fluid	Pore fluid pressure Fluid velocity distribution
Mass or Chemical Transport (steady flow and isothermal)	Fluid velocity distribution Chemical boundary conditions Parameters Porosity Dispersivity Chemical Reactions Identify Kinetics	Composition of fluids and rocks
Heat Transport (steady flow and no change in fluid chemistry)	Fluid velocity distribution Heat boundary conditions Parameters Porosity Thermal conductivity Thermal dispersivity Enthalpies	Temperature
Stress and Strain (elastic)	State of stress Fluid pressure Temperature Parameters Elastic constants	Strain and failure (fracture)

number of x's are a subjective attempt to indicate where the parameters can be investigated, either in the field or the laboratory, and which are most important in obtaining a realistic solution. Two parameters and their changes during flow turn out to be critical in the transport process: (1) the **permeability** and (2) the **chemical reactions** and their **kinetics**.

The pathways for flow and therefore the nature of permeability is critical to any study of fluid flow and transport. What are the flow pathways that exist in the crust? Brace (1980) summarized both the laboratory and in situ large-scale permeability data for crystalline rocks (Figure 2). He concluded that the laboratory permeability of most of these rocks, measured on small samples is usually several orders of magnitude less than the in situ permeability measured at a larger scale in the Earth's crust. Bredehoeft *et al.* (1983) made a similar observation for the Cretaceous Pierre Shale in South Dakota. This difference presumably reflects the dominant effect of fluid flow along widely spaced fractures in the crust. As fractures become less prominent at deeper crustal levels where pressures and temperatures are higher, the mechanism of dominant flow becomes ambiguous. Whether fluids move via grain boundary migration, fluid overpressure/hydrofracturing, or diffusion along defects remains unknown. Changes in grain boundary geometry occur when fluid compositions change, resulting in significant variation in permeability and porosity. The extent to which such changes are important in controlling fluid migration is unclear. Migrating fluids can encourage recrystallization of the rocks through which they pass; if this occurs, changes in grain geometry can lead to variations in the local porosity. Devel

TABLE 3 Relative Need of System Properties Important to Understanding Coupled Fluid Properties and Dynamics (the greater the need, the greater number of x's)

	Scale Effects		Time-Dependent Effects	
System Properties	Lab	Field	Lab	Field
Basic Geometry (structure and stratigraphy)		xxx		x
Boundary Conditions				
Topography		xxx		x
Heat Sources		xxx		xxx
Tectonic Effects (rock mechanics)	xx	xx	xx	
Sources and Sinks		xxx		xxx
Parameter Distribution				
Permeability				
Porous media	xx	xxxx		xxx
Fracture media		xxxxx		xxx
Porosity	xx	xx		xx
Fluid Density	x		x	
Viscosity	xx		x	
Chemical Reactions				
Identification	xx	xxxx		
Kinetics			x	xxxx
Dispersivity				
Mass		xxx		x
Heat		xx		x
Thermal Conductivity	xxx			x
Elastic Constants	xxx			
Plastic Behavior			xxx	xx

Note: Also shown is the type of investigations needed relative to the field or laboratory.

opment of preferential, or channelized, flow regimes may occur where local conditions or processes enhance permeability. The extent to which porous media versus channelized flow occurs has yet to be determined.

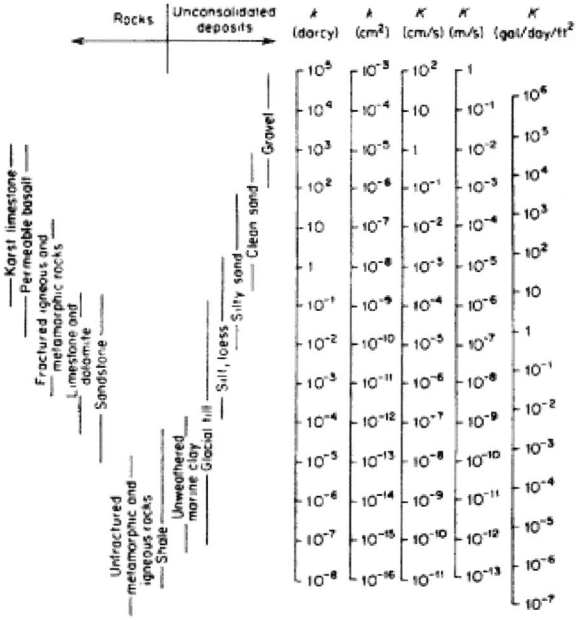

Figure 2
Range of in situ permeability measurements for crystalline and argillaceous rocks from various sites around the world (after Brace, 1980).

Mineral-Fluid Reactions

The advection of chemical components by fluid flow and dispersive fluxes from one chemical environment to another causes chemical reactions between the minerals and fluids. The processes of dissolution, precipitation, ion exchange, and sorption can all be represented by the general equations for the conservation of mass.

The general problem of chemical mass transport by groundwater is one of considering advective mass transport through a region in which reversible chemical reactions must be considered in terms of overall disequilibrium, yet with local equilibrium among some of the minerals and the aqueous phase. This condition has long been recognized in the study of natural weathering and hydrothermal processes and more recently in engineering studies of contaminant transport in groundwater. Although the analysis of chemical contaminants in shallow groundwater systems does not involve the long periods typical for geologic problems, both situations rely on a similar formulation. Equilibrium is simply a special case in the more general formulation.

The **irreversible formulation** requires that we understand the **kinetics of the reactions** of interest. The equilibrium between minerals and the fluid require a set of thermodynamic relations to describe their activity-composition relationships, and the minerals out of equilibrium require a kinetic-rate law consistent with the thermodynamic standard state to describe their rate of change. Also, a single equation must be written for each of the basic chemical components. In typical problems related to ore deposition, from 15 to 20 equations have nonlinear features that make them difficult to integrate numerically; realistic problems require large computer capacity, often supercomputers. For contaminant transport problems, as many as five simultaneous concentration equations have been used to analyze specific problems.

COUPLING

How are thermal, mechanical, hydrological, and chemical processes coupled? The closed nature of many mineralized fractures illustrates the important interaction between the chemistry and physical transport properties of the system. In general, fluid flow transports both heat and solute mass. This transport can perturb a system in chemical equilibrium into a disequilibrium state, and consequently, chemical reactions between the fluid and rock will occur. These reactions dissolve and precipitate solid material, changing the hydraulic characteristics of the fluid pathways and inducing heterogeneities in permeability and dispersivities; these changes in turn modify fluid flow patterns and mechanical properties. In multiphase fluid systems, heat transfer can also modify in situ fluid saturations through boiling, evaporation, and condensation. Saturation changes can also modify transport properties by changing relative permeabilities.

Chemical reactions within the system are strongly dependent on the solvation properties of the aqueous phase, because this phase controls the dissolution and deposition of minerals, and hence the porosity distribution. Diffusion and advection processes are controlled by porosity and permeability. Ultimately, fluid transport is coupled to solution and precipitation. If rock strain is considered to be dependent on changes in total stress, as well as temperature, then pore pressure is strongly coupled to heat transport. In some contexts it is possible to ignore the coupling and thus simplify the appropriate mathematics.

The equations for pressure, chemical composition, temperature, and Darcy's Law are coupled and nonlinear. An additional coupling occurs because of the dependence of both fluid density and viscosity on pressure, temperature, and concentration. The proportionality constant in Darcy's Law includes properties of the fluid, both density and viscosity. Density is influenced by pressure, temperature, and chemical concentration, whereas for most problems, viscosity is strongly influenced only by the temperature.

To fully understand the role of fluids in crustal processes, it will be necessary to unravel the complex coupling between thermal, chemical, mechanical, and hydrological processes. Each of the processes, properties, and driving forces that are currently included in the basic theory is the consequence of existing experimental data, theoretical analysis, and computer simulation. Few of these linkages have been directly observed in the field, yet it is likely that it is the interaction of these processes that ultimately controls, to a large extent, the movement and transport by fluids in the Earth's crust. Understanding these complex coupled processes is not only necessary for unraveling the nature and origin of mineral deposits, hydrothermal systems, and fluid-dependent crustal changes, but it is also needed to rationally address many societal issues, including groundwater contamination, radioactive and hazardous waste disposal, enhanced oil recovery, and exploitation of geothermal heat. Obtaining clearer answers to these questions will provide the means to establish the extent to which fluids have contributed to the distribution of heat and resources in the crust. Furthermore, these answers will give us a better understanding of the role of fluids in initiating or influencing tectonic events.

DISCUSSION: GEOLOGIC PHENOMENA

It is obvious from the above discussion that fluids within the crust are intimately involved in many processes of interest to geologists. For example, Hubbert and Rubey (1959) pointed out the mechanical problems associated with large overthrust sheets such as those observed in the Alps. They argue that one possible mechanical solution to large-scale overthrusting would be for associated pore fluids to have pressures approaching the lithostatic pressure exerted by the overlying rock. Under such conditions, the frictional resistance to sliding becomes negligible; overthrusting can occur as a gravitational process (sliding). While the original thought was to apply this idea to thrust sheets, the theory has been extended to the general problem of frictional failure.

Indeed, the more we observe dynamic processes operating in the active tectonic areas of the Earth, the more we see additional evidence suggesting that the pore pressure in many, if not most, of these regions is high, well above hydrostatic. Unfortunately, most of the evidence is indirect, often only inferences from other geophysical observations. However, **the current weight of the circumstantial evidence is such that a good case can be made for a rather general condition of high pore pressure in tectonically active areas**.

DRIVING FORCES

The flow of fluid in response to a force is the central mechanism in the theory for single-phase nonisothermal, reactive transport in a porous medium. The potential driving forces provide considerable insight into the associated geologic phenomena. The force fields that drive flow are caused by the following mechanisms:

1. **topographic relief**;
2. **tectonic dilation and compression**;
3. **diagenesis**;
4. **heat**; and
5. **fluid source**.

Although there are other potential driving forces, such as chemical concentrations across clay membranes (osmosis) and electrical potential, they are small in relation to those listed above.

The generation of pore pressure is a rate-controlled phenomenon; some mechanism, or set of mechanisms, operates to generate a hydraulic gradient. Pore pressure is dissipated by fluid flow outward from the source. The amount of the pore pressure increase depends on the ease with which flow can occur, which in turn depends on the hydraulic conductivity (permeability) of the host rock. As shown in Figure 2, observed permeabilities can range over 15 orders of magnitude within the crust.

All of the major mechanisms that drive fluid flow (given above) have also been suggested to create high pore pressure in active tectonic regimes of the crust.

Topographic Relief

On the tectonically stable portions of the continents, most, if not all, groundwater flow is the result of topographic relief. Tóth (1963) pointed out that the water table is the upper boundary for saturated groundwater flow, and that this boundary is usually closely approximated by the land surface. While the water table may fluctuate seasonally and from year to year because of wet and dry seasons, it does not fluctuate a great deal. Under most climatic regimes the water table can be demonstrated as approximately stationary over time.

Forces associated with the groundwater table surface are important enough in understanding groundwater flow that they should be stated another way. If we install a piezometer into the groundwater flow system within the Earth, we generally find the hydraulic head to be within 50 m of the land surface, certainly in most instances within 100 m. Hubbert and Rubey (1959) referred to this as the "normal" condition. This condition implies that the land surface is approximately the upper boundary condition for the groundwater flow system. The lower boundary for the system is no flow at some depth. The flow system is driven by topographic relief on the water table, the upper boundary for the saturated groundwater system.

Lateral boundaries are formed by drainage divides that separate the flow system into discrete cells. Tóth (1963) showed that the local topography provides perturbation on the regional flow system (Figure 3). Because of significant scale effects with small flow systems superimposed on larger systems the appropriate scale for the problem of interest needs to be determined.

The patterns of groundwater flow produced by normal conditions are determined by the topography, and changes in permeability only distort the pattern in flow. In the normal situation, topography drives groundwater flow. Groundwater flows downhill, just like surface water. Under normal conditions the maximum topographic elevation places an upper bound on the hydraulic head, while the lowest point imposes a lower bound. The total difference in elevation forms a constraint on the hydraulic gradient. Thus under conditions in which elevation differences are the driving force for the groundwater flow system, the total drive on the system is limited by the available topographic relief.

Flow in most large, mature sedimentary basins within the stable continental craton is generally thought to be driven by topographic relief. Early in the development of sedimentary basins both compaction and diagenesis may also be important driving mechanisms. In most sedimentary basins, significant quantities of fluids move vertically across confining layers (cap rocks). Low-permeability layers retard flow and sometimes cause large decreases in hydraulic head. Some changes in permeability are compensated for by the large areas through which flow, especially vertical flow, can occur in a regional flow system (Figure 4). In South Dakota, for example, approximately 50 percent of the recharge and discharge to and from the Dakota Sandstone aquifer flow through the overlying Pierre Shale, a shale that is often thought to be impermeable. This phenomenon is typical of flow in many, if not most, sedimentary basins.

High fluid pressures may have been generated by topographically driven flow across the Appalachian Basin as a consequence of uplift of the core of the Appalachians early in the Alleghanian Orogeny (Engelder, Chapter 9). The high fluid pressures accompanying this topographically driven flow later facilitated the development of first-order structures in the Valley and Ridge Province. Subsequent joint sets, which are not correlated across the Allegheny Front, are likely to be a consequence of fluid pressure pulses developed during local tectonic compaction.

Tectonic Dilation and Compression

Changes in tectonic stress affect the pore pressure. The degree to which the pore pressure is increased or decreased by changes in stress depends on the rate at which flow can dissipate the pressure.

Various tectonic strains lead to a volume decrease of the host rock. Usually the volume strain rate is small. Currently along the San Andreas Fault in central California, the rate of shear strain is on the order of 10^{-6} per year. This rate is accompanied by a volume strain of approximately 10^{-8} per year, two orders of magnitude smaller but capable of producing high pore pressure where the permeability of the rocks is small enough. Unfortunately, in

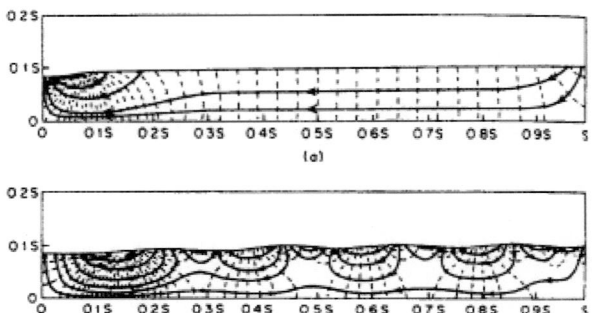

Figure 3
Cross section illustrating the role of topographic relief in groundwater flow (after Tóth, 1963).

active tectonic areas such as along the San Andreas Fault, we have no direct observations of pore pressure below 2 to 3 km. While it is clear from theoretical considerations that active rock deformation causes pore pressure changes, the data to demonstrate these effects are limited.

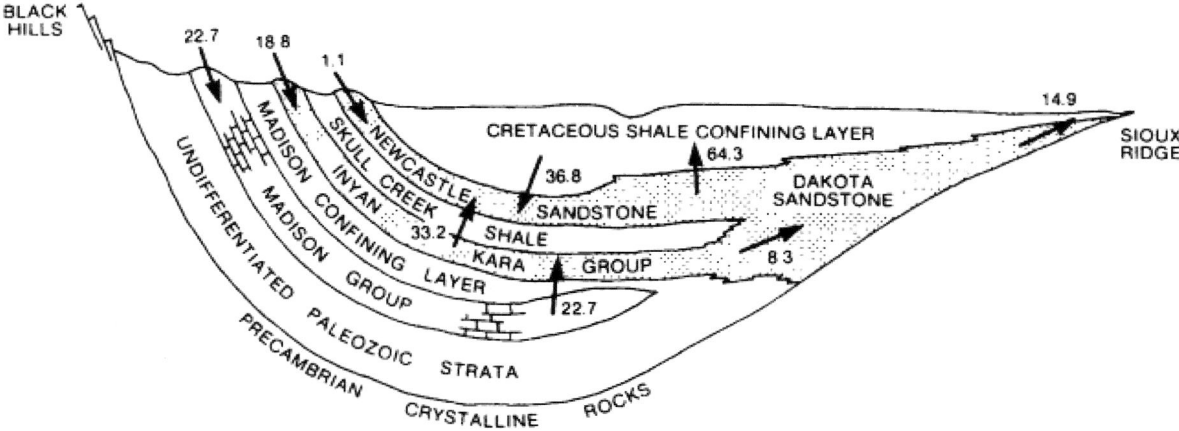

Figure 4
Idealized cross section of the Dakota flow system in South Dakota in which the quantities of flow through various layers are indicated. The numbers are integrated over the entire state of South Dakota (from Bredehoeft et al., 1983).

Once tectonic strain ceases, any pore pressure change caused by the strain becomes a transient phenomenon that will be dissipated by flow over time. However, if the rocks are of low permeability, such transient effects will not dissipate rapidly; indeed, in some instances of very low permeability material, these transient effects may persist for periods of geologic interest.

Diagenesis

The simplest model for the hydraulics of the crust is based on the assumption that porosity and, consequently permeability, remain unchanged with time. Under that assumption, permeability inferred from in situ phenomena directly provides an estimate of the ambient crustal permeability. As pointed out by Brace (1980), pore pressure in a crust with these rock permeabilities will generally have to be close to hydrostatic. This implies that water at depth is sufficiently connected to the free surface of the crust so that the pressure at any depth is simply the weight per unit area of a column of water reaching to the Earth's surface. There is, however, a wide range of direct and indirect observations that indicates fluid pressures can build-up locally to approach the lithostatic pressure in a variety of geological environments. To further complicate the problem, laboratory experiments and geological evidence suggest that rocks in situ rapidly seal hydraulically.

To reconcile these competing ideas of high crustal permeabilities and associated hydrostatic fluid pressures with evidence for hydraulically-sealed low permeability rocks with elevated fluid pressures, Nur and Walder (Chapter 7) have proposed a time-dependent process to relate fluid pressure, flow pathways, and fluid volumes. In their model crustal porosity, permeability and hence fluid pressures are in general timedependent because of the gradual closure of crustal pore space via heating, sealing, and inelastic deformation. Under certain circumstances, this process will lead to the local drying out of the crust. Under other conditions in which the pore fluid cannot escape fast enough, pore pressure will build-up leading perhaps to natural hydraulic fracturing, fluid release, pore pressure drop, and resealing of the system. If there is an adequate supply of fluid, this process could repeat leading to the intriguing possibility of cyclic episodes of pore pressure buildup and natural hydraulic fracturing. This model, involving coupling of hydrological and mechanical processes, needs to be rigorously evaluated with in situ observations and measurements.

A number of young geologic basins that are actively receiving sediments have anomalously high pore pressures that approach the total weight of the overburden. The simplest mechanism to generate such high pore pressure is one in which low permeability sediments are deposited in the basin at rates sufficiently high that consolidation cannot keep pace. If the rate of loading, caused by sedimentation in the basin, exceeds the rate at which the pore pressure can dissipate through the expulsion of fluid, then pore pressures that approach the lithostatic load will result (Figure 5). Under these conditions there is little or no effective stress, and little or no consolidation takes place.

The simple model of rapid sedimentation can account for abnormally high pore pressures such as that which exists in the Gulf Coast Basin. There are, however, other complicating processes at work in the Gulf. Montmorillonite changes to illite with a release of free water from the clay structure at approximately the same depth as the first occurrence of the anomalous pore pressure. In addition, there may be a thermal generation of high pore pressure caused by high thermal gradients.

In contrast to the Gulf Coast Basin, the Caspian Basin is subsiding at a rate of 1,100 m per million years and has a maximum thickness of accumulated sediments of 25 km. The basin has a very low thermal gradient, 16°C/km; temperatures of approximately 110°C exist at 6,000 m depth. There is no change of montmorillonite to illite at depths of 6 km but pore pressures are well above lithostatic. The South Caspian Basin then is at least one instance where simple mechanical loading at a high rate can account for the high pore pressures.

The extent to which the simple sedimentation-consolidation model is the appropriate mechanism for generating high pore pressure depends on the loading of the overburden. High pore pressure in a subsiding basin is a dynamic process that requires active loading

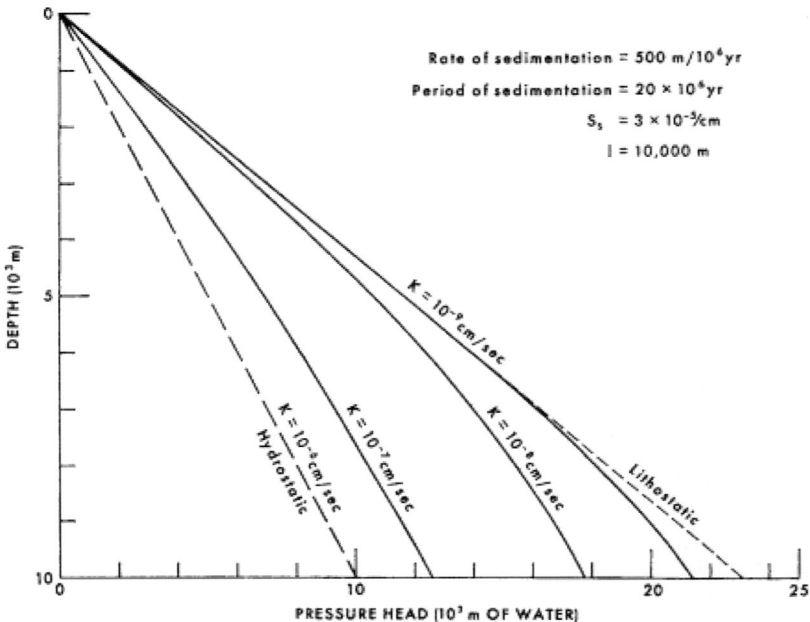

Figure 5
Calculated pore pressure versus depth profiles for a subsiding basin such as the Gulf Coast. The model used for these calculations considers only purely mechanical compaction (from Bredehoeft and Hanshaw, 1968).

to be maintained. Once subsidence and sedimentation cease, the high pore pressure becomes a transient within the system that will dissipate with time at a rate that depends on the hydraulic diffusivity of the deposits and the boundary conditions.

Heat

The introduction into the crust of a source of heat such as an igneous intrusive body creates a heat engine for the groundwater system, causing convective flow around the intrusive body due to fluid-rock interactions. Of interest is how the convecting groundwater leaches metals from the surrounding country rocks and crystallized portions of the intrusion and redistributes them. (See Norton, Chapter 2, for additional details.) The convective flow and fluid-rock interactions around plutons can result in base and precious metal deposits, whereas flow of formational waters out of basins (due to both topographic relief and heating) can yield deposits of zinc, lead, and barium.

The pore pressure is also strongly affected by changes in temperature and pore volume caused by the deposition or dissolution of mineral phases. The magnitude of the change in pore pressure induced by such changes depends on a function of the rates of these processes relative to the rate at which groundwater flow dissipates the pressure increases. The rate of flow, in turn, depends on the permeability of the material. If the permeability is sufficiently high, flow occurs away from the density perturbation and mitigates the pore pressure increase. On the other hand, if the permeability of the material is very low, pore pressure can increase until it exceeds the effective strength of the pore wall and an hydraulic fracture is formed.

How much heat and mass are transported by migrating fluids? Studies of hydrothermal systems have demonstrated that large quantities of heat and dissolved materials can be transported for hundreds to thousands of meters in short time spans (few thousand years) near the Earth's surface. A thorough analysis has been conducted for the Skaergaard Intrusion (Norton and Taylor, 1979; Norton et al., 1984) (see Figure 6). Their computer analyses and field observations suggest the development of a complex system of fractures throughout the long hydrothermal history of the system. The extent to which similar systems can exist at intermediate and deep crustal levels has yet to be established. Since most hydrothermal systems occur in the vicinity of cooling magma bodies, will similar fluid flow systems develop when magmas are trapped deeper in the crust? Conversely, can large circulation systems be established in the intermediate and deep crust in the absence of a cooling magma body and, if so, what are the conditions that allow such flow?

The change in pore pressure associated with heating is potentially enormous. Large pore pressure pulses can be generated by heating (e.g., igneous intrusions). Palciauskas and Domenico (1982) examined the problem taking into account the compressibility of the rocks; they suggest that these pulses in pressure can range from 5 to 10 bars/°C in low-permeability rocks.

Pore pressure increases are limited by the stress state at the point where the rock fails. Failure can occur either as shear or tension. If the failure occurs in tension it is a hydraulic fracture. At the point at which the rock fails, a fracture is created. Fractures increase the permeability of the rock mass and allow high pore pressure to dissipate rapidly. Many ore deposits consist of mineral-filled fractures. Near the critical point, water has a very steep pressure-density relationship. The rapid changes of density with pressure and temperature of the fluids as they move through the critical region is reflected in dramatic shifts in mineral solubilities, a factor of major importance in the formation of some types of hydrothermal mineralization. (See Titley, Chapter 3, for further discussion.)

Moving groundwater can be an efficient mechanism to transport heat introduced by an igneous body. Convective heat transport is much more efficient than by the transport of heat by pure conduction. The rate of cooling of the igneous body strongly depends on the rate of heat dissipation by circulating groundwater. In the lower and middle crust, the amount of groundwater that can be convected is poorly known.

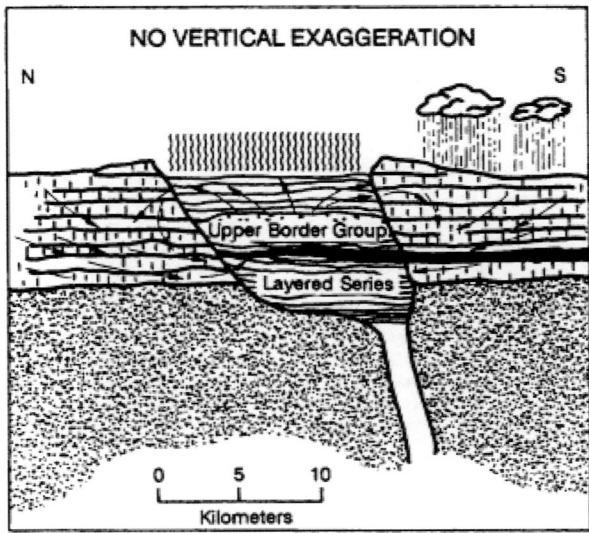

Figure 6
Schematic geologic cross section of the Skaergaard intrusion, showing the large-scale meteoric-hydrothermal circulation pattern in the plateau lavas above the granite gneiss basement (from Norton and Taylor, 1979).

Fossil geothermal systems constitute several classes of hydrothermal ore deposits. In these systems, ore has been emplaced by the moving fluids. It seems possible that many vein-type ore deposits are formed in hydraulic fractures held open by the pore pressure during mineral deposition. Most sills and dikes seem also to be simple hydraulic fractures into which magma was injected.

Geothermal power development exploits a concentrated heat source in the crust and depends on flowing groundwater (or steam). Whether or not commercial development of such a heat source is feasible depends almost entirely on whether the permeability of the hot rock is high enough to allow enough groundwater flow to replenish the heat efficiently. Geothermal reservoir engineering today involves numerical simulation of the reservoir's performance, which requires that the coupled partial differential equations be solved for the boundary conditions of interest.

Source of Fluids

What are the sources of fluid in the crust? Although fluids within a few thousand meters of the Earth's surface are likely to be derived primarily from precipitation, the source for fluids at deeper crustal levels is problematic. Fluids released from crystallizing magmas must contribute some fluid to the crustal fluid reservoir. Rocks that are heated as they are buried to deeper levels, or that are in the vicinity of intruding magma bodies, will also contribute fluids as they recrystallize to new mineral assemblages that are more compatible with their new thermal environment. (See Walther, Chapter 4, for additional discussion.) Finally, fluids may be released directly from the mantle to the overlying crust during "degassing" events. The relative proportions of these sources in any particular region have not been well established, although some isotopic studies have been undertaken to evaluate the relative roles of these processes. (See Taylor, Chapter 5, and Bredehoeft and Ingebritsen, Chapter 11, for additional discussion.)

Hydration or dehydration of minerals changes the fluid mass in the pores and will change the pore pressure. Depending on the volume change associated with the dehydration, such a source of fluids can increase the pore pressure. Perhaps the most commonly discussed example is the montmorillonite-illite transformation that Burst (1969) first suggested as the explanation for high pore pressure in the Gulf Coast. Various higher temperature metamorphic reactions release water. Most of these dehydration reactions are endothermic (i.e., absorbing heat), which slightly reduces the potential pore pressure build-up from the release of water.

The magnitude of the fluid-pressure change that accompanies such a reaction is rate dependent. If fluid is generated at such a rate that it cannot flow readily away from the source, then fluid-pressure will build up. If the permeability of the surrounding host rock is sufficiently small, the buildup in fluid pressure may be quite large.

One other potential source of fluid flow into the crust is mantle degassing. Rubey (1951) presented the case that the volatiles associated with the Earth's surface were not present early in the life of the planet and have accumulated with time. Perhaps the most commonly accepted hypothesis is that the volatiles have been degassed from the mantle, Figure 7. The role of mantle degassing as a source of crustal fluids is reviewed in Chapter 11 by Bredehoeft and Ingebritsen.

RECOMMENDED RESEARCH

There are numerous important research topics involving the role of fluids in the crust. The list below is not all inclusive; it reflects those topics that we felt deserved special consideration. Each item is discussed in more detail following the list.

1. There should be continued studies of **field areas** where the role of fluids in geologic processes can be well documented, especially rock assemblages from deep in the crust. Of particular interest are the hot active areas—the geothermal areas and the mid-ocean ridges.
2. Research is needed on the continued quantification of fluid transport in **fractured rocks**. Such research is important in understanding both crustal fluid flow and contaminant transport in shallower groundwater systems.
3. There needs to be a greater understanding of the **nature of permeability**, for example, the processes of diagenesis in altering permeability and the relationship of permeability to dispersion or mixing.
4. A greater understanding is needed of fluid flow through **rocks of low permeability**, for delineating the history of flow through sedimentary basins and for questions of the disposal of hazardous materials in the shallow crust.
5. Continued research is necessary on the **kinetics of mineralogic reactions** in the presence of fluid transport. Such reactions are important in understanding the rate of fluid phase generation. Not only should the classic mineralogical reactions be considered, but a greater emphasis should be placed on understanding the kinetics of organic reactions involving microbes with the mineralogical matrix.
6. There needs to be greater study of **pore fluids in tectonically disturbed geologic regimes**, both fossil and active regimes, particularly in situ sampling.
7. Research on the role of fluids in **crustal deformation** should be encouraged to help understand the time-dependent behavior of deformation and the role of fluids in rock mechanics.
8. Advances are required in the **mathematical simulation of coupled flow** problems, which are now addressable using very large-scale computational facilities and could supply an increased understanding of the physical processes and their prediction.

1. Field Studies

Much of the theory concerning mass and energy transport by moving fluids is derived from laboratory and near surface investigations. The application of these ideas to the deep crust is speculative. New investigations often provide additional insights.

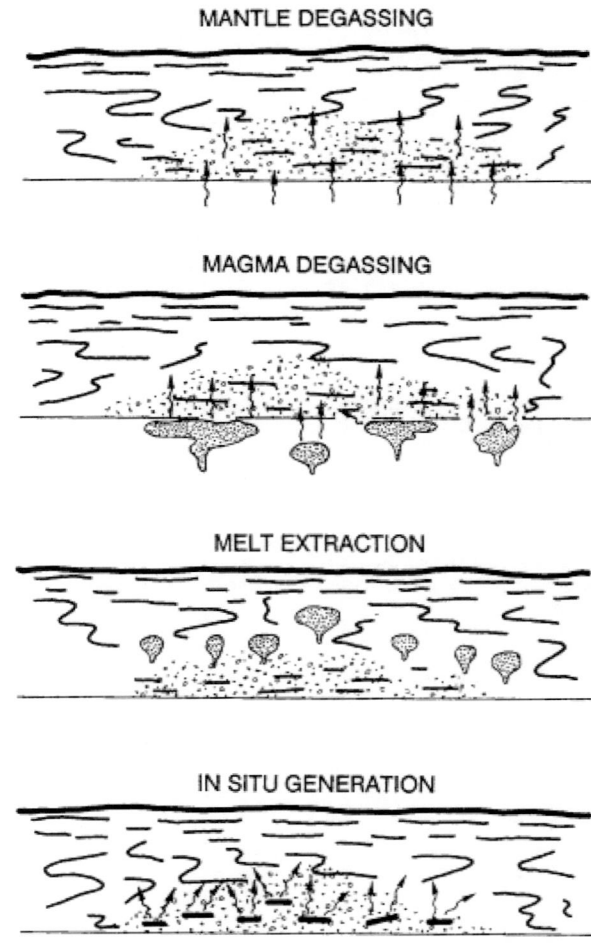

Figure 7
Schematic illustration of the possible sources for deep crustal fluids.

The application of the theory to real field areas is difficult and laborious. As Table 3 indicates, a large set of data must be assembled and evaluated quantitatively if one is to apply the theory rigorously. Commonly there is a lack of data to understand a wide range of fluid flow problems; some critical parameters must be evaluated by analogy to other areas where they are known. Often *inverse methods are* applied to determine the range of critical parameters. Because of the demands of data few attempts to apply mass and energy transport theory mathematically to analyze real field problems have been attempted. Even the attempts to simulate basin-scale fluid movements are limited in number.

In dealing with the deep crust, knowledge of the fluids will be largely indirect. Our knowledge of the fluids will depend on our understanding of the petrology of the rocks and the role the fluids must have played. A number of studies of exposed deep metamorphic rocks indicate large fluxes of fluids; often these studies indicate that fluids several times the "pore volumes" circulated through the rock. Often the only remaining clues are isotopic signatures and sometimes fluid inclusions in the rock. It is from this scant indirect evidence that the properties and budgets of fluids and their effects must be inferred.

In the active areas one can study the processes operating currently. It is for this reason that the active geothermal areas are interesting. In these areas one can observe heat and mass transport properties directly along with concurrent chemical reactions. Much work has gone into modeling heat and fluid (including steam) movement in active geothermal areas. These reservoir simulations were partly initiated to evaluate power development; however, the principles are being applied to study geologic processes as well (see, for example, Bredehoeft and Ingebritsen, Chapter 11). To date the geochemistry of the geothermal systems has not been integrated formally into an irreversible thermodynamic construct (see Norton, Chapter 2). Ore deposits are often fossil geothermal systems.

The discovery of "black smokers" at the mid-ocean ridge with 350°C sulphide-rich water is evidence for large-scale geothermal circulation at active spreading centers through young oceanic crust. This hydrothermal circulation is almost certainly driven by high temperatures below the ridge. Evidence suggests that the depths of fluid circulation may have been 1 to 2 km. There also may be mantle-derived volatile fluids as indicated by ^{3}He contents of the fluids coming up along the mid-ocean ridges.

The more actual field studies in which the fluids and their role are recognized and analyzed, the more confident earth scientists can be in their understanding of the role of fluids. The application of the theory to the Earth is the experimental test. Often a careful field investigation provides new insight.

2. Fractured Rocks

Numerous studies have indicated that the laboratory measurements of permeability do not characterize the flow properties of rock in situ. Often, in situ measurements indicated that the presence of fractures dominates the flow regime at a field scale. These fractures have a spacing larger than those in typical laboratory specimens. This is especially true for older well lithified sedimentary rocks as well as crystalline rocks.

How to handle the problems of fluid migration and transport in a fractured medium is still an open question for both hydrologists and reservoir engineers. Two schools of approach have developed. **The first school holds** that the problem is tractable, provided one can describe the geometry of the **fractures and the fracture network** (see Figure 8). This requires that the orientation, extent, and aperture of each fracture be described. Clearly, the problem of describing the fracture geometry is formidable, especially when this must be done at depth in the subsurface. Mathematically treating the fractures as fractals holds promise for greater characterization. **The second school** views the **fractured rock as essentially a porous medium** in which the pores are more widely spaced. The second group tends to describe the continuum in more classical terms as an anisotropic porous medium. Other characterizations include the so-called dual porosity medium in which flow in the interior blocks is characterized as flow in a porous medium while flow in the fractures is characterized by a much higher permeability, presumably representing the network of fractures.

No consensus currently exists within the scientific community on how to deal with problems of fracture flow. Much more field laboratory and theoretical research must be done before our understanding will permit us to confidently treat this important problem. Much recent research has gone into studying fractures and attempting to characterize their distributions.

3. Permeability

The rates of fluid migration depend on the permeability of the rock. Neglecting for the moment the role of fractures in permeability, discussed above, there are important research questions associated with the permeability of classical porous media.

Permeability before diagenesis is controlled by the lithology of the rocks. There are surprisingly few studies of the lithologic control on permeability. This is especially true when one looks at rocks of low permeability, the "confining layers" for the hydrologist, the "cap rock" of reservoir engineering.

Figure 8
An example of a computer-generated model of a fracture network. Fractures are represented as disc-shaped features of variable size.

One major phenomenon that occurs in transport of chemical constituents in a porous medium is hydrodynamic dispersion, a process of physical mixing of the fluid through the matrix. This process has been shown to result from local small-scale variations in the fluid velocity field. Velocity variations in turn depend on the local variations in permeability. It is important, therefore, to understand the structure of permeability including its variability. Unfortunately, detailed data describing permeability, especially in situ permeability, is lacking. Research into the nature of permeability and its relationship to lithology, including the diagenesis of rocks, is badly needed. Geostatistics is being applied both to the study of permeability and to dispersion; permeability distribution is also amenable to study by the fractal geometry. The application of these methods may provide new insights into how permeability varies.

4. Low-Permeability Rocks

The role of low-permeability or "tight" rocks in controlling subsurface fluid movement is not well understood. Careful measurements need to be made in situ, but the measurement in rocks with such low permeability is difficult. For example, a borehole represents a potential short circuit for flow in the system, and this simple act of making the measurement might significantly disturb the system. Measurements in such a low permeability regime are at the level Heisenberg uncertainty.

Bedded salt was thought by many to be impermeable because it is plastic and at some places it has existed since the Paleozoic era. Recent investigations at the Waste Isolation Pilot Plant (WIPP), which involve a mined nuclear repository in Permian Salado Formation salt, indicate that the salt is saturated with brine. One hypothesis for movement of the brine is that the salt behaves as any other porous material, except in this instance with very low permeability (10^{-8} to 10^{-9} Darcy).

Shales also have very low permeability when measured in the laboratory. However, when the permeability is measured in situ where the scale of measurement is of the order of kilometers or larger, the permeability has been shown to be larger by two or more orders of magnitude. This suggests that more conductive fractures exist, which may be spaced kilometers apart.

Low-permeability formations isolate deeper formations from the surface and make possible lateral flow in deeper sedimentary basins. Without these tight formations, fluids would circulate down and back to the surface within short distances. Understanding and measuring the physical properties of the low permeability rocks is of great importance in gaining knowledge about regional subsurface fluid movement, which is critical for applications such as the isolation of toxic wastes within the Earth's crust.

5. Chemical Reactions

Moving fluids that transport both heat and mass facilitate the precipitation of minerals, dissolution of the rock matrix, ion exchange, and a host of potential chemical reactions. The total system is dynamic with reactions occurring in the moving fluids. The nature of the problem dictates that it is not enough to know which chemical reactions will occur (a problem in itself), but one must also understand the reaction kinetics if one is to make a quantitative analysis. Sometimes the flow is sufficiently large that assuming a local chemical equilibrium is adequate. In other instances, the rate of reaction is the controlling parameter.

In the past the emphasis in geochemistry has been on understanding the inorganic compounds, the rock and mineral deposit forming minerals. It is increasingly clear that much of society's interest is in the organic compounds, especially hydrocarbons and a variety of contaminants. There are many unknowns about the role of organic geochemistry; even more poorly known are the catalytic properties of mineral surfaces and their effects on organic reactions. In addition, some ore-forming minerals may be transported as organic complexes in aqueous systems.

Hydrologists are becoming increasingly aware that microbes exist in the subsurface environment. Most of the chemical reactions associated with organic groundwater contamination are controlled biologically. The microbes appear to exist in the subsurface in a state of near starvation. Any potential food source is immediately seized upon. It appears that a rapid biologic adaption is possible to accommodate the potential food source. The microbes have been found in the groundwater-saturated subsurface to depths approaching a kilometer; but the maximum depth at which they can exist is not clear.

The entire area of geochemistry, especially organic geochemistry, in a moving subsurface fluid system is ripe for exciting new research. At shallow depths, organic reactions are clearly biologically controlled. It is not clear to what depths the biota play a significant role.

6. Pore Fluids in Active Tectonic Areas

Pore pressures are observed to be very anomalous in a number of petroleum provinces. However, empirical observations in tectonically active areas where oil and gas are not known to be present are limited to very shallow depths, usually less than 300 m (approximately 1000 feet). Indirect methods suggest that many of the tectonically active areas may have pore pressures approaching lithostatic. However, one needs empirical observations to corroborate geophysical and other indirect data.

A program of deep drilling for scientific purposes would greatly add to our empirical knowledge concerning pore fluids, and, in turn, their role in active tectonic processes. The few drillholes that have sampled fluids at depth—e.g., at the Kola Peninsula in the USSR and at Cajon Pass in California—have provided some intriguing data. However, in many areas of great scientific and societal interest there is no direct information at depth. Pore fluid may play an important role in failure of the Earth materials. The fluid state may be most important in understanding the mechanics of deformation including earthquakes.

7. Crustal Deformation

Hubbert and Rubey (1959) indicated that fluids can play an important role in the faulting process. Subsequent investigations into controlling earthquakes at Rangely, Colorado, demonstrated the applicability of the Hubbert and Rubey hypothesis. This hypothesis has only been applied to a linear elastic theory; earth materials are known to be viscoelastic. Recent development of constitutive laws for rocks undergoing failure include viscous effects. What we know of the earthquake failure mechanism is best explained by a viscoelastic failure behavior.

The role of pore fluids in the newer models for viscoelastic rock behavior is an area for continued research. The exact rock behavior prior to and during an earthquake is not well known. "Is there dilatancy; and what is its effect on failure?" is still the subject of debate and investigation. There may need to be observations of pore pressure within the focal zone during an earthquake to provide conclusive information—a formidable task of both predicting an earthquake and making the measurements.

This area of the role of fluids in the deformation and failure process of rocks is one of continuing investigation and interest. Results from the earthquake prediction experiment at Parkfield, California, may provide new insight.

8. Mathematical Analysis

Modeling of the roles of fluids in crustal processes and prediction of how fluids (especially those containing contaminants) move is a broad goal for the earth scientists. The set of partial differential equations referred to above and described in detail in Chapter 1 incorporate our fundamental understanding of the physics and chemistry associated with the role of fluids in the crust. Solutions to the basic equations provide insight into how the systems operate under a variety of complex geologic settings. Only in the simplest cases are analytical solutions to a problem of interest feasible. However, the digital computer has made it feasible to solve realistic problems of interest. As the computers become more powerful the problems addressed have also become more realistic. The numerical methods commonly in use—e.g., finite difference, finite element, and method of characteristics— have become household words in the geosciences. Investigators now solve transient-flow problems in two- and three-space dimensions routinely, in both petroleum engineering and groundwater.

There are, however, problems of interest that tax even the largest computers. The approach to mass and heat transport in the geosciences has adopted methods of nonequilibrium thermodynamics developed for chemical engineering; this approach, outlined in Table 2, involves writing a separate equation for fluid pressure and temperature and for each chemical constituent of interest. A set of partial differential equations is then solved simultaneously. For the time-dependent problem in three-space dimensions, this becomes a very large computational problem. Such problems can involve thousands of CPU hours on even the largest available super computers. **There is an obvious need in the earth sciences for very large-scale computing facilities.**

A critical part of the analytical methods used in geohydrology is the comparison of field data to analytical results; in petroleum reservoir engineering this comparison is referred to as "**history matching**"—perhaps its best description (other disciplines, such as hydrology, use other terms to describe this comparison such as calibration or parameter estimation, which give the comparison an aura of more reality than it deserves). The history match is all important for real problems. It addresses the question, can the system as the investigator thinks he understands it, produce this set of observations?

Most real geologic systems are sufficiently unconstrained by information so that no unique solution is possible. One is usually led to ask the question, is there a set of reasonable parameters that describe the system and yield something close to the observed result? In actual fact, one is usually left to play with the analytical model of the system to see which parameters the system behavior is sensitive to, this procedure is often referred to as **sensitivity analysis**. Sensitivity analysis can be performed by trial and error, or some more rigorous procedure. In the end, the investigator gains a feeling for just how the system responds. Since the information is incomplete with respect to uniqueness, one is often left to make a professional judgment as to the actual nature of (1) fundamental relationships that describe the system, (2) boundary and initial conditions, and (3) the parameter distribution within the system.

The parameter distribution of most geologic problems is described by a very limited sample. For problems at depth within the Earth, **drilling and sampling, while expensive, are a critical endeavor to permit the state of parameter estimation and history matching**.

Since the system behavior depends on the parameter distribution within the system, one would like to look at the effect upon the system of variations in parameter distribution. Past investigations have generally tested what was thought to be the best **single** interpretation of the parameters of interest; for example, the permeability distribution. Given this distribution, the system behavior was computed. However, one would like to place a confidence band about the computed system behavior. Given confidence bands about system response, one is in a better position to judge the "goodness" of the history match. However, in the end the "goodness" of the history match is a judgment.

One advantage of using a time-dependent mathematical analysis is that one can make a quantitative prediction of system behavior. This is particularly useful for environmental problems. For example, one would like to know how a plume of contaminated groundwater is moving in an aquifer, in order to take action. **Future research in the geosciences must be directed toward placing confidence bands about the output of analytical models, including predictions**.

References

Brace, W. F. (1980). Permeability of crystalline and argillaceous rocks: Status and problems, *International Journal of Rock Mechanics in Mineral Sciences and Geomechanical Abstracts 17*, 876-893.

Bredehoeft, J. D., and B. B. Hanshaw (1968). On the maintenance of anomalous fluid pressure, I. Thick sedimentary sequences, *Geological Society of America Bulletin 79*, 1097-1106.

Bredehoeft, J. D., C. E. Neuzil, and P. C. D. Milly (1983). Regional flow in the Dakota aquifer, *U.S. Geol. Survey Water Supply Paper 2237*, pp. 1-45.

Burst, J. F. (1969). Diagenesis of Gulf Coast clayey sediments and its possible relation to petroleum migration, *American Association of Petroleum Geologists Bulletin 53*, 73-79.

Fyfe, W. S., N.J. Price, and A. B. Thompson (1978). *Fluids in the Earth's Crust*, Elsevier, Amsterdam, 383 pp.

Hubbert, M. K., and W. W. Rubey (1959). Role of fluid pressure in mechanics of overthrust faulting, *Geological Society of America Bulletin 70*, 115-166.

Norton, D. (1984). A theory of hydrothermal systems, *Annual Reviews of Earth and Planetary Sciences 12*, 155-177.

Norton, D., and H. P. Taylor, Jr. (1979). Quantitative simulation of the hydrothermal systems of crystallizing magmas on the basis of transport theory and oxygen isotope data: An analysis of the Skaergaard intrusion, *Journal of Petrology 20*, 421-486.

Norton, D., H. P. Taylor, Jr., and D. K. Bird (1984). The geometry and high-temperature brittle deformation of the Skaergaard intrusion, *Journal of Geophysical Research 89*, 10,178-10,192.

Palciauskas, V. V., and P. A. Domenico (1982). Characterization of drained and undrained response of thermally loaded repository rocks, *Water Resources Research 18*, 281-290.

Rubey, W. W. (1951). Geologic history of the sea, *Geological Society of America Bulletin 87*, 1111-1148.

Tôth, J. (1963). A theoretical analysis of ground-water flow in small drainage basins, *Journal of Geophysical Research 68*, 4795-4812.

BACKGROUND

1

Mass and Energy Transport in an Deforming Earth's Crust

JOHN D. BREDEHOEFT
U.S. Geological Survey, Menlo Park
DENIS L. NORTON
University of Arizona

INTRODUCTION

Groundwater is ubiquitous throughout the crust to depths of at least 15 to 20 km, perhaps deeper in some places. Because geologists have traditionally focused on minerals and lithologic units in their study of Earth processes, the importance and distribution of the fluid phase have been overlooked. In fact, fluids are suspected to have a strong influence over both local state conditions and the redistribution of mass and energy, and the question arises as to whether studies that have not taken into account the presence of water have uncovered the correct mechanisms.

The field and theoretical evidence brought forth by the contributions to this volume demonstrate that groundwater plays a major role in crustal processes. The following are but a few of the phenomena for which fluids clearly play a central role: migration and entrapment of oil and gas, formation of ore deposits, metasomatic alteration of large volumes of rock, changes in the state of stress, failure of rocks in shear and tension, formation and emplacement of magmas, formation of geothermal systems, triggering of earthquakes, cooling of magmas, diagenesis of sedimentary rocks, movement of landslides, composition of the oceans, transport of contaminants, and distribution of heat. Consequently, understanding the geology of the crust requires that the distribution and behavior of groundwater be considered.

One example of how H_2O-rich fluid strongly affects the evolution of conditions in the crust is the central role it plays in hydrothermal systems. One can show that a rise in temperature of 1°C can cause a pressure increase of several bars in the local fluid pressure. Consequently, the introduction of a large body of magma into the crust will inevitably set up a convecting groundwater system around the magma body, which transports heat away from the magma at supercritical conditions. Because supercritical fluid is efficient as a heat-transporting medium, the rates of groundwater flow will control the cooling rates of the igneous body. The same convective process also redistributes chemical components from the host to the pluton environments, including the ore-forming constituents. Therefore, groundwater flow controls the distribution and grade of many ore deposits.

The magnitude of groundwater convection and consequently the extent of mineral alteration around a magma depend on the permeability of rocks. However, even if the rocks were initially impermeable, the pore pressure generated by heat dispersed into the host rocks is high enough to create hydraulic fractures. Thus, the system generates its own permeability by making fractures. This fracturing process tends to be episodic because of the intrinsic strength of the rock. Therefore, repeated sequences of fracturing followed by mineral deposition in the fractures are typical of magma environments. As the fractures gradually fill with minerals, permeability decreases and fluid flow is mitigated. This cycle of processes forms an intricately interconnected feedback system (Figure 1.1) in which the properties of the fluid phase exert primary control on the evolution of the system.

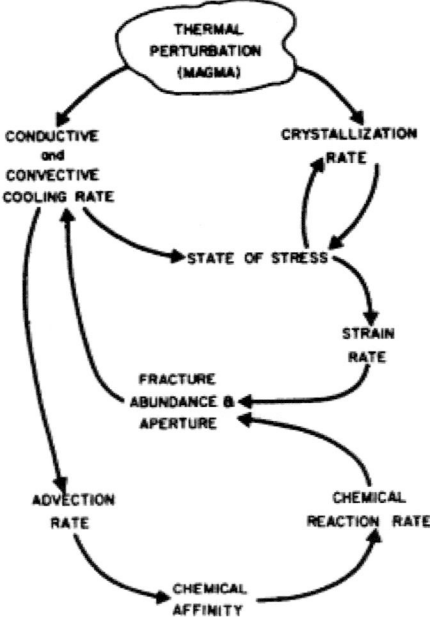

Figure 1.1
Systematic relationships between principal transport rates and their products. Arrows depict directions of energy, mass flow, and feedback effects of state conditions on rates (from Norton, 1984).

Because each of the contributions to this volume discusses ways in which the properties of H_2O-rich fluids determine how processes evolve, this chapter summarizes some of the general-process dynamics. The symbolic formalism of mass and energy transport is used in an attempt to sketch a coherent picture of processes and to show the usefulness of the equations in revealing the role of the fluid phase in mass and energy transport within a deformable medium. The equations presented below are not original, so the reader is encouraged to explore in greater depth the transport theories such as those used in engineering (e.g., Bird *et al.*, 1960; Slattery, 1972) to solve geologic problems.

CONSERVATION EQUATIONS

The concept of conservation provides a basis for writing a set of equations that symbolically represent the transport processes involving water in the Earth's crust. For each of the mass, energy, and momentum quantities that participate in the transport, an equation is formulated that expresses the conservation of the quantity with respect to the local system. These equations describe the rate of change in quantities with respect to time and a representative volume of the system.

During the 1800s some carefully conducted experiments produced a set of empirical laws that express the flux of mass and energy and in terms of a driving force and a proportionality constant that incorporates a medium's properties. These flux laws form the basis for discussing transport; they include Fick's Law of Diffusion, Fourier's Law of Heat Conduction, DeDonder's Law of Affinity, and Darcy's Law. Each has the general form of the flux of a quantity, where the flux is proportional to the gradient in a field parameter. The conservation equations for thermal, mechanical, and chemical energy discussed below all derive from one or more of these flux laws.

Darcy's Law

Fundamental to the transport processes in rocks is the conservation of fluid momentum that derives from the experiments by Henry Darcy (Darcy, 1856). In his experiments Darcy investigated the flow of water through a sand filter and suggested a general relationship for the rate of flow versus the drop in hydraulic head. In a classic paper on the theory of flow in porous media, Hubbert (1940) showed that Darcy's Law is equivalent to the Navier-Stokes equation for fluid momentum and can be stated as follows:

$$v_i = - \frac{k_{ij}}{\mu \phi} \frac{\partial}{\partial x_j} (\rho \hat{g} h), \qquad (1.1)$$

where v_i is the mean velocity in the ith direction, k_{ij} is the permeability, μ is the fluid viscosity, is the porosity through which fluid flows, g is the acceleration of gravity, and h is the hydraulic head. While this is the simplest form of Darcy's Law, Hubbert went on to show that when fluid density (ρ) varies in space, a more general form stated in terms of fluid pressure (p) is necessary:

$$v_i = - \frac{K_{ij}}{\mu \phi} \left(\frac{\partial p}{\partial x_j} + \rho \hat{g} \frac{dz}{dx} \right). \qquad (1.2)$$

Although the hydraulic conductivity, K_{ij}, is commonly used as the proportionality constant m Darcy's Law, it is a function of properties of both the rock and the fluid. In subsurface conditions the medium and fluid properties not only vary independently but also range over extreme values; they must be independently expressed in the equations. Hydraulic conductivity is related to permeability as follows:

$$K_{ij} = \frac{\rho}{\mu} k_{ij} . \quad (1.3)$$

Permeability is a property of the medium itself that is nonhomogeneous and anisotropic and is a symmetric tensor.

Darcy's Law is useful to geologic processes because it is a macroscopic expression that involves averaging over some "representative elementary volume" of the porous medium (see Bear, 1972).

Flow Equation

The conservation of mass for a single-phase fluid is

$$\frac{\partial}{\partial x_i}(\phi \rho v_i) = \frac{\partial \rho \phi}{\partial t} , \quad (1.4)$$

where ρ is the fluid density within the interconnected pore space whose fraction of the system is . The pore pressure and rock stress are coupled; the coupling is discussed more fully below under somewhat restrictive assumptions of coupling. Theis (1935) and Jacob (1940) showed that by substituting Darcy's Law for the velocity, v_i in Eq. (1.4), and combining the compressibility of both the rock and the water, we obtain an equation to describe isothermal flows in compressible media:

$$\frac{\partial}{\partial x_i}\left(\frac{\rho k_{ij}}{\mu}\right)\left(\frac{\partial h}{\partial x_j}\right) = S_s \frac{\partial h}{\partial t} , \quad (1.5)$$

where S_s is a coefficient that expresses the effective vertical compressibility of the coupled rock and fluid, called the "specific storage" by Jacob (1940). The flow equation can also be stated in terms of the fluid pressure:

$$\frac{\partial}{\partial x_i}\left[\frac{\rho k_{ij}}{\mu}\left(\frac{\partial p}{\partial x_j} + \rho \hat{g} \frac{\partial z}{\partial x_j}\right)\right] = \frac{S_s}{\rho \hat{g}} \frac{\partial p}{\partial t} . \quad (1.6)$$

As shown below, the flow equation becomes more complicated as we account for changes in tectonic stress and nonisothermal conditions. The fluid velocity derived from these relations forms a basis for the transport of other quantities in the system.

Dispersion-Diffusion-Advection

When we observe the movement of chemical components through a porous medium, a combination of dispersion, diffusion, and advection is apparent. This combination of transport mechanisms can be used to explain variations in velocities of chemical components. The phenomenon of dispersion is depicted in Figure 1.2. It occurs both at the microscopic level, because of the differing velocities of flow through the pores, and at the macroscopic level, where heterogeneities in geologic materials greatly increase the magnitude of the dispersive process.

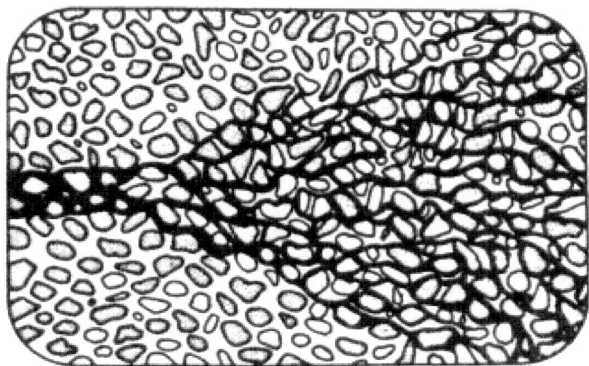

Figure 1.2
Microscopic dispersion (after Freeze and Cherry, 1979).

Diffusion laterally away from the flow channels into fracture-controlled matrix blocks was recognized early in the study of hydrothermal ore deposits as the mechanism by which alteration haloes and high assays form in the rock matrix adjacent to fossil flow channels called veins. Although it is a distinctly different mechanism from the dispersion of components, its effects are difficult to separate and are generally treated as a combined process.

The mass flux caused by dispersion and diffusion is formulated as an expression analogous to Fick's Law of Diffusion:

$$j_{ij} = D_{ij}^* \frac{\partial C}{\partial x_j} , \quad (1.7)$$

where D_{ij}^* is the coefficient of dispersion, which is the sum of the hydrodynamic dispersion coefficient and the diffusion coefficient

$$D_{ij}^* = D_{ij} + D_d . \quad (1.8)$$

Anderson (1984, p. 39) presents a clear discussion of dispersion:

... D_{ij} is the coefficient of mechanical dispersion and D_d is the coefficient of molecular diffusion. An effective diffusion coefficient is generally taken to be equal to the diffusion coefficient of the ion in water (D_d) times a tortuosity factor. The tortuosity factor has a value less than 1 and is needed to correct for the obstructing effect of the porous medium. Effective diffusion coefficients are generally around 10^{-6} cm^2/sec, although a range of 10^{-5} to 10^{-7} cm^2/sec is not inconceivable (Grisak and Pickens, 1981). Except for systems in which groundwater velocities are very low, the coefficient of mechanical dispersion generally will be one or more orders of magnitude larger than D_d. Therefore, in many practical applications the effects of molecular diffusion may be neglected ($D_d = 0$). The coefficient of mechanical dispersion is routinely taken to be the product of the magnitude of the velocity times a parameter known as dispersivity, which is commonly and somewhat vaguely referred to as a characteristic mixing length.

Advective transport is caused by fluid flow from one chemical environment to another. In this situation the fluid carries along chemical components that alter the local chemical conditions and shift local equilibrium to far from equilibrium conditions. The long-distance transport of components and the imposition of one lithologic composition onto another over broad regions occur as a consequence of the carrying-along capacity of the flowing water.

The general form of the advection-dispersion equation is

$$\frac{\partial}{\partial x_i}\left(D_{ij}^* \frac{\partial C}{\partial x_j}\right) - \frac{\partial}{\partial x_i}(Cv_i) + \sum_{k=1}^{s} R_k = \frac{\partial C}{\partial t}, \quad (1.9)$$

where C is the concentration of a particular chemical solute of interest, and R_k is the rate of production of the solute in the kth reaction. In the case where more than one solute is of interest, an advection-dispersion equation is written for each.

Heat Transport

A general expression for heat transport by moving groundwater can be derived from a statement about the dispersive as well as the conductive flux of heat, much as was done for the dispersion-diffusion of mass above. The dispersive-conductive flux of heat follows directly from Fourier's Law of Heat Conduction, written here in terms of temperature:

$$q_i = L_{ij}^* \frac{\partial T}{\partial x_j}, \quad (1.10)$$

where L_{ij} is the coefficient of heat dispersion. The coefficient of heat dispersion is

$$L_{ij}^* = L_{ij} = K_T, \quad (1.11)$$

where L is the thermal dispersion coefficient and K_T is the thermal conductivity of the porous medium (fluid and rock). The usual assumption is that the rock and fluid are in thermal equilibrium, and consequently the temperature in Eq. (1.10) is that of the fluid and rock.

The convective movement of heat associated with the flowing fluid is described by the flux equation:

$$q_{convection} = v_i (\rho c_p T)_{heat\ content\ of\ fluid}, \quad (1.12)$$

where c_p is the isobaric heat capacity and T is the temperature of the fluid phase flowing at velocity v_i.

Combining the convective, diffusive, and dispersive fluxes and transforming them into rates of change lead to the expression for single-phase energy transport in terms of the temperature and heat capacity of the fluid (Domenico, 1977):

$$\underbrace{\frac{\partial}{\partial x_i}\left(L_{ij}^* \frac{\partial T}{\partial x_j}\right)}_{Conduction} + \underbrace{\frac{\partial}{\partial x_i}(v_i \rho c_p T)}_{Advection} = \rho c_p \frac{\partial T}{\partial t}, \quad (1.13)$$

where the fluid velocity is defined by Eq. (1.2) and the velocity of the rock matrix is presumed to be zero.

Although this set of partial differential equations—the flow equations for pressure and the advection-dispersion equations for chemical solutes and energy, along with Darcy's Law—form a complete conservation statement, the effects of stress and strain must also be considered.

Stress and Strain

In the 1920s, in investigating consolidation, Terzaghi treated soils as a water saturated porous medium. In a series of papers Biot (1941, 1955, 1956) introduced a set of constitutive relationships for stress and strain in a fluid-filled porous media. Several authors have extended Biot's work. Following Nut and Byerlee (1971), Palciauskas and Domenico (1982) extended the theory to nonisothermal conditions. They suggested that the constitutive relationship, which includes thermal effects, is

$$\varepsilon_{ij} = \frac{1}{2\mu_s}\left(\sigma_{ij} - \frac{1}{3}\sigma_{kk}\delta_{ij}\right) + \frac{1}{9K_b}(\sigma_{kk}\delta_{ij})$$

$$- \frac{1}{3H} p \delta_{ij} - \alpha_T (T - T_0), \quad (1.14)$$

where ε_{ij} is the strain, μ_s is the stress, σ_{kk} is the sum of the normal stresses, δ is the Kroenecker delta, K_b is the bulk modulus of elasticity, H is a coefficient introduced by Biot, p is the pore pressure, α_T is a coefficient of linear thermal expansion, and T is the temperature. Pore pressure changes are usually thought to be the result of volumetric strain, :

$$\theta = \varepsilon_{11} + \varepsilon_{22} + \varepsilon_{33} = \frac{1}{3K}(\sigma_{11} + \sigma_{22} + \sigma_{33})$$

$$- \frac{1}{H}(p - 3\alpha_t)(T - T_0). \quad (1.15)$$

For many problems in rock mechanics it is convenient to think of the total stress as the sum of an effective stress plus the pore pressure. Nur and Byerlee (1971) defined a relationship for effective stress:

$$\hat{\sigma}_{ij} = \sigma_{ij} - \hat{\alpha} P \delta_{ij}. \quad (1.16)$$

In this expression α is defined for an isotropic elastic medium as

$$\hat{\alpha} = 1 - \frac{K}{K_s}, \quad (1.17)$$

where K_s is the bulk modulus of the grain alone (bulk modulus of the minerals). Biot's coefficient H can be defined as (Nur and Byerlee, 1971):

$$H = 1 - \frac{K}{\hat{\alpha}}. \quad (1.18)$$

Terzaghi (1943) observed that for the most porous materials, $\alpha \approx 1$. The constitutive relationship can be stated in terms of effective stress:

$$\epsilon_{ij} = \frac{1}{2\mu_s}\left(\hat{\sigma}_{ij} - \frac{1}{3}\hat{\sigma}_{kk}\delta_{ij}\right)$$

$$+ \frac{1}{9K_b}(\hat{\sigma}_{kk}\delta_{ij} - \alpha_T)(T - T_0). \quad (1.19)$$

One can introduce into the flow equation the effects that would be produced by changes in total stress, nonelastic deformation, and thermally induced rock deformation. These effects are in addition to the deformation due to changing pore pressure, considered above, Eqs. (1.5) and (1.6). Palciauskas and Domenico (1980) suggested a more general, fully coupled flow equation (for the case where the grains are incompressible, i.e., $\alpha = 1$):

$$\frac{\partial}{\partial x_i}\left[K_{ij}\left(\frac{\partial p}{\partial x_j} + \rho \hat{g}\frac{\partial z}{\partial x_j}\right)\right]$$

$$\overset{\text{Pore}}{\underset{\text{Pressure}}{}} \quad \overset{\text{Normal}}{\underset{\text{Stress}}{}} \quad \overset{\text{Dilatancy}}{} \quad \overset{\text{Thermal}}{\underset{\text{Elasticity}}{}}$$

$$= \hat{S}_s \frac{\partial P}{\partial t} - \left(\frac{\hat{S}_s B}{3}\right)\left(\frac{\partial \sigma_{kk}}{\partial t}\right) - \hat{D}\frac{\partial \tau}{\partial t} - \alpha_T \phi \frac{\partial T}{\partial t}$$

$$\overset{\text{Reaction}}{}$$

$$+ \rho \hat{g} \frac{\partial \phi}{\partial t}, \quad (1.20)$$

where S_s is a three-dimensional specific storage defined by Van der Kamp and Gale (1983):

$$\hat{S}_s = \rho \hat{g}\left[\left(\frac{1}{K} - \frac{1}{K_s}\right) + \phi\left(\frac{1}{K_f} - \frac{1}{K_s}\right)\right]. \quad (1.21)$$

where K_f is the bulk modulus of the fluid.

B is a coefficient defined by Skempton (1954), generally referred to as "Skempton's B coefficient." B relates the change in pore pressure to a change in mean stress in the absence of flow, referred to as the undrained state, that is,

$$dp = Bd\left(\frac{\sigma_{11} + \sigma_{22} + \sigma_{33}}{3}\right). \quad (1.22)$$

where B is defined as

$$B = \frac{\left(\frac{1}{K} - \frac{1}{K_s}\right)}{\left(\frac{1}{K} - \frac{1}{K_s}\right) + \phi\left(\frac{1}{K_f} - \frac{1}{K_s}\right)}. \quad (1.23)$$

In this formulation Palciauskas and Domenico (1980) express the inelastic volume change at the porous medium (dilatancy of the material):

$$\left(\frac{1}{1 - \phi}\right)d\phi = -\hat{D}d\tau, \quad (1.24)$$

where τ is the maximum deviatoric stress. In the words of Palciauskas and Domenico (1980):

D is a dilatation coefficient that can be positive or negative, depending upon the material.... The coefficient is not likely to be a constant for a given material, but will be somewhat strain dependent. When D is negative, the inelastic volume change acts to increase porosity with increasing shear stress.

Mineral-Fluid Reactions

The advection of chemical components by fluid flow and dispersive fluxes from one chemical environment to another causes chemical reactions between the minerals and fluids. The processes of dissolution, precipitation, ion exchange, and sorption can all be represented by the general equations for the conservation of mass.

The general problem of chemical mass transport by groundwater is one of considering advective mass transport through a region in which both reversible and irreversible reactions are occurring. Although in many instances fluid-flow rates are slow enough that local equilibrium prevails in the aqueous and mineral phases, in all systems where rock metasomatism has occurred the system must be considered in terms of overall disequilibrium with local equilibrium among some of the minerals and the aqueous phase. This condition has long been recognized in the study of natural weathering and hydrothermal processes (Helgeson et al., 1970) and more recently in engineering studies of contaminant transport in groundwater (Cherry et al., 1984).

Although the analysis of chemical contaminants in shallow groundwater systems does not involve the long periods typical for geologic problems, both situations rely on a similar formulation. Equilibrium is simply a special case in the more general formulation. The irreversible formulation requires that we understand the kinetics of the reactions of interest (Aagaard and Helgeson, 1982).

The geochemical basis for describing reactions among minerals and fluids in natural systems derives from the work of Helgeson *et al.* (1970, 1979). Once the rigorous description of mass transfer in systems that are only locally in equilibrium was placed on a thermodynamic basis, the rate of the irreversible reactions could be incorporated into the theory (Aagaard and Helgeson, 1982). More recently Helgeson's formulation has been extended to provide for the diffusion and advection of chemical components (Lichtner, 1986).

In the geochemical theory conservation of each of the basis components,

$$\Psi_i, \quad i = 1, 2, 3 \ldots \hat{I}, \qquad (1.25)$$

where Ψ_i is the concentration of the component in the respective phase, is described in terms of a transport equation of the form:

$$\underbrace{\frac{\partial(\phi_f \Psi_f)}{\partial t}}_{Fluid} + \underbrace{\sum_p \frac{\partial(\phi_p \Psi_p)}{\partial t}}_{Products} + \underbrace{\sum_r \frac{\partial(\phi_r \Psi_r)}{\partial t}}_{Reactants}$$
$$+ \underbrace{v_f \cdot \nabla \Psi_f}_{Advection} - \underbrace{\kappa \phi_f \rho_f \Psi_f}_{Diffusion} = 0. \qquad (1.26)$$

where the mineral products, $p = 1, 2, 3 \ldots P$, are those minerals in local equilibrium with the fluid and the mineral reactants, $r = 1, 2, 3 \ldots R$, are those phases that react irreversibly with the fluid. Eq. (1.26) is derived from the general mass transport equation, Eq. (1.9), by assuming the flow is steady and incompressible, that is,

$$\nabla \cdot v_f = 0 \qquad (1.27)$$

and that the dispersion is negligible. Although this assumption is questionable, the practical issue of how to determine the magnitude of the dispersion for either metamorphism or ore deposits has not been solved. The equilibrium minerals and the fluid require a set of thermodynamic relations to describe their activity-composition relationships, and the minerals out of equilibrium require a kinetic-rate law consistent with the thermodynamic standard state to describe their rate of change. Also, an equation of the form of Eq. (1.26) must be written for each of the basis components.

In typical problems related to ore deposition, 15 to 20 equations in the form of Eq. (1.26) are necessary. These equations have nonlinear and stiff features that make them difficult to integrate numerically--realistic problems require large computer capacity, often super computers. For contaminant transport problems as many as five simultaneous concentration equations have been used to analyze specific problems.

COUPLING

Chemical reactions within the system that appear as a source term in Eq. (1.9) are strongly dependent on the solvation properties of the aqueous phase. Because this phase controls the dissolution and deposition of minerals, and hence the porosity distribution, the diffusion and advection processes are controlled by these properties of the solvent. The argument can be made that changes in pressure associated with the flow of fluid through tortuous pore space can cause substantial changes in the mass of material deposited in the channels.

If rock strain is considered to be dependent on changes in total stress as well as temperature [Eq. (1.20)], then pore pressure is even more strongly coupled to heat transport [Eq. (1.13)]. In some contexts it is possible to ignore the coupling and thus simplify the appropriate mathematics.

The equations for pressure [Eq. (1.5) and (1.20)], concentration [Eq. (1.9)], temperature [Eq. (1.13)], and Darcy's Law [Eqs. (1.1) and (1.2)] are coupled and nonlinear. The coupling occurs because of the dependence of both fluid density and viscosity on pressure, temperature, and concentration. The proportionality constant in Darcy's Law includes properties of the fluid, both density and viscosity. Density is influenced by pressure, temperature, and chemical concentration; that is,

$$\rho = f(P, T, X_i). \qquad (1.28)$$

Viscosity is also influenced by all three independent variables: pressure, temperature, and concentration. However, for most problems only the temperature effects need to be considered:

$$\mu = f(T). \qquad (1.29)$$

DISCUSSION: GEOLOGIC PHENOMENA

It is obvious from the discussion above that fluids within the Earth's crust are intimately involved in many processes of interest to geologists. Rather than attempt to deal with all of these processes, we have chosen to focus on one particular problem, the pore pressure in the active tectonic areas of the Earth.

Hubbert and Rubey (1959) have pointed out the mechanical problems associated with large overthrust sheets such as those observed in the Alps. They argue that one possible mechanical solution to large-scale overthrusting would be for associated pore fluids to have pressures approaching the lithostatic weight of the overlying rock. Under such conditions the frictional resistance to sliding becomes negligible; overthrusting can occur as a gravitational process (sliding). While the original thought was to apply this idea to thrust sheets, numerous earth scientists have extended the idea to the general problem of frictional failure.

Indeed, the more we observe dynamic processes operating in the active tectonic areas of the Earth, the more we see additional evidence suggesting that the pore pressure in many, if not most, of these regions is high, well above hydrostatic. Unfortunately, most of the evidence is indirect, often only inferences from other geophysical observations. However, the current weight of the circumstantial evidence is such that a good case can be made for a rather general condition of high pore pressure.

In stating the case of high pore pressure in the Earth's active tectonic belts, we could address a number of issues and, indeed, ramifications of those issues. Yet the purpose of this volume is not to provide an exhaustive treatise but rather to indicate the importance of considering the pore fluids. We believe the chapters that follow illustrate our case.

HOW PERMEABLE IS THE CRUST?

In considering geologic materials in the upper part of the crust, everything must be considered permeable. Geologists have known this since at least the early 1900s, and more recent quantitative analyses of regional groundwater systems demonstrate that flow through lithologies previously considered to be impermeable is often significant, even through confining beds such as the Cretaceous Pierre Shale that include layers of bedded bentonite (Bredehoeft et al., 1983). Permeability and resistivity are the two physical parameters associated with earth materials with the widest range of possible values; permeability is observed in nature to vary over at least 15 orders of magnitude (see Figure 1.3).

The depth to which fluids can circulate within the crust has been a subject of continued investigation by earth scientists. There are several lines of study. Perhaps the most direct information comes from earthquakes generated by fluid injection at the Rocky Mountain Arsenal near Denver, Colorado. Earthquakes that occurred at the arsenal to a depth of between 7 and 8 km were rather clearly triggered by the injection (Hsieh and Bredehoeft, 1981). Although other studies of fluid-induced earthquakes show similar depths, in most instances much less is known about the nature of the permeability at depth than what is known at the Rocky Mountain Arsenal.

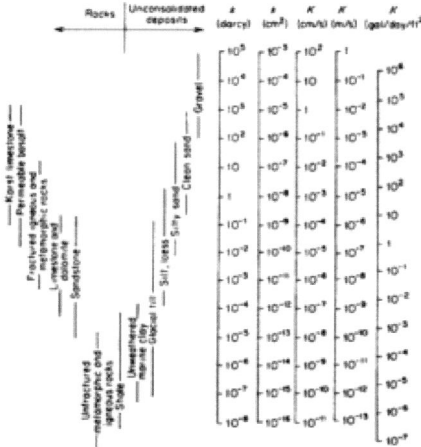

Figure 1.3
Ranges of permeability and hydraulic conductivity for a variety of rocks.

Deep resistivity measurements of the Earth's crust also indicate reasonable porosity to rather substantial depths. Brace (1980) summarized the information and suggested that some permeability must exist within the crust to depths of between 13 and 20 km.

There is additional evidence for deep circulation of fluids within the crust from metamorphic and igneous petrology. Taylor (Chapter 5, this volume) presents $^{18}O/^{16}O$ isotope data from a number of major intrusive bodies indicating circulation of groundwater to 10 to 15 km. Walther (Chapter 4, this volume) estimates that fluid flow has occurred as deep as 20 km, and Wickham and Taylor (Chapter 6, this volume) come to a similar conclusion after studying regional metamorphism in the Pyrenees. Clearly, the pressure exerted by the pore fluids plays a fundamental role in the motion of fluid through the crust.

Field evidence and theoretical relations indicate that permeabilities in excess of 10^{-14} cm^2 are common in magma environments (Taylor, 1974, 1977; Norton and Taylor, 1979). The measured mass transfer of chemical components in ore deposits, layered intrusions, and batholiths requires that advection was a dominant process. Norton and Knight (1977) demonstrated that for the driving forces encountered in the vicinity of magmas the threshold permeability at which convective exceeds conductive heat transport occurs at these permeability values.

MECHANISMS THAT CREATE HIGH PORE PRESSURE

The theory for single-phase, nonisothermal, reactive transport in a porous medium contains as the central mechanism the flow of fluid in response to a force. The force fields that drive this flow are caused by the following mechanisms:

1. topographic relief,
2. tectonic dilation and compression,
3. diagenesis,
4. geothermal systems, and
5. fluid source.

Although there are other potential driving forces, such as chemical concentrations across membranes (osmosis) and electrical potential, they are small relative to those listed above.

The generation of high pore pressure is a rate-controlled phenomenon; some mechanism, or set of mechanisms, operates to generate higher-than-hydrostatic pore pressure. This pore pressure is dissipated by fluid flow outward from the source. The amount that the pore pressure increases depends on the ease with which flow can occur, which in turn depends on the hydraulic conductivity (permeability) of the host rock. As suggested above, observed permeabilities range over 15 orders of magnitude within the crust.

All of the major mechanisms indicated above that drive fluid flow are also suggested in this volume to create high pore pressure in active tectonic regimes of the crust. We discuss examples of each.

Topographic Relief

On the tectonically stable portions of the continents most if not all groundwater flow is the result of topographic relief. The driving force or head, h, is composed of the pressure and elevation terms in Darcy's Law [Eq. (1.2)]. The relation of these quantities can be stated as

$$h = \frac{p}{\rho g} + z, \qquad (1.30)$$

where h is the height above some arbitrary datum to which fluid would rise in a manometer. The gradient in the head generates the field of force that moves groundwater.

In this situation the water table replicates the land surface. Tôth (1963) pointed out that the water table is the upper boundary for saturated groundwater flow and that this boundary is usually closely approximated by the land surface. While the water table may fluctuate seasonally and from year to year because of wet and dry seasons, it does not fluctuate a great deal. Under most climatic regimes the water table can be demonstrated as approximately stationary over time.

Forces associated with the groundwater table surface are important enough in understanding groundwater flow that they should be stated another way. If we install a piezometer into the groundwater flow system within the Earth, we generally find the water level, the hydraulic head, to be within 50 m of the land surface, certainly in most instances within 100 m. Hubbert and Rubey (1959) referred to this as the "normal" condition. This condition implies that the land surface is approximately the upper boundary condition for the groundwater flow system. The lower boundary for the system is no flow at some depth. The flow system is driven by topographic relief on the water table, the upper boundary for the saturated groundwater system.

Lateral boundaries are formed by drainage divides that separate the flow system into differing cells. Tôth (1963) showed that the local topography provides perturbations on the regional flow system (Figure 1.4). Because of the obvious scale effects, with small slow systems superimposed on larger systems we have to determine the appropriate scale for the problem of interest.

Low-permeability layers retard flow and sometimes cause large decreases in hydraulic head. Some changes in permeability are compensated for by the large areas through which flow, especially vertical flow, can occur in a regional flow system (Figure 1.5).

The patterns of groundwater flow produced by normal conditions are determined by the topography, and changes in permeability only distort the pattern in flow. A simple flow net analysis will convince us that the pattern of flow is established by the boundary conditions. In the normal situation topography drives groundwater flow. Groundwater flows downhill, just like surface water. Under

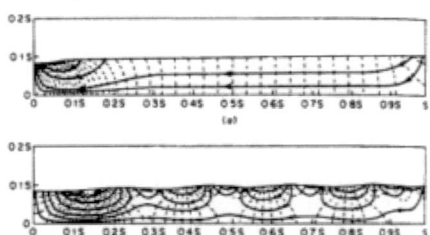

Figure 1.4
Cross section illustrating the role of topographic relief in groundwater flow (after Tôth, 1963).

normal conditions the maximum topographic elevation places an upper bound on the hydraulic head, while the lowest point imposes a lower bound. The total difference in elevation forms a constraint on the hydraulic gradient. Under conditions in which elevation differences are the driving force for the groundwater flow system, the total drive on the system is limited by the available topographic relief.

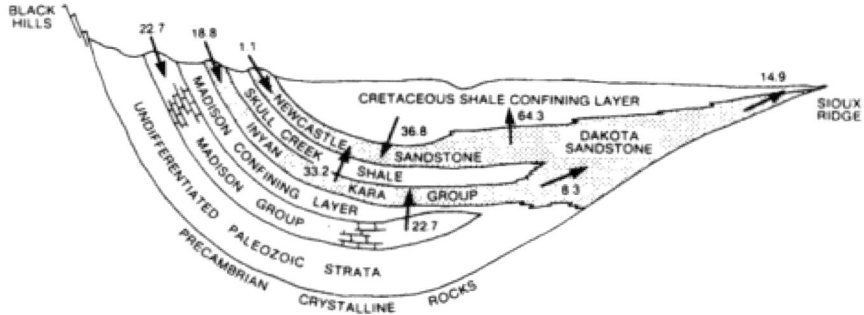

Figure 1.5
Idealized cross section of the Dakota flow system in South Dakota in which the quantities of flow through various layers are indicated. The numbers are integrated over the entire state of South Dakota (from Bredehoeft et al., 1983).

Flow in most large mature sedimentary basins within the stable continental craton is generally thought to be driven by topographic relief. Early in the development of sedimentary basins both compaction and diagenesis may also be important driving mechanisms. In most sedimentary basins significant quantities of flow move vertically across confining layers (cap rocks). Low-permeability layers retard flow and sometimes cause large decreases in hydraulic head. Some changes in permeability are compensated for by the large areas through which flow, especially vertical flow, can occur in a regional flow system. In South Dakota, for example, approximately 50 percent of the recharge and discharge to and from the Dakota Sandstone aquifer flows through the overlying Pierre Shale, a shale that is often thought to be "impermeable." This phenomenon is typical of flow in many if not most sedimentary basins.

If one were to raise the pore pressure to levels at which hydraulic fractures of the rock are possible by simple topographic relief, there must be very high local or regional relief. Engelder (Chapter 9, this volume) examined fractures he believes are generated hydraulically in the Devonian Catskill sands of the northern Appalachians. Engelder suggests that the most plausible mechanism to create the necessary pore pressure is high mountainous relief in the Appalachians to the east during later Paleozoic time. This mechanism requires approximately 5 km of relief to generate the appropriate pore pressure but is often dismissed because the necessary relief is so high.

Tectonic Dilation and Compression

Changes in tectonic stress change the pore pressure. The effects are included in the more general form of the flow equation, Eq. (1.20), by the second and third terms on the right-hand side, which take into account changes in normal as well as shear stresses. The degree to which the pore pressure is increased by changes in stress depends on the rate at which flow can dissipate the pressure.

Tectonic dilation, as we have defined it, takes a number of forms. Perhaps the most universal is pressure solution, which gradually over time reduces the porosity of the rock and compresses the fluids within the pore space. Nur and Walder (Chapter 7, this volume) argue that this is a universal phenomenon that can lead to high pore pressure when the process operates. Pressure solution is most commonly observed in quartz sandstones, but it is not restricted in rock type.

Various tectonic strains lead to a volume decrease of the host rock. Usually the volume strain rate is small. Currently along the San Andreas Fault in central California the rate of shear strain is on the order of 10^{-6} per year. This rate is accompanied by a volume strain of approximately 10^{-8} per year, two orders of magnitude smaller but capable of producing high pore pressure as long as the permeability of the rocks is small enough. Unfortunately, in active tectonic areas such as along the San Andreas Fault, we have no direct observations of pore pressure below 2 to 3 km. While it is clear from theoretical considerations that active rock deformation causes pore pressure changes, the data to demonstrate these effects are limited.

Oliver (Chapter 8, this volume), in perhaps the most speculative paper of this volume, suggests that volume strain associated with tectonic deformation between plates has caused fluid hydrocarbons to flow laterally into reservoirs within more stable areas of the continents. Vrolijk and Myers (Chapter 10, this volume) investigated the Kodiak accretionary complex and suggest that the high pore pressure there is the result of active subduction.

Once tectonic strain ceases, any pore pressure change caused by the strain becomes a transient phenomenon that will be dissipated by flow over time. However, if the rocks are of low permeability, such transient effects will not dissipate rapidly; indeed, in some instances of very low permeability material, these transient effects may persist for periods of geologic interest (Figure 1.6) (see Bredehoeft and Hanshaw, 1968; Hanshaw and Bredehoeft, 1968).

Diagenesis

To reconcile the competing ideas of high crustal permeabilities and associated hydrostatic fluid pressures with evidence for hydraulically sealed low-permeability rocks with elevated fluid pressures, Nur and Walder (Chapter 7, this volume) propose a time-dependent process to relate fluid pressure, flow pathways, and fluid volumes. In their model crustal porosity, permeability, and hence fluid pressures are in general time-dependent due to the gradual closure of crustal pore space via heating, sealing, and inelastic deformation. Under certain circumstances this process will lead to local drying out of the crust. Under other circumstances in which the pore fluid cannot escape fast enough, pore pressure will build-up, leading perhaps to natural hydraulic fracturing, fluid release, pore pressure drop, and resealing of the system. If there is an adequate supply of fluid, this process could repeat leading to the intriguing possibility of cyclic episodes of pore pressure buildup and natural hydraulic fracturing. This model involving coupling of hydrological and mechanical processes needs to be rigorously evaluated with in situ observations and measurements.

A number of young geologic basins that are actively receiving sediments have anomalously high pore pressures. Several such basins—the Gulf Coast Basin in the United States (Dickinson, 1953) and the Caspian Basin in the Soviet Union, to name two—have pore pressures that approach the total weight of the overburden (Figure 1.7).

The simplest mechanism to rate such high pore pressure is one in which low-permeability sediments are deposited in the basin at rates sufficiently high that consolidation cannot keep pace. The effective stress law, Eq. (1.17), implies that the total stress can be decomposed into two components—the effective stress (the so-called grain-to-grain stress) and the pore pressure. If the rate of loading

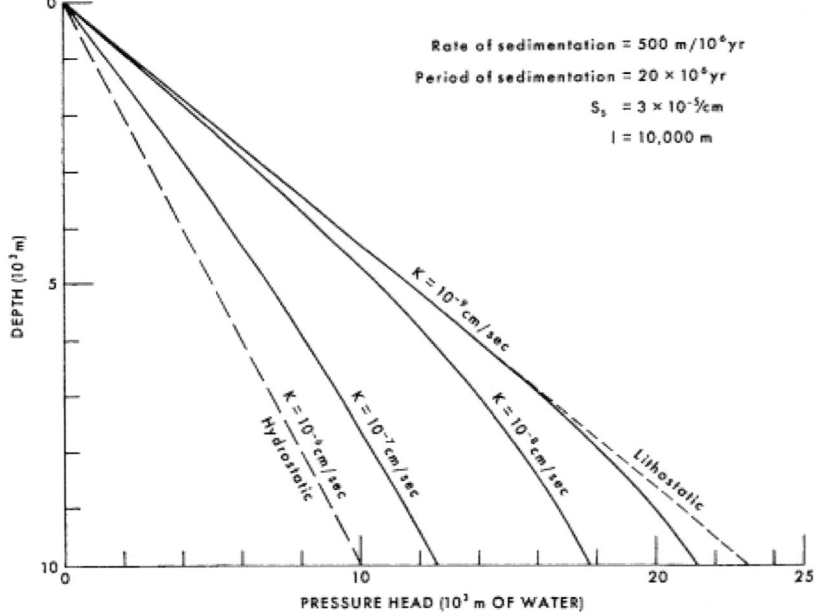

Figure 1.6
Calculated pore pressure versus depth profiles for a subsiding basin such as the Gulf Coast. The model used for these calculations considers only purely mechanical compaction (from Bredehoeft and Hanshaw, 1968).

caused by sedimentation in the basin exceeds the rate at which the pore pressure can dissipate through the expulsion of fluid, then pore pressures that approach the lithostatic load will result [this process is described mathematically by the flow equation, Eq. (1.20)]. Under these conditions there is little or no effective stress, and little or no consolidation takes place.

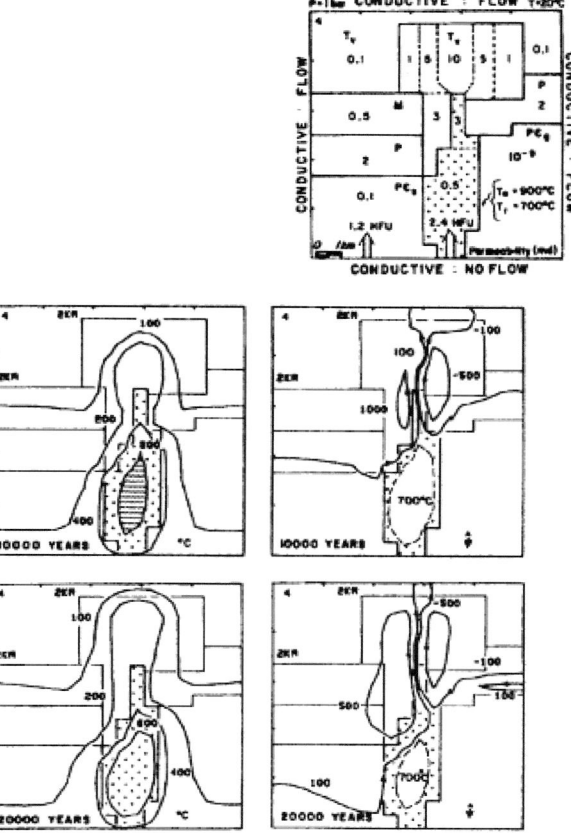

Figure 1.7
Convective cells established in the groundwater around a hot igneous intrusion (from Norton, 1982).

The simple model of rapid sedimentation can account for abnormally high pore pressures such as exist in the Gulf Coast Basin (Bredehoeft and Hanshaw, 1968). There are, however, other complicating processes at work here. Montmorillonite changes to illite with a release of free water from the clay structure at approximately the same depth as the first occurrence of the anomalous pore pressure (Burst, 1969). In addition, there may be a thermal generation of high pore pressure caused by somewhat higher thermal gradients (Sharp and Domenico, 1976).

In contrast to the Gulf Coast Basin, the Caspian Basin is subsiding at a rate of 1100 m per million years. The maximum thickness of accumulated sediments is 25 km. The basin has a very low thermal gradient, 16°C/km; temperatures are only approximately 110°C at 6000 m. There is no change of montmorillonite to illite to depths of 6 km, but there are substantial overpressures. The south Caspian Basin is at least one instance where simple mechanical loading at a high rate can account for high pore pressures.

The extent to which the simple sedimentation-consolidation model is the appropriate mechanism for generating high pore pressure depends on the weight of the overburden. High pore pressure in a subsiding basin is a dynamic process that requires active loading to be maintained. Once subsidence and sedimentation cease, the high pore pressure becomes a transient within the system that will dissipate with time. The rate at which the anomalous pore pressure dissipates depends on the hydraulic diffusivity of the deposits and the boundary conditions (Hanshaw and Bredehoeft, 1968).

Geothermal Systems

Fluid density is a function of temperature, pressure, and concentration of solute. Perturbations in any of these fields induce a buoyancy effect on the fluid and consequently cause the fluid to circulate, but the most common cause is temperature changes. Increases in temperature also reduce the fluid viscosity, making it easier for the fluid to flow [see Darcy's Law, Eqs. (1.1) and (1.2)].

Fluid density changes are incorporated into the flow equations through the Boussinesq approximation (Boussinesq, 1903; Rayleigh, 1916) in which the density is considered to be a function of the ambient, ρ_0, and a small density perturbation, dr:

$$\rho = \rho_0 + \delta\rho, \qquad (1.31)$$

where the perturtbation density is in turn a function of temperature, pressure, and solute content. This linear approximation to the buoyancy force permits the highly nonlinear variations in the density of supercritical fluid to be incorporated into the driving force field that generates convective flows.

As Eq. (1.24) indicates, the pore pressure is also strongly affected by changes in temperature and pore volume caused by the deposition or dissolution of mineral phases. The magnitude of the change in pore pressure induced by such changes is a function of the relative rate of change with respect to the rate at which groundwater flow dissipates the pressure increases. The rate of flow, in turn, depends on the permeability of the material. If the permeability is sufficiently high, flow occurs away from the density perturbation and mitigates the pore pressure increase. On the other hand, if the permeability of the material is very low, pore pressure can increase until it exceeds the effective strength of the pore wall.

The introduction into the crust of a source of heat such as an igneous intrusive body creates a heat engine for the groundwater system, causing flow. Several investigators have looked at fluid circulation associated with igneous plutons; more recent attempts to analyze the flow system have resulted in numerical models that solve the appropriate coupled partial differential equations (Cathles, 1977; Norton and Knight, 1977; Faust and Mercer, 1979a,b; Norton, 1984). Norton (Chapter 2, this volume) summarizes results associated with fluid variation in the near-field region around magma bodies. Convection cells are set up in the groundwater flow around the intrusive body. Of interest is how the convecting groundwater might redistribute metals and ore-forming fluids emanating from the igneous body. In other instances the convecting groundwater leaches metals from the surrounding country rocks and crystallized portions of the intrusion and redistributes them.

Large pore pressure pulses can be generated by the introduction of hot, or molten, intrusions. Knapp and Knight (1977) showed that it is possible to get pore pressure increases as large as 20 bars/°C even in regions of rather low temperature and pressure simply as a consequence of the properties of water. Palciauskas and Domenico (1982) reexamined the problem taking into account the compressibility of the rocks; they suggest that these pulses in pressure can range from 5 to 10 bars/°C in low-permeability rocks. In either case the change in pore pressure associated with heating is potentially enormous.

Pore pressure increases are limited by the stress at which failure is the hydraulic fracture in which the pore pressure increases to the point where the rock fails. Failure can occur either as shear or tension. If the failure occurs in tension, it is an "hydraulic fracture" (Hubbert and Willis, 1957). At the point at which the rock fails, a fracture is created. The energy release associated with these fractures accounts for the microseismic noise noted in active geothermal areas (Knapp and Knight, 1977; Palciauskas and Domenico, 1980). Thermal shock may also account for the seismic noise.

Fractures increase the permeability of the rock mass and allow high pore pressure to dissipate rapidly. The stress conditions at which failure will occur limit the maximum pore pressure that can be sustained. Many of the fractures are associated with vein-forming ore deposits. Titley (Chapter 3, this volume) discusses the sequences of fracturing associated with ore deposits at Sierrita, Arizona. Near the critical point, water has a very steep pressure-density relationship. Whether fractures fill with veinforming crystalline deposits depends on this thermodynamic behavior in the region near the critical point.

Moving groundwater is an efficient mechanism to transport heat introduced by an igneous body. Convective heat transport is much more efficient in transporting heat than pure conduction [see Eq. (1.13)]. The rate of cooling of the igneous body strongly depends on the rate of heat dissipation by circulating groundwater (Cathles, 1977; Norton and Knight, 1977).

Fossil geothermal systems often constitute hydrothermal ore deposits. In these systems ore has been emplaced by the moving fluids (Norton, 1984). It seems possible that many vein-type ore deposits are formed in hydraulic fractures held open by the pore pressure during mineral deposition. Most sills and dikes seem also to be simple hydraulic fractures into which molten igneous rock was injected. Hubbert and Willis (1957) discuss the basic physics of hydraulic fracturing; their explanation is still the seminal work on the problem.

Geothermal power development exploits a concentrated heat source in the crust. Whether commercial development of such a heat source is feasible depends almost entirely on whether the permeability of the hot rock is high enough to allow enough groundwater flow to recover the heat efficiently (Donaldson, 1982). Heat transport in a geothermal system depends on flowing groundwater (or steam). Geothermal reservoir engineering today involves numerical simulation of the reservoir's performance, which requires that the coupled partial differential equations be solved for the boundary conditions of interest.

Source of Fluids

What are the sources of fluids in the crust? Although fluids within a few thousand meters of the Earth's surface are likely to be derived from precipitation, the source of fluids at deeper crustal levels is problematic. Fluids released from crystallizing magmas must contribute some fluid to the crustal fluid reservoir. Rocks that are heated as they are buried to deeper levels, or that are in the vicinity of intruding magma bodies, will also contribute fluids as they recrystallize to new mineral assemblages that are more compatible with their new thermal and pressure environment. Finally, fluids may be directly released from the mantle to the overlying crust during "degassing" events. The relative proportions of these sources in any particular region have not been well established, although some isotopic studies have been undertaken to evaluate the relative roles of these processes.

Hydration or dehydration of minerals changes the fluid mass in the pores and will change the pore pressure. Such an addition or removal of fluid is represented as an appropriate source of sink terms in the general set of equations given above. Depending on the volume change associated with the dehydration, a source of fluids can increase the pore pressure. Perhaps the most commonly discussed example is the montmorillonite-illite transformation that Burst (1969) first suggested as the explanation for high pore pressure in the Gulf Coast. Various higher temperature metamorphic reactions release free water. Walther (Chapter 4, this volume) discusses fluid dynamics during progressive metamorphism.

The magnitude of the fluid-pressure change that accompanies such a reaction is rate dependent. If fluid is generated at such a rate that it cannot flow readily away from the source, then fluid-pressure will build. If the permeability of the surrounding host rock is sufficiently small, the buildup in fluid pressure may be quite large.

One other potential source of fluid flow into the crust is mantle degassing. Rubey (1951) presented the case that the volatiles associated with the Earth's surface were not present early in the life of the planet and have accumulated with time. Perhaps the most commonly accepted hypothesis is that the volatiles have been degassed from the mantle. There are various arguments about this topic. Gold (1979) and Gold and Soter (1985) argue that the basic composition is methane; numerous other workers believe that the principal carbon-containing volatile is carbon dioxide. Barnes *et al.* (1978) systematically mapped the distribution of carbon dioxide springs around the Earth and showed that they are almost exclusively associated with active tectonic areas. Irwin and Barnes (1975) suggest that carbon dioxide could give rise to high pore pressures. Bredehoeft and Ingebritsen (Chapter 11, this volume) examine the question of whether the current suggested rates of carbon dioxide outgassing could give rise to high pore pressure and conclude that it is possible; however, the permeabilities of the host rocks would have to be quite low.

SUMMARY

The central role of H_2O-rich fluids in determining the dynamic conditions in the Earth's crust is apparent in the repeated occurrence of fluid properties in all of the transport equations. This symbolic depiction of the processes shows not only the influence of processes on one another but also that this coupling condition is a consequence of the presence of an often sparse but essential occurrence of water in the systems. Each chapter that follows demonstrates that water is an active agent of the mechanical, chemical, and thermal processes that control the tectonic regimes of the crust.

Recognition of the role of water as the material that controls the extent of coupling among processes shows promise of reducing the magnitude of the analytical problem to one that focuses on the controlling links in the system.

References

Aagaard, P., and H. C. Helgeson (1982). Thermodynamic and kinetic constraints on reaction rates among minerals and aqueous solutions. I. Theoretical considerations, *American Journal of Science 282*, 237-285.

Anderson, M.P. (1984). Movement of contaminants in groundwater: Groundwater transport—advection and dispersion, in *Groundwater Contamination*, Studies in Geophysics, National Research Council, National Academy Press, Washington, D.C., pp. 37-45.

Barnes, I., W. P. Irwin, and D. E. White (1978). Global distribution of carbon dioxide discharges and major zones of seismicity, *U.S. Geological Survey Water Resources Investigation 78-39*, Open File Report, 12 pp.

Bear, J. (1972). *Dynamics of Fluids in Porous Media*, Elsevier, New York, 764 pp.

Biot, M. A. (1955). Theory of elasticity and consolidation for a porous anisotropic solid, *Journal of Applied Physics 26*, 182-185.

Biot, M. A. (1956). Thermoelasticity and irreversible thermodynamics, *Journal of Applied Physics 27*, 240-253.

Bird, R. B., W. E. Stewart, and E. N. Lightfoot (1960). *Transport Phenomena*, John Wiley & Son, Inc., New York, 780 pp.

Boussinesq (1903). *Theorie Analytique de la Chaleur*, t. ii.

Brace, W. F. (1971). Resistivity of saturated crustal rocks to 40 km based on laboratory measurements, in *The Structure and Physical Properties of the Earth's Crust*, L G. Heacock, ed., American Geophysical Union Monograph 14, Washington, D.C., pp. 243-255.

Brace, W. F. (1980). Permeability of crystalline and argillaceous rocks: Status and problems, *International Journal of Rock Mechanics in Mineral Science and Geomechanical Abstracts 17*, 876-893.

Bredehoeft, J. D., and B. B. Hanshaw (1968). On the maintenance of anomalous fluid pressure. I. Thick sedimentary sequences, *Geological Society of America Bulletin 79*, 1097-1106.

Bredehoeft, J. D., C. E. Neuzil, and P. C. D. Milly (1983). Regional flow in the Dakota aquifer, *U.S. Geological Survey Water Supply Paper 2237*, pp. 1-45.

Burst, J. F. (1969). Diagenesis of Gulf Coast clayey sediments and its possible relation to petroleum migration, *American Association of Petroleum Geologists Bulletin 53*, 73-79.

Cathles, L. M. (1977). An analysis of the cooling intrusives by ground-water convection which includes boiling, *Economic Geology 72*, 804-826.

Cherry, J. A., R. W. Gillham, and J. F. Barker (1984). Contaminants in groundwater: Chemical processes, in *Groundwater Contamination*, Studies in Geophysics, National Research Council, National Academy Press, Washington, D.C., pp. 46-64.

Darcy, H. (1856). Determination of the laws of the flow of water through sand (translated from the French, 1983), in *Physical*

Hydrogeology. R. A. Freeze and W. Back, eds., Hutchinson Ross, Stroudsburg, Pa., pp. 14-19.

Dickinson, G. (1953). Geological aspects of abnormal reservoir pressures in Gulf Coast Louisiana, *American Association of Petroleum Geologists Bulletin 37*, 410-432.

Domenico, P. A. (1977). Transport phenomena in chemical rate processes in sediments, *Annual Reviews of Earth and Planetary Sciences 5*, 287-317.

Donaldson, I. G. (1982). Heat and mass circulation in geothermal processes in sediments, *Annual Reviews of Earth and Planetary Sciences 10*, 155-164.

Faust, C. R., and J. W. Mercer (1979a). Geothermal reservoir simulation: I. Mathematical models for liquid- and vapor-dominated hydrothermal systems, *Water Resources Research 15*, 23-30.

Faust, C. R., and J. W. Mercer (1979b). Geothermal reservoir simulation: II. Numerical solution techniques for liquid- and vapor-dominated hydrothermal systems, *Water Resources Research 15*, 31-46.

Freeze, R. A., and J. A. Cherry (1979). *Groundwater*, Prentice-Hall, Inc., Englewood Cliffs, N.J., 604 pp.

Freeze, R. A., and P. A. Witherspoon (1967). Theoretical analysis of regional groundwater flow: Effect of water-table configuration and subsurface permeability variation, *Water Resources Research 3*, 623-634.

Gold, T. (1979). Terrestrial sources of carbon and earthquake outgassing, *Journal of Petroleum Geology 1*, 3-19.

Gold, T., and S. Soter (1985). Fluid ascent through the lithosphere and its relation to earthquakes, *PAGEOPH*, 492-530.

Grisak, G. E., and J. F. Pickens (1981). An analytic solution for solute transport through fractured media with matrix diffusion, *Journal of Hydrology 52*, 47-57.

Hanshaw, B. B., and J. D. Bredehoeft (1968). On the maintenance of anomalous fluid pressures: II. Source layer at depth, *Geological Society of America Bulletin 77*, 1107-1122.

Helgeson, H. C. (1979). Mass transfer among minerals and hydrothermal solutions, in *Geochemistry of Hydrothermal Ore Deposits*, 2nd ed., H. L. Barnes, ed., John Wiley & Sons, New York, pp. 568-610.

Helgeson, H. C., and D. H. Kirkham (1974a). Theoretical prediction of the thermodynamic behavior of aqueous electrolytes at high pressures and temperatures: I. Summary of the thermodynamic/electrostatic properties of the solvent, *American Journal of Science 274*, 1089-1198.

Helgeson, H. C., and D. H. Kirkham (1974b). Theoretical prediction of the thermodynamic behavior of aqueous electrolytes at high pressures and temperatures: II. Debye-Huckel parameters for activity coefficients and relative partial molar properties, *American Journal of Science 274*, 1198-1261.

Helgeson, H. C., T. H. Brown, A. Nigrini, and T. A. Jones (1970). Calculation of mass transfer in geochemical processes involving aqueous solutions, *Geochimica et Cosmochimica Acta 34*, 569-592.

Hsieh, P. A., and J. D. Bredehoeft (1981). A reservoir analysis of the Denver earthquakes: A case of induced seismicity, *Journal of Geophysical Research 86*, 903-920.

Hubbert, M. K. (1940). The theory of ground-water motion, *Journal of Geology 48*, 785-944.

Hubbert, M. K., and W. W. Rubey (1959). Role of fluid pressure in mechanics of overthrust faulting, *Geological Society of America Bulletin 70*, 115-166.

Hubbert, M. K., and D. G. Willis (1957). Mechanics of hydraulic fracturing, *Transactions of the AIME 210*, 153-166.

Irwin, W. P., and I. Barnes (1975). Effect of geologic structure and metamorphic fluids on seismic behavior of the San Andreas Fault system in central and northern California, *Geology 1*, 713-716.

Jacob, C. E. (1940). On the flow of water in an elastic artesian aquifer, *EOS Transactions American Geophysical Union 21*, 574-586.

Knapp, R. B., and J. E. Knight (1977). Differential thermal expansion of pore fluids: Fracture propagation and microearthquake production in hot pluton environments, *Journal of Geophysical Research 82*, 2515-2522.

Lichtner, P. C. (1986). Continuum model for the simultaneous chemical reactions and mass transport in hydrothermal systems, *Geochimica et Cosmochimica Acta 49*, 779-800.

Norton, D. (1982). Fluid and heat transport phenomena typical of copper-bearing pluton environments, southeastern Arizona, in *Advances in Geology of Porphryr Copper Deposits*, Southwestern North America, S. R. Titley, ed., University of Arizona Press, Tucson, pp. 59-72.

Norton, D. (1984). A theory of hydrothermal systems, *Annual Reviews of Earth and Planetary Sciences 12*, 155-177.

Norton, D., and J. Knight (1977). Transport phenomena in hydrothermal systems cooling plutons, *American Journal of Science 277*, 937-981.

Norton, D., and H. P. Taylor, Jr. (1979). Quantitative simulation of the hydrothermal systems of crystallizing magmas on the basis of transport theory and oxygen isotope data: An analysis of the Skaergaard Intrusion, *Journal of Petrology 20*, 421-486.

Nut, A., and J. D. Byerlee (1971). An exact effective stress law for elastic deformation of rocks with fluids, *Journal of Geophysical Research 76*, 6414-6419.

Palciauskas, V. V., and P. A. Domenico (1980). Microfracture development in compacting sediments: Relations to hydrocarbon-maturation kinetics, *American Association of Petroleum Geologists Bulletin 64*, 927-937.

Palciauskas, V. V., and P. A. Domenico (1982). Characterization of drained and undrained response of thermally loaded repository rocks, *Water Resources Research 18*, 281-290.

Rayleigh, Lord (1916). On convection currents in a horizontal layer of fluid, when the higher temperature is on the underside, *Physical Magazine Series 6(32)*, 529-545.

Rubey, W. W. (1951). Geologic history of the sea, *Geological Society of America Bulletin 87*, 1111-1148.

Sharp, J. M., Jr., and P. A. Domenico (1976). Energy transport in thick sequences of compacting sediment, *Geological Society of America Bulletin 87*, 390-400.

Skempton, A. W. (1954). The pore pressure coefficients A and B, *Geotechnique 4*, 143-147.

Slattery, J. C. (1972). *Momentum, Energy, and Mass Transfer in Continua*, McGraw-Hill, New York, 679 pp.

Taylor, H. P., Jr. (1974). The application of oxygen and hydrogen isotope studies to problems of hydrothermal alteration and ore deposition, *Economic Geology 69*, 843-883.

Taylor, H. P., Jr. (1977). Water/rock interactions and the origin of H$_2$O in granitic batholiths, *Journal of the Geological Society of London 133*, 509-558.

Terzaghi, K. (1943). *Theoretical Soil Mechanics*, Wiley, New York, 510 pp.

Theis, C. V. (1935). The relation between the lowering of the piezometric surface and the rate and duration of discharge of a well using groundwater storage, *EOS Transactions American Geophysical Union 16*, 519-524.

Tóth, J. (1963). A theoretical analysis of ground-water flow in small drainage basins, *Journal of Geophysical Research 68*, 4795-4812.

Van der Kamp, G., and J. E. Gale (1983). Theory of Earth tide and barometric effects in porous formations with compressible grains, *Water Resources Research 19*, 538-544.

2

Pore Fluid Pressure Near Magma Chambers

DENIS L. NORTON
University of Arizona

INTRODUCTION

Fluid pressure in the Earth's crust is a function of variations in the regional stress, temperature, and composition of H_2O-rich fluids. Whereas in relatively quiescent geologic environments the effects of fluid pressure are subtle, its effects in magma environments can be easily identified (Knapp and Norton, 1981; Burton and Helgeson, 1983; Lantz, 1984; Norton, 1984). In the near-field regions of magmas the occurrence of high fluid pressures is evident from the functional form of the equation of state for the fluids; the distribution and geometry of fluid-filled pores; and relationships among the thermal, chemical, and mechanical processes. This chapter discusses processes in magma-hydrothermal systems to demonstrate the mechanisms involved and to show that in these environments fluid pressure plays an essential role in the generation and maintenance of percolation networks for hydrothermal fluid flow.

Extreme variations in fluid pressure in the near-field region of magmas are caused by sparse but significant amounts of H_2O-rich fluids that are ubiquitous in the host rocks and common in the magmas. Pore fluids typically found in the host rocks have large positive values of the isochoric coefficient of thermal pressure, whereas those in the magma have large negative values. Therefore, fluid pressure increases during the dissipation of thermal energy from magmas are a natural consequence of the cooling process, where temperature increases in the host and concurrently decreases in the magma. The resultant pressure increase in both domains generates large local stresses. Once the rock fails, fracture networks form that are continuous between the domains on either side of the pluton wall. These networks allow fluid flow and convective transport of thermal energy, thereby increasing the rate of pressure change in both environments.

Because the evidence for these conditions derives from both field observation and theory, the following discussion first reviews the conditions of ambient pressure in crustal environments and then examines the consequences of perturbing these conditions in the context of the distribution and geometry of pores.

PRESSURE AT DEPTH

Pressure conditions that exist prior to the imposition of a thermal perturbation on the crust are critical to predicting the magnitude and consequences of the subsequent fluid pressure increases. Pressure at the base of a column of rock that is composed of minerals and fluid-filled pore spaces is given by

$$P_{\text{total}} = P_0 + \hat{g} \int_{z_0}^{z} \rho_r (T, P, X) \, dz , \quad (2.1)$$

where $\hat{g}$ is the magnitude of the gravitational vector and $\rho_r(T, P, X)$ is the volume-averaged density of the rock in the column. The term *rock* refers to minerals plus pore space, and all pore space is assumed to be filled with fluid. Because density is a function of the material composition and of temperature and pressure, an explicit function is needed to integrate Eq. (2.1). It expanded into a statement that independently accounts for the density of solid and fluid phases:

$$P_{total} = P_0 + \hat{g} \int_{z_0}^{z} \left[\phi_m \rho_m + \phi_f \rho_f (T, P, X) \right] dz , \quad (2.2)$$

where ϕ_m is the total volume fraction of mineral phases, ρ_m is the average density of the mineral assemblage, ϕ_f is the volume fraction of the fluid phase, and $\rho_f(T, P, X)$ is the temperature-, pressure-, and composition-dependent density of the phase. Mineral densities are nearly constant relative to fluid density over the state conditions commonly encountered in the crust. Consequently, the pressure contribution from the minerals can be removed from the integral:

$$P_{total} = P_0 + \hat{g} \left[\phi_m \rho_m + \int_{z_0}^{z} \phi_f \rho_f (T, P, X) \right] dz . \quad (2.3)$$

Total pressure can therefore be determined by expressing the density function as an equation of state using relations like those reported by Helgeson and Kirkham (1977) and Johnson and Norton (1989).

However, the crust is generally in a state of heterogeneous stress that can be represented as a stress tensor, $\ddot{T}$:

$$\ddot{T} = \begin{vmatrix} \sigma_{xx} & \sigma_{xy} & \sigma_{xz} \\ \sigma_{yx} & \sigma_{yy} & \sigma_{yz} \\ \sigma_{zx} & \sigma_{zy} & \sigma_{zz} \end{vmatrix} , \quad (2.4)$$

where σ_{ij} refers to the stress exerted along the ith axis normal to the jth direction. The resultant pressure caused by these stresses is equivalent to the trace of the stress tensor and is called the mean pressure, P_m:

$$P_m = \frac{\sigma_{xx} + \sigma_{yy} + \sigma_{zz}}{3} . \quad (2.5)$$

The mean (P_m) and total (P_{total}) pressures from Eq. (2.3) are equal only in conditions where $\sigma_{xx} = \sigma_{yy} = \sigma_{zz}$. To demonstrate the role of fluid pressure in deformation, we will assume that the stress field is homogeneous.

Pressure conditions actually encountered in the subsurface are unlikely to correlate closely with those values derived from the integrations using extreme values of ϕ_m in Eq. (2.3). Neither (ϕ_m 0 for $0 < z < z_{max}$) nor hydrostatic ($\phi_m > 0$; $\phi_f = 1$ for $0 < z < z$) pressure conditions are likely because pressure generally depends on the geologic history of the lithologic units.

As an example of the effect of dynamic loading history that can alter the pressure from the conditions just mentioned, computations that describe the burial of a fluid filled pore are reviewed in this chapter; the details of this experiment can be found in Knapp and Knight (1977). The loading process can be thought of as diagenesis of a fluid packet, first isolated from the other pores in the rock by near-surface compaction and then subjected to increases in temperature and confining pressure as it subsides to greater depths in the basin.

Even though the fluid in this situation may constitute a small fraction of the rock, typically less than 10 wt.%, its density variation in response to changes in state conditions can have a large effect on local pressure conditions. Because temperature and pressure are functions of depth, the total differential of fluid density with respect to depth is

$$\frac{d\rho_f}{dz} = \left(\frac{\partial \rho_f}{\partial T} \right)_{PX} \frac{dT}{dz} + \left(\frac{\partial \rho_f}{\partial P} \right)_{TX} \frac{dP}{dz}$$

$$+ \sum_i^I \left(\frac{\partial \rho_f}{\partial X_i} \right)_{TPX_i} \frac{dX_i}{dz} , \quad (2.6)$$

where the partial derivatives are derived from the equation of state for the fluid, $f(P, T, a, X) = 0$, and the derivatives of pressure and temperature with depth are independent quantities that must be defined from solutions to the heat transfer equations. The partial derivative of composition, on the far right of Eq. (2.6), is significant in all natural environments. However, an important aspect of fluid pressure variations can be demonstrated with a single-component fluid. Therefore, the following discussion focuses on the single-component H_2O system and omits the compositional variation in Eq. (2.6).

Density variation with depth for a fluid in the pure system is given by

$$\frac{d\rho_f}{dz} = \left(\frac{\partial \rho_f}{\partial T} \right)_P \frac{dT}{dz} + \left(\frac{\partial \rho_f}{\partial P} \right)_T \frac{dP}{dz} , \quad (2.7)$$

where the dependent partial derivatives can be replaced with the intrinsic properties of the fluid—the isobaric coefficient of thermal expansion, α_f:

$$\alpha_f = -\frac{1}{\rho_f} \left(\frac{\partial \rho}{\partial T} \right)_P , \quad (2.8)$$

and the isothermal coefficient of compressibility, β_f:

$$\beta_f = -\frac{1}{\rho_f} \left(\frac{\partial \rho}{\partial P} \right)_T . \quad (2.9)$$

Substitution of these quantities into Eq. (2.7) gives

$$\frac{d\rho_f}{dz} = -\rho_f \alpha_f \frac{dT}{dz} + \rho_f \beta_f \frac{dP}{dz}. \quad (2.10)$$

If the fluid-filled pore is assumed to remain at a constant volume, the variation of pressure within the pore is only a function of the expansivity-to-compressibility ratio and the thermal gradient:

$$\frac{dP}{dz} = \frac{\alpha_f}{\beta_f}\frac{dT}{dz}, \quad (2.11)$$

where α_f/β_f is the isochoric coefficient of thermal pressure. It varies from 5 to 20 bars/°C for ranges in state conditions commonly found in subsiding sedimentary basins (Figure 2.1).

Knapp and Knight (1977) found pressure increases within the pore to be large at shallow depths even for modest thermal gradients. They also found that only for thermal gradients of less than 10°C/km will the pressure within the constant-volume pore remain less than the mean confining pressure, P_m, as subsidence occurs; for larger thermal gradients the fluid pressure increases to values much greater than the confining stress. This overpressure occurs at depths that depend primarily on the magnitude of the thermal gradient. Knapp and Knight's computation showed clearly that fluid pressure can vary with depth if the pore geometry is poorly interconnected and that the variation is a function of the thermal pressure coefficient of the fluid. The upper limit of the fluid pressure is a function of the pore wall strength and failure process.

Prior to examining the consequences of failure on fluid pressure, the effect of a thermal perturbation imposed on the same pore space as described above is examined.

CHANGE IN FLUID PRESSURE

Thermal perturbations imposed on a porous rock generate differential stresses because of large differences in the constitutive properties between minerals and fluids. The physical significance of this difference in response to thermal changes is that small thermal changes in the crust can cause large differences between values of mean pressure and fluid pressure. For a particular thermal state in the crust the hydrostatic and mean or total confining pressure can be defined from Eqs. (2.3) and (2.5). Perturbations from this state caused by changes in the thermal flux are now examined.

Consider the pressure changes that occur within an isolated fluid-filled pore. Assume that the fluid pressure exerted on the pore wall is initially in equilibrium with the mean confining pressure:

$$P_f = P_m. \quad (2.12)$$

The time derivative of fluid pressure increase as a result of temperature and pore volume changes is

$$\frac{dP_f}{dt} = \left(\frac{\partial P_f}{\partial T}\right)_V \frac{dT}{dt} + \left(\frac{\partial P_f}{\partial V}\right)_T \frac{dV}{dt}, \quad (2.13)$$

where V is the pore volume. Because the pores are assumed to be filled with a single-phase fluid, this volume is equivalent to $1/\alpha_f$. Again, consider a constant volume process, $dV/dt = 0$, consistent with regional strain rates that are much smaller than those caused by the local change in temperature and with the situation in which a local increase in fluid pressure does not dilate the pore at a significant rate. Under this condition the change in internal pressure as a function of time is given by

$$\frac{dP_f}{dt} = \left(\frac{\partial P_f}{\partial T}\right)_V \frac{dT}{dt}, \quad (2.14)$$

where $(\partial P_f/\partial T)_V$ is the ratio of the isobaric coefficient of thermal expansion to the isothermal compressibility, α_f/β_f (Figure 2.1). Notice that the time derivative of the pressure is analogous to its spatial derivative in Eq. (2.12). As a consequence of α_f/β_f pressure increases of several bars per degree centigrade can occur over relatively short times

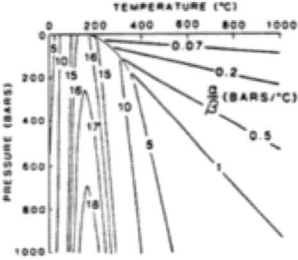

Figure 2.1
The isochoric coefficient of thermal pressure, β_f/β_f, computed with the equation of state for the H_2O-system (from Johnson and Norton, 1989).

in the near-field region of magma bodies, where the change in temperature with time is rapid.

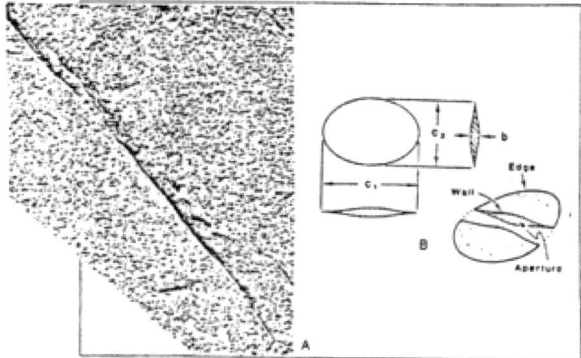

Figure 2.2
(A) Intersect of fracture with topographic surface. Slit-like form is typical of fracture topology at all scales of observation (Norton, 1987). (B) Idealized fracture form based on dislocation theory (from Mavko and Nut, 1978; Norton, unpublished field data).

The magnitude of pressure increases associated with this situation and the burial computation in the previous section are likely to exceed the strength of the walls of intergranular pore space or of the mineral grains themselves. The conditions for failure depend on the mode of occurrence of the fluid in the rocks, particularly on the shape of the pore space.

FLUIDS IN ROCKS

Fluids in rocks are seldom directly observed, but their presence is inferred from the presence of pore space. Although the mode of occurrence of the fluid phase is traditionally expressed in terms of the volume or porosity, the geometry of the fluid and its distribution with respect to the mineral phases are of equal importance to considerations of transport processes, particularly the consequences of fluid pressure changes. Studies of the mode of occurrence of fluid have revealed that the shape of the pore space is determined by properties of the fluid that fills them and by the local conditions of stress.

The total porosity of crystalline igneous rocks ranges from a few percent to less than a fraction of a percent, and of this total pore space the ever-present fractures in these rocks contribute only a very small fraction of pore space to the total, circa 10^{-3} (Snow, 1970; Norton and Knapp, 1977). Igneous rocks commonly have a relatively large permeability, $>10^{-14}$ cm^2, during their active thermal history, but only a small porosity, or total fluid fraction, is associated with the flow channels. This is because the flow channels are elongate fractures with apertures on the order of a few hundred microns (Norton and Villas, 1977). The redistribution of only a small fraction of the fluid contained in the total pore space of such rocks into fractures can significantly affect the rock permeability and the mechanism of transport through it.

The slit-like nature of fluid-filled pores (Figure 2.2) is caused by the active deformation of their host rock within a heterogeneous stress field and the redistribution of a portion of the fluid phase into fractures whose orientation is a function of the stress trajectories. Once formed, these fractures are extremely sensitive to the differential between the mean confining pressure and the fluid pressure within the fracture. They are therefore delicate indicators of fluid pressure conditions because the mean confining pressure is relatively constant over long times, whereas the fluid within the fractures is sensitive to local changes in temperature.

FAILURE CRITERIA

The small finite strength of the pore wall implies that for the initial condition of $P_f = P_m$ only a small pressure increase is necessary to reach the yield strength of the wall. The strain rate associated with this deformation is strictly a function of the local time derivative of the temperature. The duration of the elastic deformation may range from a few months to years. But the failure of the wall and consequent fracture propagation is an instantaneous process, in which fluid expands irreversibly against a fixed pressure.

Extensive commentary on failure criteria exists in the

literature; the findings by Berbabe (1987) indicate that a simple failure law matches experimental data on fracture formation in crystalline rocks. This law describes the failure of the pore walls when the sum of the confining pressure and wall strength is exceeded by the internal fluid pressure:

$$P_c + \tau = P_f, \quad (2.15)$$

where τ is the tensile strength of the rock. This failure criterion oversimplifies the mechanical problem considerably but highlights the forces involved and accurately describes the transitions from one pore configuration to another.

The generation of pressures in excess of the mean confining pressure requires material strengths that will withstand the excess pressure. Because on a regional scale a rock has no effective strength, pressures in excess of P_m are not likely to develop over large portions of the crust. However, if the initial condition were one in which fluid in equilibrium with the local value of hydrostatic pressure was sealed into an isolated pore and then heated, the fluid pressure might increase several tens of bars up to the local confining pressure before failure would occur. Therefore, the following two failure conditions should be considered:

1. The condition in which the initial condition was $P_f \approx P_m$. A fluid-filled pore fails and the fluid expands against a pressure only slightly lower than the condition of failure.
2. The condition in which the initial condition was $P_f \approx P_{hydro}$. Failure may not occur until pressure increases to P_m or even slightly greater depending on the rock strength. Expansion may then occur against P_{hydro} if the newly formed fracture intersects a regime of hydrostatic pressure.

The main difference between the two situations is the amount of energy required to attain failure. The second situation requires more energy simply because the pressure increase required to attain failure is greater. In either of the above situations the volume increases associated with failure can be examined in terms of irreversible expansion against a constant pressure value, P_b.

VOLUME CHANGE CAUSED BY FAILURE

The volume of the newly formed fracture is a function of the expansivity of the fluid and the energy lost by viscose flow. Because fluid flows away from the breached pore wall a relatively short distance and for a limited time, viscose dissipation of energy is likely to be small. The rate of volume change in a fluid subjected to differential changes in temperature and pressure is given by the total differential of volume with respect to time:

$$\frac{dV}{dt} = \left(\frac{\partial V}{\partial T}\right)_P \frac{dT}{dt} + \left(\frac{\partial V}{\partial P}\right)_T \frac{dP}{dt}, \quad (2.16)$$

where the dependent partial derivatives can be abbreviated as

$$\alpha \equiv \frac{1}{V}\left(\frac{\partial V}{\partial T}\right)_P \quad (2.17)$$

and

$$\beta \equiv -\frac{1}{V}\left(\frac{\partial V}{\partial P}\right)_T. \quad (2.18)$$

Substituting these quantities into Eq. (2.16) gives an expression for volume change in terms of the fluid properties and the change in temperature and pressure:

$$\frac{dV}{dt} = V\alpha \frac{dT}{dt} - V\beta \frac{dP}{dt}. \quad (2.19)$$

Both the expansivity and the compressibility of an H_2O-rich fluid phase are much greater than those of minerals. Therefore, Eq. (2.19) can be written only in terms of the fluid properties without introducing significant errors in the analysis (Moskowitz and Norton, 1977):

$$\frac{dV}{dt} = V_f \alpha_f \frac{dT}{dt} - V_f \beta_f \frac{dP}{dt}. \quad (2.20)$$

The pressure differential is relatively small where the failure occurs at pressure slightly greater than the confining pressure, but where expansion is against hydrostatic pressure the change is substantial. Integration of Eq. (2.20) for a given pressure over fixed temperature limits gives

$$\int_{V_f^0}^{V_f} \frac{dV_f}{V_f} = \int_{T_0}^{T} \alpha_f(T)\,dT + \int_{P_0}^{P} \beta_f(P)\,dP \quad (2.21)$$

and

$$V_f = V_f^0 \exp\left[\int_{T_0}^{T} \alpha_f(T)\,dT + \int_{P_0}^{P} \beta_f(P)\,dP\right]. \quad (2.22)$$

The concomitant propagation of the fracture and the thermal expansion of the fluid as it is released by the pore wall failure result in a net increase in porosity that potentially augments the flow porosity. The porosity increase that results from the failure of pressurized pores can be computed from the equation of state of the fluid and the change in fluid volume defined by Eq. (2.18). The porosity, , contributed by the isolated pores in a region is the initial volume fraction of fluid in those pores:

$$\phi_f = \frac{V_f}{V_{total}} = \frac{V_f}{V_m + V_f} . \quad (2.23)$$

The initial fluid volume is a function of the initial porosity,

$$V_f = \phi V_{total} , \quad (2.24)$$

and the total volume of minerals in the rock is

$$V_f = (1 - \phi) V_{total} . \quad (2.25)$$

Therefore, the fluid porosity at a time following a thermal change can be expressed as a combination of Eqs. (2.23) and (2.17):

$$\phi_f = \frac{\phi_f^0 F(T)}{(1 - \phi_f^0) + \phi_f^0 F(T)} , \quad (2.26)$$

where the function, $F(T)$, in Eq. (2.26) is

$$F(T) = \exp\left[\int_{T_0}^{T} \alpha_f(T) \, dT + \int_{P_0}^{P} \beta_f(P) \, dP\right] . \quad (2.27)$$

Porosity increases in a super-exponential manner because of the exponential term and because α_f increases exponentially in the critical region of the H_2O system. Under near-critical conditions the porosity doubles in response to only a few degree increase in temperature (Figure 2.3).

Because a large portion of the total porosity in a crystalline rock is in the form of isolated porosity, circa 0.01 to 0.1, the thermal expansion of pore fluid could be quite effective in increasing the interconnected pore space, which constitutes the flow network in a rock and therefore directly affects the value of permeability. Furthermore, the flow porosity required for moderately large values of permeability is on the order of 10^{-4} to 10^{-5}. Conversion of even a small percentage of the isolated pores into oriented fractures can lead to enormous increases in permeability.

ENERGY REQUIRED FOR FRACTURE

The total energy available from a volatile-rich magmatic body is ≈200 cal/g. In regions where the alteration process produces hydrous phases, the exothermic heat can add an additional amount of up to 40 cal/g. Mafic and volatile-poor magmas contain more thermal energy than felsic magmas because of the large 100 cal/g heat of crystallization. The thermal energy converted to mechanical energy during the failure process described is only a few calories per gram (Knapp and Knight, 1977).

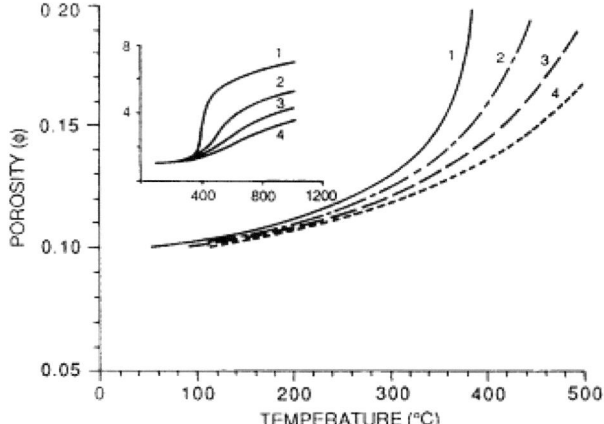

Figure 2.3
Maximum change in porosity as a function of temperature change for an isolated fluid-filled pore that propagates through the rock matrix on failure (Moskowitz and Norton, 1977).

Although the energy expended in a typical situation amounts to only a few calories per gram, the increase in permeability that can result from the failure is significant. This increase allows a small fraction of the pressure to be dissipated by viscose flow. However, the generation of an interconnected network over a broad region also disperses the fluid force field generated by the buoyancy forces. Therefore, the increase in permeability results in an increase in fluid velocity and consequently an increase in the advective transport of chemical components and heat.

ADVECTION AND PORE PRESSURE

Increases in rock permeability cause proportionate increases in fluid velocity. The fluid flow is driven by the pervasive buoyancy force field, which is an inevitable consequence of changes in heat flux, particularly those changes caused by the infiltration of magma. Therefore, when the synchronous propagation of fractures produces an interconnected network of flow channels, fluid motion ensues and augments the flux of heat and chemical components. Localized flow associated with the individual failure events also augments the flux but is less effective than the buoyancy force driven flow.

If local equilibrium between the fluid in an isolated pore and the minerals that form the pore wall is assumed, then at the time of pore failure and propagation of a fracture a gradient in fluid composition will be established along the interconnected fracture network because of gradients in state conditions. This compositional gradient in the fluid composition and the coinciding fluid velocity advect chemical components from one environment to another.

As an example of this process, consider a system in which the lithologic units can be represented by the system SiO_2-H_2O; they contain only quartz and an aqueous fluid locally in equilibrium with quartz. Chemical changes caused by advective flow in a system, where temperature, pressure, and composition gradients occur and where equilibrium prevails locally between quartz and fluid, are symbolically depicted by

$$\underset{\text{Fluid}}{\frac{\partial (\phi \Psi)_f^{SiO_2}}{\partial t}} + \underset{\text{Quartz}}{\frac{\partial (\phi \Psi)_{qtz}^{SiO_2}}{\partial t}} + \underset{\text{Advection}}{v_f \cdot \Psi_f^{SiO_2}} = 0, \quad (2.28)$$

where the relationship between the rate of change in quartz and the concentration of aqueous silica can be defined by the equilibrium equation:

$$SiO_2 = SiO_{2\,aqueous}. \quad (2.29)$$

The partial derivatives of $\Psi_f^{SiO_2}$ in the fluid phase with respect to time, t, and distance, l, can then be defined in terms of the standard state partial molal enthalpy, ΔH_r^0, and partial molal volume, ΔV_r^0, of the equilibrium reaction between quartz and aqueous silica, Eq. (2.29):

$$\frac{\partial \Psi_f^{SiO_2}}{\partial t} = \Psi_f^{SiO_2} \, 0.026 \left(\frac{\Delta H_r^0}{RT^2} \frac{dT}{dt} - \frac{\Delta V_r^0}{RT} \frac{dP}{dt} \right) \quad (2.30)$$

$$\frac{\partial \Psi_f^{SiO_2}}{\partial l} = \Psi_f^{SiO_2} \, 0.026 \left(\frac{\Delta H_r^0}{RT^2} \frac{dT}{dl} - \frac{\Delta V_r^0}{RT} \frac{dP}{dl} \right), \quad (2.31)$$

where the assumption has been made that the activity of aqueous silica is equal to its molality, and the concentration of silica in Eqs. (2.30) and (2.31) is in grams per cubic centimeter. The distance operator in the advection term in Eq. (2.31) is equated to the distance along the flow path, l. Therefore, the rate of change in the amount of quartz within a fracture is a function of both the thermodynamic properties of the reaction and the temporal and spatial derivatives of temperature and fluid pressure:

$$\underset{}{\frac{\partial \Psi_f^{SiO_2}}{\partial t}} = \underset{\text{Advection}}{-v_f \cdot \Psi_f^{SiO_2} \, 0.026} \left(\frac{\Delta H_r^0}{RT^2} \frac{dT}{dl} - \frac{\Delta V_r^0}{RT} \frac{dP}{dl} \right)$$

$$\underset{\text{Rate of change in fluid}}{} - \Psi_f^{SiO_2} \, 0.026 \left(\frac{\Delta H_r^0}{RT^2} \frac{dT}{dt} - \frac{\Delta V_r^0}{RT} \frac{dP}{dt} \right). \quad (2.32)$$

This relation demonstrates the coupling among the thermal, pressure, and fluid flow fields in which the deposition or dissolution of a mineral phase in the flow network can effectively change its continuity.

The manner in which the mineral actually blocks the flow channel is a consequence of detailed interaction between the local flow field within the fracture (Brown, 1987) and the local gradients in temperature and pressure. These interactions are unfortunately not predictable from the transport theory. However, there is ample evidence in the textures of veins to indicate that sealing of the flow channel actually occurs numerous times during the history of the thermal event.

SUMMARY

Field relations indicate that the processes that lead to the formation of fracture-controlled percolation networks and that fill these networks with vein minerals are related through an episodic series of events. The fractures that constitute the networks are slit-like features of limited extent; they occur in interconnected sets, and in any one environment there may be many such sets of different orientation and distinct chronology (see Titley, Chapter 3, this volume). The vein material within a fracture is pervaded with discordant contacts that suggest many discrete events of fracture and vein fill.

Transport equations demonstrate that the functional relations among the field variables form a coupled set of differential equations in which the principal transport mechanism has a hyperbolic form. In all but the simplest of systems in which hyperbolic functions depict the mechanism of transport, the system evolves in a chaotic manner, particularly if other nonlinearities are present. In the situation discussed in this chapter, highly nonlinear properties of H_2O-rich fluids have long been recognized as exerting primary control on the evolution of temperature, pressure, and fluid velocity (Norton and Knight, 1977).

Independent lines of evidence—one from transport theory and the other from the geometric properties of veins and fracture systems—point to anomalous fluid pressure as the force that not only generates systematic fractures and causes them to interconnect and form percolation networks but that also retards flow through mineral deposition.

References

Berbabe, Y. (1987). The effective pressure law for permeability during pore pressure and confining pressure cycling of several crystalline rocks, *Journal of Geophysical Research* 92, 649-657.

Brown, S. R. (1987). Fluid flow through rock joints: The effect of surface roughness, *Journal of Geophysical Research* 92, 1337-1347.

Burton, C. J., and H. C. Helgeson (1983). Calculation of the chemical and thermodynamic consequences of differences between fluid and geostatic pressure in hydrothermal systems, *American Journal of Science 283-A*, 540-548.

Helgeson, H. C., and D. H. Kirkham (1974). Theoretical prediction of the thermodynamic behavior of aqueous electrolytes at high pressures and temperatures. I. Summary of the thermodynamic/electrostatic properties of the solvent, *American Journal of Science 274*, 1089-1198.

Hubbert, M. K., and D. G. Willis (1957). Mechanics of hydraulic fracturing, *Transactions, Society of Petroleum Engineering, AIME 210*, 153-166.

Johnson, J. W., and D. Norton (1989). Critical phenomena in magma-hydrothermal systems: I. State, thermodynamic, transport, and electrostatic properties of H_2O in the critical region, *American Journal of Science*, in press.

Knapp, R. B., and J. E. Knight (1977). Differential thermal expansion of pore fluids: Fracture propagation and microearthquake production in hot pluton environments, *Journal of Geophysical Research 82*, 2515-2522.

Knapp, R. B., and D. Norton (1981). Preliminary numerical analysis of processes related to magma crystallization and stress evolution in cooling pluton environments, *American Journal of Science 281*, 35-68.

Lantz, R. (1984). The influence of the geometry of the pluton-host rock interface on the orientations of thermally induced hydrofractures at the Cochise Stronghold pluton, Cochise County, Arizona, M.S. thesis, University of Arizona, Tucson.

Mavko, G., and A. Nur (1978). The effect of nonelliptical cracks on the compressibility of rocks, *Journal of Geophysical Research 83*, 4459-4468.

Moskowitz, B. M., and D. Norton (1977). A preliminary analysis of intrinsic fluid and rock resistivity in active hydrothermal systems, *Journal of Geophysical Research 82*, 5787-5795.

Norton, D. (1984). A theory of hydrothermal systems, *Annual Reviews of Earth and Planetary Sciences 12*, 155-177.

Norton, D., and R. Knapp (1977). Transport phenomena in hydrothermal systems: The nature of porosity, *American Journal of Science 277*, 913-936.

Norton, D., and J. E. Knight (1977). Transport phenomena in hydrothermal systems: Cooling plutons, *American Journal of Science 277*, 937-981.

Norton, D., and R. N. Villas (1977). Irreversible mass transfer between circulating hydrothermal fluids and the Mayflower stock, *Economic Geology 72*, 1471-1504.

Snow, D. T. (1970). The frequency and aperture of fractures in rocks, *Journal of Rock Mechanics 7*, 23-40.

3

Evolution and Style of Fracture Permeability in Intrusion-Centered Hydrothermal Systems

SPENCER R. TITLEY
University of Arizona

ABSTRACT

Deep and tall columns of permeable crustal rocks evolve as a consequence of fracturing during shallow emplacement and rapid cooling of water-beating felsic magma. Rapid cooling of the magma and sudden pressure drops result in both the development of the characteristic porphyritic texture of rocks in these systems and overpressures of magmatic water; sudden release of the hydraulic energy built up in the melt is believed to have resulted in development of stockwork-style (reticulate) fracturing of the magma and its wallrock. Early stages of cooling of magmas considered here takes place through conduction, but subsequent cooling is dominated by development of the stockwork and consequent convective flow; rapid transfer of heat to wallrock is focused by the stockwork system and results in rapid local rise of temperatures and consequently pressures of pore water, resulting in further rock failure in the walls. The dominant character of flow porosity that evolves is that of an intricately interconnected three-dimensional network of planar fractures with large length-to-aperture ratios and continuities that range from microns to hundreds of meters.

Field study and analysis of these systems reveal that they form a complex process characterized by episodic rock failure and a consistent and predictable evolutionary course of hydrothermal reaction with wallrocks, usually in a thermal regime of declining wallrock temperatures. Studies of waters suggest changing provenances with magmatic waters dominating early hydrothermal systems, subsequent mixing with waters of different sources including pores, and, ultimately, dominance by meteoric water.

INTRODUCTION

Emplacement and cooling of magma at shallow crustal levels are attended by thermal-mechanical stresses resulting in the formation of extensive interconnected networks of closely spaced (centimeter) joints in large rock volumes. These volumes of densely fractured rocks with reticulate fracture patterns, informally referred to as stock-works, represent secondary permeability imposed on crustal rocks that underlie areas of up to hundreds of square kilometers, centered on the locus of magma emplacement. In most instances a volcanic center may be inferred as such a locus from regional or areal geology. These centers of areally small (kilometers), subvolcanic intrusions are the object of discussion in this chapter.

The shape of the rock volume affected, which includes the pluton complex and many times its volume of wall-rocks, is crudely cylindrical and may have had a height that extended from a permeable volcanic superstructure to depths corresponding to the height of the cooled magma column. This height is believed to have been as great as 7 km at Yerington, Nevada (Dilles, 1987), and at least 3 km at San Manuel, Arizona, as scaled from Lowell (1968). The permeability imposed on these systems in the form of joints provides a means for localized rapid cooling of magmas by convection of fluids of magmatic and meteoric provenances. The movement of these fluids from crustal wallrocks and magmatic sources results in the formation of metal-sulfur deposits and some centers of modem geothermal activity. The passage of fluids through joints centered on the igneous system also results in mineralogical modification (hydrothermal alteration) of joint walls and filling of open space in the joint system by precipitated minerals.

This chapter presents the results of detailed field and laboratory analyses of the geometrical and mineralogical characteristics of these joints and their host rocks from a broad sampling of sites. These studies reveal information concerning the development of flow porosity in a specific but widespread geological phenomenon of modem as well as old tectonic regimes. Intrusive centers have been studied in both island arc and continental settings, where they have been found to reveal generally consistent styles and histories of joint evolution as well as a consistent pattern of evolution of hydrothermal minerals; these phenomena are, in turn, consistently interpretable in a context that allows assessment of some aspects of geochemical and thermal evolution of the hydrothermal systems.

This chapter addresses some relevant geological aspects of the occurrence of these intrusive centers, relevant aspects of petrology, an overview of the characteristics of fracture-joint evolution, and through a synthesis, some of the implications of the results of these studies.

Geological Setting and Environments

Volcanic and seismic activity are phenomena of modem orogenic regions. The existence of volcanic rocks and young epizonal intrusions within orogenic belts attests to historical volcanic activity; the presence of the eroded tops of swarms or chains of epizonal intrusions, as well as the presence of erosional remnants of coeval volcanic rocks, trace and reveal the position of still older orogens. In Figure 3.1 regions of both older (Mesozoic and Cenozoic) orogenic belts and modem orogenic regions are shown. The belts of current activity may be seen to transgress both oceanic and continental lithosphere and are marked by the lines of seismic activity and volcanism. These settings are sites of plate convergence, where, apparently, subduction-related processes result in generation of magma and transfer of heat to the shallow crust.

Widespread geological field evidence from the circum-

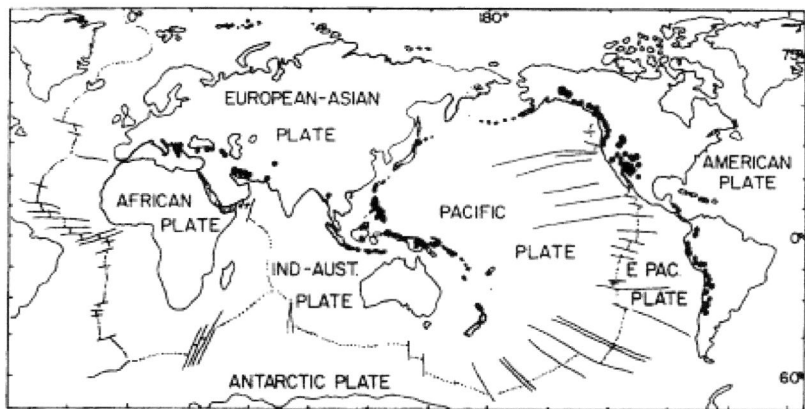

Figure 3.1
Map of major plates and continental masses of the world, showing (black circles) important centers of studied Mesozoic and younger pluton-centered hydrothermal activity. Shaded areas trace presently active regions of seismic and volcanic activity where orogenic processes are taking place. Modified from Stanton (1978).

Pacific orogenic belts indicates that extmsive volcanic activity and epizonal intrusion are related processes. Magmas rise in regions where subduction-related stress has resulted in localized sites of crust weakened by faulting, folding, or thinning or where extensive magma generation at depth has resulted in deep regions of buoyant silicate melt. Within the cores of volcanic systems, bodies of magma are emplaced and ultimately cool.

Some important characteristics of these systems merit comment. They occur in clusters of a few to many tens of intrusion centers in areas of 10^4 to 10^5 km^2, and many were apparently formed at times of high subduction-related compression in both island arcs (Titley and Heidrick, 1978) and continental settings (Heidrick and Titley, 1982). Individual pluton centers or clusters give rise to chains that trace the edges of cratons or lie on or adjacent to island arcs. In some locations pluton centers occur within crust that has been strongly deformed, and their settings have been described as mobile regions. Viewed at regional scale, the generally circular intrusion centers ordinarily range from a kilometer in diameter to systems up to, but rarely more than, about 5 km across, but they have been localized beneath volcanic rocks covering areas an order Of magnitude greater.

In young Tertiary systems, volcanic rocks coeval with intrusions are commonly, although not invariably, closely adjacent; in older Tertiary or Mesozoic systems, similarly coeval volcanic rocks are present but distant, which is interpreted as representing remnants after weathering of a larger volcanic superstructure. The intrusive rocks comprise porphyritic phases and are ordinarily members of felsic rock clans, ranging from quartz diorite through granite. The suites of many intrusive centers are accurately described as complex, such as at Ajo, Arizona (Wadsworth, 1968), and Ray, Arizona (Banks *et al.*, 1972). Volcanic rocks are seldom more mafic than andesite and range through progressively more felsic rocks to rhyolite. Whereas many different textural varieties of igneous rocks are present in the complex rocks of a volcanic system, a young mass of porphyritic igneous rock is invariably at the core of the volume of fractured rock.

NATURE OF SECONDARY POROSITY IN INTRUSION-CENTERED SYSTEMS

Stockworks associated with porphyry intrusions are vertical columns of fractured rocks, kilometers in diameter and height. This volume of fractured rock includes not only rocks of the intrusion complex but also volumes of wallrock surrounding porphyry that are many times greater than the intrusions. Within the stockwork, the fractures that compose the porous volume vary widely in scale and style; they range from through-going veins with continuity of kilometers, through intersecting veins and veinlets of meters to centimeters of continuity, visible to the unaided eye (the mesoscopic veins of this chapter), to veinlets whose scale ranges downward in size to those resolvable only with special optics. Evidence from the field, based on the volume and distribution of hydrothermal alteration products, suggests that it is the fractures (joints) at mesoscopic scale that dominate the properties of flow porosity (permeability) of these systems.

The style of mesoscopic joints in the stockwork is shown in Figure 3.2. There it can be seen that the style of openings in the stockwork is dominated by a three-dimensional mesh of intersecting planar to curviplanar joints. The development of domains of variably fractured rock within stockworks is an important characteristic wherein, for reasons that are uncertain and that are not obviously related to the mechanical properties of lithology, some volumes cf rock manifest many more fractures than contiguous volumes; the flow porosity is thus anisotropic at the scale of tens of meters.

Field studies reveal that there has been little or no rotation or transport of rock following the fracturing process in most porphyry-centered systems. Such a generalization excludes volumetrically insignificant masses of localized breccia that occur in some systems and clearly

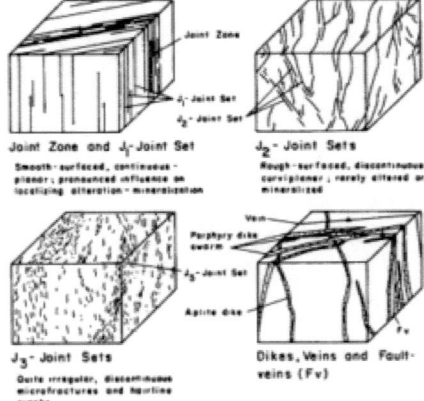

Figure 3.2
Schematic block diagrams illustrating typical and characteristic styles of fractures with stockwork systems. The several kinds of fractures successively overprint each other and result from episodic events. Reprinted from Heidrick and Titley (1982) with permission, University of Arizona Press, Tucson.

excludes some systems such as Cerro Verde (Peru) and Cananea (Mexico), where large breccia masses are major constituents of the fractured rock volumes. Whereas the mass of hydrothermally derived minerals that occur within and adjacent to the joints may aggregate up to 10 percent of the total volume of fractured rock, field evidence does not indicate that this volume was produced instantaneously. The walls of both steep and flat veins of stock-works are parallel over meters of length, regardless of vein thickness, revealing negligible rotation. Evidence of offset of old veins by younger fractures is extremely rare and where present is at the scale of centimeters. Fracture asperity may locally result in pinching and swelling of vein walls, but such apparent roughness cannot be unequivocally related to movement. Consequently, transport of fractured rock has been considered trivial, the dominant motion or displacement only that of true joints, normal to the fracture wall. The only reasonable explanation of the relatively high volume of crack-related alteration minerals is that fractures evolved episodically and appear to have been filled continuously through a time of cooling and consequent shrinking of the hot rocks.

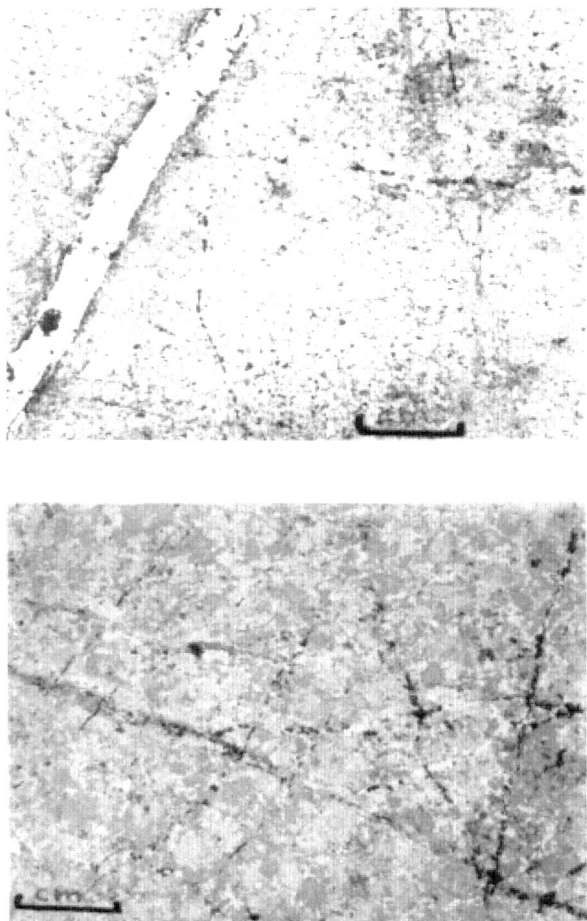

Figure 3.3
Photographs of thinned sections showing habit of microveins as used in this paper. Scale bar is 1 cm. The upper diagram shows apparently early microveins in relationship to thicker 0.5-cm quartz veins. In both the upper and lower diagrams an apparent systematic orientation or set of orientations is shown that may be interpreted in the context of oriented stresses that control fracture directions, even down to this scale.

Within the domains of mesoscopic joints, smaller domains of microscopic fractures are present. These fractures, defined here as features less than 0.1 mm in width, are also interconnected but are discontinuous with persistence ordinarily on the scale of centimeters (Figure 3.3). Whereas they contribute significantly to the porosity of the rocks, by virtue of their discontinuous nature and apparent small widths, they do not appear to have contributed to the permeability of the stockwork; filled with fluid nonetheless, they may be representative of fluid-rock reactions controlled more by diffusion than flow.

ROLE OF MAGMATIC PROCESSES IN FORMATION OF JOINTS

The occurrence of porphyry plutons associated with volcanic centers is significant. The intricately interconnected, closely spaced joints that compose the common and typical stockwork evolve in association with rocks of porphyritic textures; stockworks are not widely, if at all, recognized as having evolved in the course of the formation of igneous intrusions with phaneritic textures.

The evolution of porphyritic igneous rocks and their associated stockwork fractures appears to be integrally related, and the evolution of porphyries continues as a subject of fundamental petrological study. A generalized and simplified concept concerning the origin of these rocks explains that they represent multistage cooling of felsic silicate melts. Interruptions in uniform cooling, which commenced at great but unknown depths, and chilling of melt accompanied by drops in pressure, attend the rapid rise of magma to shallow levels of the crust, a phenomenon consistent with the apparently rapid emplacement of magmas in volcanic systems and the presence there of porphyritic igneous rocks.

The characteristics of porphyry intrusions in the complexes considered here, as seen in the field, attest to this likelihood of rapid emplacement of magmas. Xenoliths of any sort are rare to absent within, at the margins, or in the caps of these porphyry plutons, and wallrocks are undeformed, the implication being that magmas were emplaced as fluids with a minimum amount of stoping. Studies of many porphyry stocks only rarely reveal the presence of recognizable foliation, flow lines, or other evidence of motion of magma along contacts. These characteristics of rock texture, foliation, and the absence of xenoliths have been ascribed to a process of "permissive" rather than "forceful" (Mutch and McGill, 1962) emplacement of the porphyry magmas. A "permissive" or unimpeded emplacement of such magmas is envisioned as possible, and even likely, in the active tectonic environment in which they occur, with tectonic stresses resulting in deeply penetrating faults and consequent channelways.

The most important result of emplacement of magmas at shallow depth is that of rapid cooling and quenching of melt that contained crystals of earlier-formed minerals. The porphyritic textures are thus formed in the walls and across some great but unknown depth of the cooled magma. Rapid cooling and crystallization of melt bring about changes in water pressure, the effects of which have been described by Burnham (1967, 1979). Briefly stated, the exclusion of water from the silicate melt by crystallization, especially at the shallow depths (3 to 5 km) considered here, results in overpressures of exsolved water that exceed the tensile strength of the rocks. Brittle failure under these conditions (hydraulic fracturing) ensues and evolves the fractures composing the stockwork.

Emplaced magmas may cool initially by conduction of heat to their walls (Norton and Knight, 1977). This process is believed to result in rock failure and to complement the highly energetic effects bringing about the hydraulic failure described above. Water contained in the wallrocks, chiefly in pores, becomes an agent of energy when heated; the effects of this heating have been described by Knapp and Knight (1977) and Knapp and Norton (1981). As wallrocks to porphyry plutons are heated by conduction or convection in the early stages of magma emplacement, contained pore water passes into the supercritical region and effective pressures exceed the tensile strength of the rock. The result is rock failure seen now in the presence of the large volumes of fractured wallrock surrounding porphyry intrusions.

NATURE OF JOINTS AND FRACTURE SYSTEMS

At regional scale the intrusive centers may be seen in some terranes to be related to, and presumably controlled by, variations in the effects of stress in different domains within regional fault systems. Whereas such control is difficult to establish or to even propose in older terranes subjected to tectonic overprinting, it may be reasonably inferred from geological features in young terranes of island arcs. As revealed in the geology of the mobile belt of Papua New Guinea (Figure 3.4), batholiths and related centers of porphyry complexes lie within or adjacent to

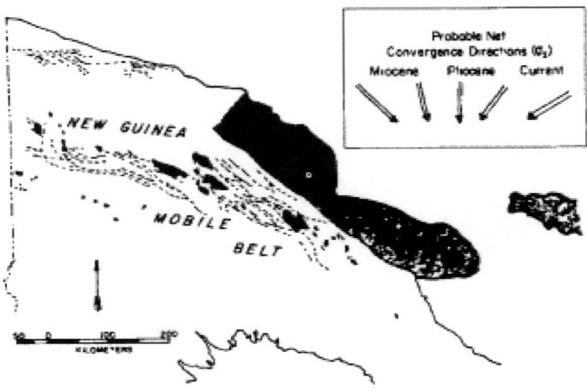

Figure 3.4
Modified geologic map of Papua New Guinea showing trace of regional faults and location of principal batholiths and centers of pluton emplacement. Black areas are those of mappable igneous bodies and centers. Modified from Titley and Heidrick (1978).

southward deflections of generally WNW-striking regional linear faults. Interpreted right-slip in this regional fault system result in areas of "low compression" at the southward deflections of their trace. Regional stresses that appear to have controlled positioning of these systems at such a scale are interpreted as having even more fundamental importance in affecting properties of joints at kilometer to meter scale.

Joint and Fracture Orientations

The orientation of mesoscopic fractures in stockworks has been traditionally described as "random," and at casual glance in outcrop the swarms of joints appear to manifest such characteristics. In detail, however, measurements of fracture orientation in numerous pluton-centered systems in cratonic and arc settings (Rehrig and Heidrick, 1976; Titley and Heldrick, 1978) reveal properties of alignment and dip that are consistent with inferred orientations of the regional or local stress field, an association that is intuitively obvious, although it is not conspicuous or prominent at outcrop scale.

Detailed studies of joint orientations in large systems reveal two extremes in the habits of orientation of mesoscopic fractures, from each of which may be deduced the orientation of principal stresses. These extremes are shown in Figure 3.5, geologic sketch maps of fracture systems in two pluton-centered stockworks in Arizona. In Figure 3.5A fractures radial and concentric to a center of intrusion reveal patterns evolved at high levels in pluton systems where the maximum principal stress is vertical and intermediate and least principal stresses are hydrostatic, equal, and horizontal. The example of Figure 3.5A views a Laramide intrusion complex very near its top with the orientations of fractures related to stress in a shallow epizonal environment. Stresses in this crustal region result in a pattern of centrosymmetric, concentric, and radial fractures. In Figure 3.5B (a map of a surface weathered into the deeper parts of a Laramide intrusion complex in southern Arizona) a pattern of joint development is present that was controlled by stresses in deeper parts of the crust. In the example of Figure 3.5B, fluid inclusion data derived from study of quartz formed in the joints are interpreted to indicate that we are viewing the system at a depth at least 2 km beneath the original surface (Preece and Beane, 1982). The fracture orientations of Figure 3.5B may be further interpreted as reflecting a horizontally oriented maximum principal stress, a roughly south to north oriented minimum principal stress and a vertical intermediate principal stress.

The analysis of geometry and distribution of joints formed near the locus of intrusions, coupled with radiometric age data that show fracturing to be essentially synchronous with intrusion, has led to interpretations of origin and orientations of regional stress. In Arizona the consistent, nearly monotonous ENE and NNW directions of fracturing seen in Laramide intrusion systems (Figure 3.6) correspond to the directions of stress inferred for the normal convergence directions of Laramide plate subduction (Rehrig and Heidrick, 1976; Heidrick and Titley, 1976, 1982). Miocene stress directions acting on eastern Papua New Guinea are seen to be consistent with the orientations of fractures surrounding centers of felsic plutons in the Eastern Highlands (Titley and Heidrick, 1978; Asami and Britten, 1980).

Field work completed in these regions, widely separated in space and in times of igneous activity, reveals without known exception joint geometries that are regionally consistent within regions with the time and inferred related tectonic style. A reasonable interpretation of the genesis of these fracture networks may be drawn from these numerous observations; the thermal and mechanical energetics of pluton emplacement and cooling bring about rock failure; the geometry of resulting joints is controlled by orientation of regional and local stresses.

Alteration, Succession, and Distribution of Altered Fractures

Three important physical and mineralogical properties of fractures that make up the volumes of stockwork have been studied in detail. The interrelationships of these features establish a basis for understanding the way in which joints and corresponding characteristics of flow porosity evolve.

The first of these properties is that of the associated hydrothermal alteration products. A second is that of the succession of veins and alteration products as revealed in cross-cutting relationships. The third is the relative abundance, in space, of different sets of veins as defined by alteration products.

Vein Alteration and Paragenesis The flow of fluids through the joint network results in alteration of the crack walls and deposition of hydrothermal minerals. Evidence from the field, from the study of many slabbed specimens, and from the results of mineralogical and fluid inclusion analyses reveals that this process is complex, takes place through many stages, and proceeds in the hydrothermal environment under declining temperatures.

Viewed at outcrop scale at the surface (Figure 3.7) and on mine bench, the mesoscopic fractures of stockworks reveal heterogeneous characteristics of style and manner of alteration and of weathering. The heterogeneity in appearance stems from the style of joint that formed at a

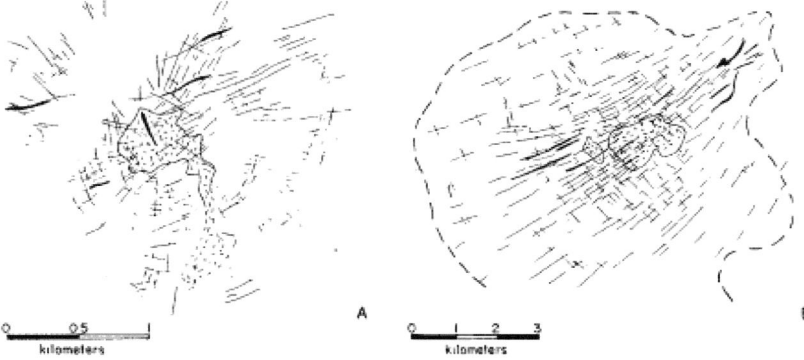

Figure 3.5
Geologic sketch maps of structures and joint-fault-dike orientations in Laramide intrusion centers in Arizona. Figure 3.5A a map of the San Juan Pluton at Safford, Arizona, modified from Heidrick and Titley (1982), manifests patterns seen at high intrusive levels where fractures trace radial and concentric patterns with respect to the center of plutons. Figure 3.5B shows fracture pattern orientations revealed at a weathered depth of about 2 km into the Sierrita, Arizona, pluton system (patterned areas). Pronounced orientation of fracture pattern believed to represent effects of region-wide compression during the Laramide.

particular time and the contrasting mineralogy formed by the hydrothermal solutions that passed through it. Within systems the habits of the altered joints that occur at specific stages in fracture evolution are broadly consistent and predictable from place to place. They differ, however, between systems.

Distinctive assemblages of alteration minerals develop within specific joint sets at different stages of the hydrothermal process. Timing of stages is revealed from the cross-cutting relationships of joints with different alteration products and styles. A typical example of this ubiquitous habit and a common succession of hydrothermal products are shown on the polished and mapped slab of Figure 3.8, wherein veins and veinlets characterized by specific alteration assemblages reveal consistent cross-cutting relationships. The habit of intersection of cracks with different vein mineralogies bespeaks a complex and protracted process of fracture development in stockwork evolution.

The development of vein alteration assemblages has been described in reports spanning many decades of work, with increased levels of understanding by numerous authors (Schwartz, 1947; Creasey, 1966; Meyer and Hemley, 1967; Carson and Jambor, 1974; Brimhall, 1979; Rose and Butt, 1979; Beane and Titley 1981; Einaudi et al., 1981; Beane, 1982; Einaudi, 1982). A widely described and generalized succession of vein alteration assemblages (Titley, 1982) from early to late in a quartz-2 feldsparmafic wallrock is as follows:

1. The assemblages quartz, quartz-biotite-orthoclase, or quartz-biotite form in and adjacent to the thermal center and in some instances may be of large areal extent, probably synchronous with assemblages 2 through 5.
2. The assemblage chlorite-epidote-carbonate-zeolites that evolves at the periphery of the system.
3. The assemblage quartz-orthoclase-(biotite)-sulfides near the thermal center.
4. The assemblage quartz-(orthoclase)-sulfides near the thermal center.
5. The assemblage quartz-sericite-pyrite developed within the system that overprints earlier stages.
6. Ultimate overprinting of stages 3, 4, and 5 by the mineralogy of stage 2 that collapses on the system as its center cools.

This common succession of alteration types in potassium silicate rocks, as also shown in the slab of Figure 3.8, reveals important characteristics of the progress of fracture and chemical evolution in the systems. Because each mineral assemblage forms in its own set of fractures at a specific time, it must be interpreted that the development of fractures is episodic. Detailed fluid inclusion data from many hydrothermal systems such as these (e.g., Preece and Beane, 1982) reveal that the succession of mineral assemblages evolves under declining temperatures of solutions. A tentative conclusion that the fractures containing the specific minerals are also opening under a corresponding lowering of temperature of the rock mass must follow. That early alteration assemblages are largely anhydrous and later ones hydrous suggests that either the

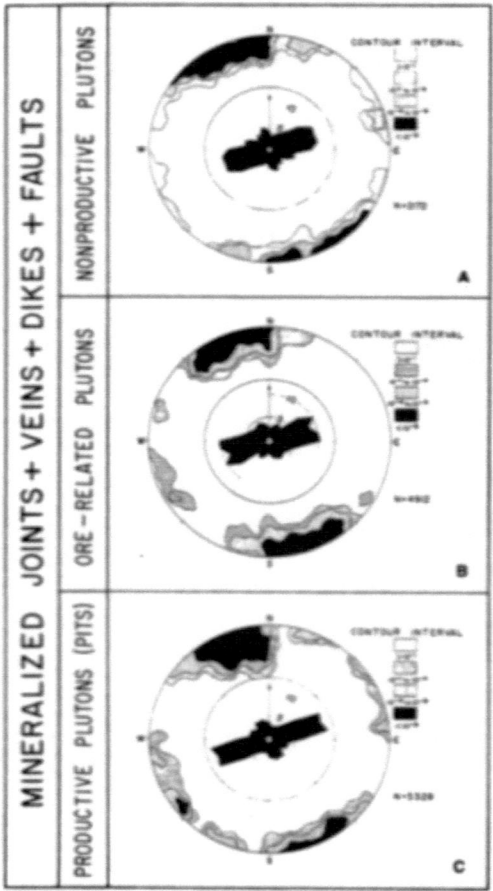

Figure 3.6 Rose diagrams of fracture patterns from stocks and wall rocks in Laramide centers in Arizona. Equal-area, pole-to-plane plots are shown for each of the fracture sets. N = number of observations. Reprinted from Heidrick and Titley (1982) with permission, University of Arizona Press, Tucson.

volume of solution increases or that certain ion activity ratios change, bringing about an effective increase in the concentration of H⁺. By stage 5 the manifestation of areally large volumes of rock, completely and pervasively converted to masses of quartz-sericite and pyrite, suggests profound H⁺ metasomatism, from which may be inferred a considerably enhanced volume of solution flow. The shift from anhydrous to hydrous phases also has been described in carbonate wallrocks where calc- and magnesium-silicate alteration minerals are formed (Einaudi, 1982; Johnson and Norton, 1985).

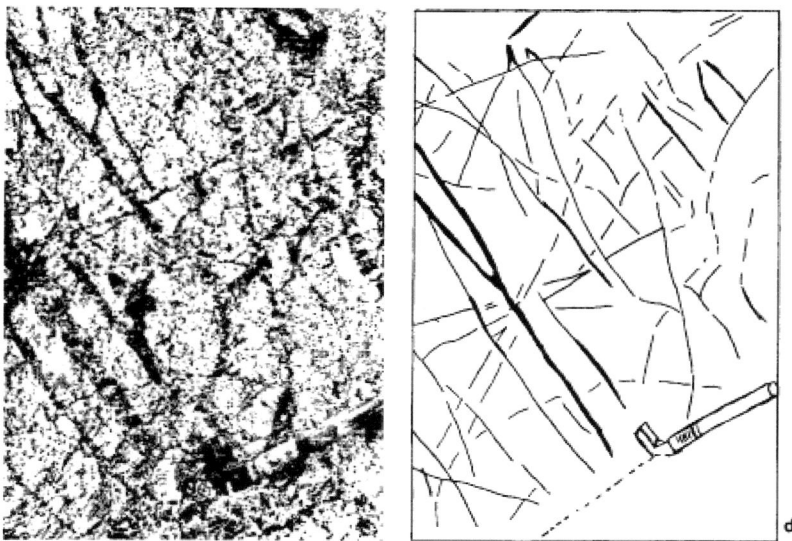

Figure 3.7
Map of weathered outcrop in granitic wallrocks of a porphyry pluton in Papua New Guinea. Photograph scale is shown by hammer near bottom. The exposure is selected for these purposes as the weathering of sulfide-bearing veins develops sufficient contrast to observe the different tones (from different alteration types) and cross-cutting nature of the differently altered joint sets. Reprinted from Titley (1982) with permission, University of Arizona Press, Tucson.

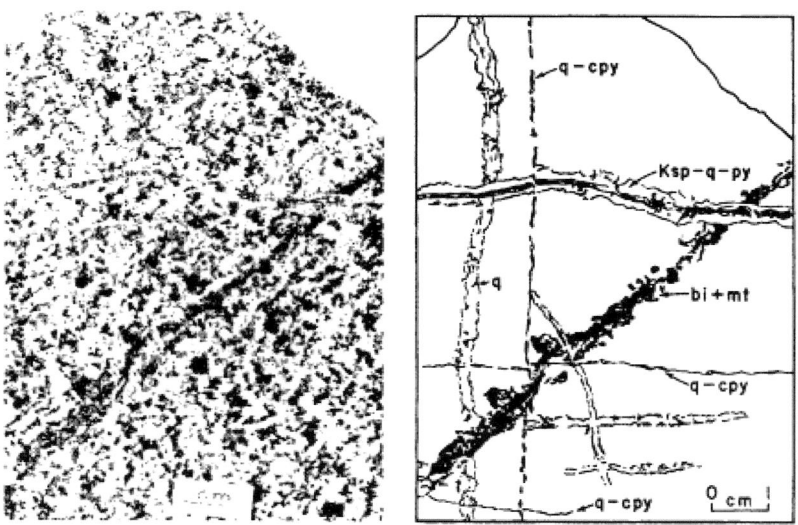

Figure 3.8 Slabbed surface of wallrock sample from Sierrita, Arizona, showing the cross-cutting habit of differently altered joints. q = quartz; cpy = chalcopyrite; bi = biotite; mt = magnetite; Ksp = orthoclase. Reprinted from Titley (1982) with permission, University of Arizona Press, Tucson.

Vein Abundance and Distribution Systematic studies of the abundance and distribution of veins have been reported from the Mayflower stock in Utah (Villas and Norton, 1977); Silver Bell, Arizona (Kanbergs, 1980; Norris, 1982); Red Mountain, Arizona (Kistner, 1984); and Sierrita, Arizona (Titley *et al.*, 1986). Results of reconnaissance studies have been reported from prospects in Papua New Guinea (Titley, 1978; Titley *et al.*, 1978).

A simple method of measurement of fracture abundance has been applied to outcrop, drill core, and selected samples in the study of the stockworks. Very simply, the method involves measuring of fracture area in rock volumes or fracture length on surfaces and then dividing that value by, respectively, either the volume or area of the sample. The value obtained is per centimeter. For the fracture abundance present in the systems studied, values were determined from length/area relationships in 2500-cm^2 circular or square sample areas. Area/volume values were determined on drill core and selected samples of slabbed rocks. The values obtained were used to determine both the distribution of vein types and their abundance (Titley, 1978; Haynes and Titley, 1980; Titley et al., 1986).

Recognition of the existence of vein successions as revealed by contrasts in alteration allows an assessment of the distribution and abundance of fractures that form at different stages in the process. Such a study for a system has been carried out in detail and complete scope only at Sierrita, Arizona. Data collected from a thousand sample sites at Sierrita, over an area of about 70 km^2, have been treated by area-averaging techniques and contoured as shown in Figure 3.9, wherein isopleths of fracture density values, representing all fractures, close on the porphyry center (from Titley et al., 1986). Separate maps showing the distribution and abundance of five different alteration types are integrated into separate curves of density versus distance (Figure 3.10).

Most of the measurements of fracture abundance at Sierrita were made on the vertical faces of benches in the pit or walls of washes or on horizontal to subhorizontal surfaces of adjoining areas. Inspection of vertical faces across a pit depth of several hundred meters, as well as inspection of vertical faces of surrounding outcrops, indicates a relatively low (much less than 10 percent) proportion of flat (less than 45°) fractures at these depths in the system. Whereas flat fractures are present at higher levels of such systems (e.g., the young systems of Papua New Guinea; see also Knapp and Norton, 1981), at the deeper (2 km?) level exposed at Sierrita they are uncommon. In these circumstances the determination of length/area relationships in fractured rocks is a close approximation of the area/volume value. Such a conversion is reasonable at Sierrita and otherwise comparable deeply exposed intrusion centers.

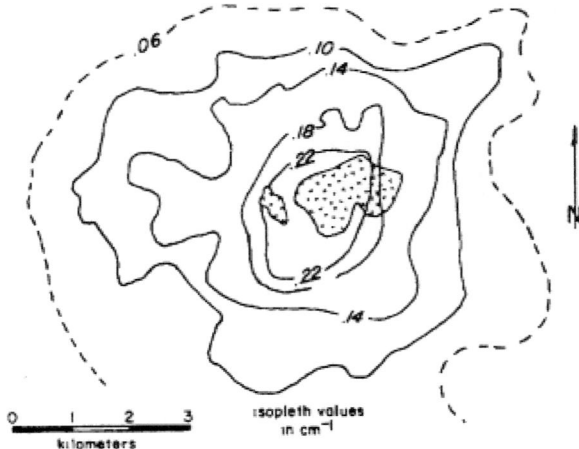

Figure 3.9
Vein/joint abundance map of the Sierrita system, Arizona. Isopleths show values for the total numbers of fractures (per centimeter) in outcrop and close on the porphyry center (patterned areas). Outer isopleth shows values nearly zero, but rare altered joints persist outward for another kilometer. Within the center of porphyry, fracture abundance values double and locally exceed values of 1.5 cm^{-1}. Figure modified from Titley et al. (1986) and data in Haynes and Titley (1980).

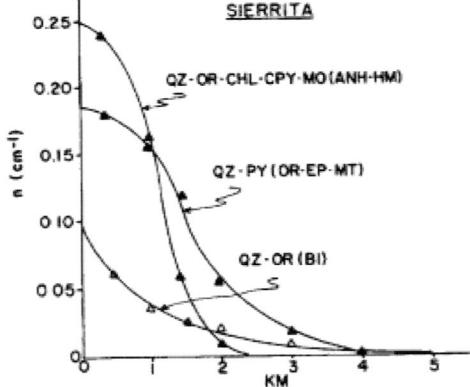

Figure 3.10
Plot of fracture abundance (vertical) as a function of vein-alteration type and distance (horizontal) from an assumed center within the porphyry mass at Sierrita. Oldest-to-youngest vein sets proceed from qz-or to the complex qz-or-chl-cpy veins that are the youngest. The values shown in Figure 3.9 include the values of the quartz-sulfide set and the quartz-orthoclase set. qz = quartz, or = orthoclase, chl = chlorite, cpy = chalcopyrite, mo = molybdenite, anh = anhydrite, hm = hematite, ep = epidote.

The information shown in Figure 3.10 is revealing in several important respects. It shows an episodic progression of fracturing that commences with an early, widespread event of relatively uniformly low numbers of fractures, followed by progressively more centrally restricted episodes of fracturing. Hydrothermal fluids flowed through fractures following each episode, and each stage of fractures is altered in a unique and characteristic way with the most dense fracturing and subsequent alteration in a stock-work closely centered on the porphyry center. Studies of fluid inclusions from vein quartz from the different vein sets reveals further that fluids depositing quartz in each subsequent event were cooler than those of each older stage of quartz formation. The most dense fracturing is the youngest, most constricted, and most centrally located of the joint sets. And where these stockworks are hosts to ore minerals such as chalcopyrite, molybdenite, or tin minerals, it is this centrally restricted set that localizes the most abundant sulfide mineralization in the porphyry-stock-work environment. From the data in Titley *et al.* (1986) and summarized and shown in Figures 3.9 and 3.10, it is possible to make estimates of rock volumes containing equal fracture areas for specific fracture sets and their measured fracture densities. Such an estimate is shown in Figure 3.11 where cylindrical volumes 1 km in height with each containing 100,000 km^2 of fracture area are schematically portrayed. At the low measured densities of fractures containing quartz and orthoclase (Figure 3.10), the relatively large volume shown is necessary, whereas at the much greater fracture densities measured in the orebody from the ore-sulfide-bearing quartz veins, a much smaller volume is required to contain the same area of fractures. Each of the fracture types shown in Figures 3.10 and 3.11 is overprinted in the central volume of rock by the younger vein stages. Viewed in two dimensions the progress appears to evolve with development of early, widespread joints, each successive fracture-forming event resulting in more closely spaced joints progressively more closely focused on the center of the system.

SOURCE OF WATER

It is appropriate to review the nature and origin of waters that have been so critically important in the evolution of stockwork and its intrinsic secondary permeability. The foregoing has underscored the complex character of these systems in the context of apparent episodic development of joints and in the context of the implied changing patterns of chemical evolution. There is no question that water in some form and likely from different sources has been an important agent of change in the evolution of these systems. Interpretations of fluid types and source stem from the data developed from analyses of fluid inclusions and from studies of oxygen and hydrogen isotopes, generalized and summarized in Figure 3.12.

Fluid Inclusion Data

Numerous workers, cited in Roedder (1984), have studied the fluid inclusions found in the hydrothermal quartz of porphyry-cored stockwork ore deposits. The solutions in these inclusions reveal both very high (e.g., 60+ wt.% NaCl equivalent) and low salinities (on the order of 2 to 10 wt.% NaCl equivalent). Although some exceptions to the generalization have been found, the high-salinity inclusions are those that are also interpreted to have been formed at high (more than 500°C) temperatures and the low salinity inclusions at lower (less than 350°C) temperatures. From these characteristics a common and widespread inference as to source has evolved—that high-temperature, high-salinity inclusions derived from magmatic water and that the low-temperature, low-salinity solutions evolved from a meteoric source. This inference is strengthened from evidence of the isotopic character of water involved.

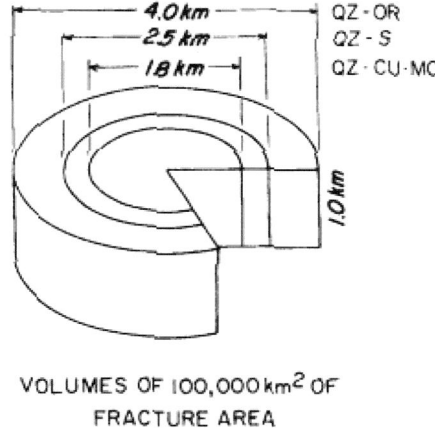

Figure 3.11
Generalized diagram, from data of Figure 3.10, showing relative cylindrical volumes of rock 1 km high, in the Sierrita system, that would contain 100,000 km^2 of fracture area. Values were estimated by planimeter of plotted and interpolated isopleths of fracture abundance in the data of Figure 3.10 and conversion to area/volume relationships. Illustration of the fracture abundance data in this way reveals the apparent focusing effect of joint-evolving events, with the passage of time, progressively closer to the intrusion center.

Data from Isotopes

Oxygen and hydrogen isotopes have been studied in a few porphyry stockwork systems and are reviewed in Sheppard *et al.* (1971) and Ohmoto (1986). Viewed independently of the data from fluid inclusions, the isotopic character of early (high-temperature, high-salinity) solutions reveals the habit of inferred magmatic solutions; the isotopic character of solutions that formed late-stage alteration minerals requires a component of meteoric water. Beyond identification of meteoric water by its isotopes, interpretations of other provenances of hydrothermal waters in these systems, such as basin, pore waters, or metamorphic waters (each likely), become clouded in uncertainty because of overlapping isotopic characteristics and uncertainties resulting from mixing of waters from different sources.

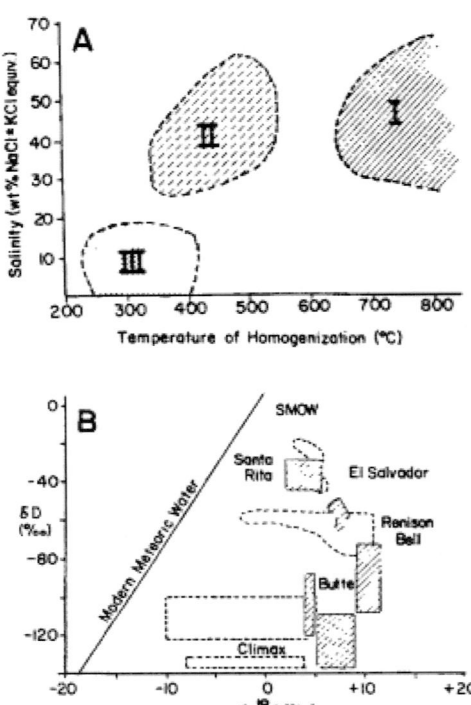

Figure 3.12
Generalized data from fluid inclusion measurements on quartz from porphyry-centered systems (Figure 3.12A), after Reynolds and Beane (1985). The diagram plots homogenization temperatures (Th) against salinities (wt.% equivalent NaCl + KCl) of fluid inclusions in hydrothermal minerals (mostly quartz) and reveals a general separation of fluids into three types, based on salinities and temperatures. Figure 3.12B shows the isotopic composition of water ($\delta^{18}O$ versus δD) from porphyry-centered systems, as modified from Ohmoto (1986). In a general way that corresponds to the data from fluid inclusions, there is a gross bimodal nature of the data revealing likely magmatic character to early waters, mixing, and ultimately meteoric character to the latest fluids.

NATURE AND EVOLUTION OF FRACTURE-RELATED PERMEABILITY

The dominating element of flow porosity or permeability in porphyry-pluton-centered systems is the fracture network. Because fracture networks are shown to evolve sequentially, however, the total number of cracks measurable in rocks is not a measurement of characteristics useful in a determination of instantaneous permeability. Earlier-formed cracks are sealed, and, as part of a succession of formation of cracks, temporally intermediate stages of joint formation are succeeded by still younger joints that are successively sealed by alteration products. It is important to emphasize, again, that interpretations of data developed from fluid filling temperatures reveals that in most instances the early history of crack formation in wallrocks (cracks that formed under increasing wallrock temperature) has not been recognized. Whereas it is likely that joints formed under these conditions, it is possible, indeed likely, that they remained closed or at least relatively impermeable through the time of rising wallrock temperatures and thermal expansion. The thermal and mechanical effects of this early stage of pluton emplacement and wallrock heating remain enigmatic and are an important object of both laboratory and field search.

Estimation of permeability may be made by evaluation of the expression

$$k = \frac{nd^3}{12}, \qquad (3.1)$$

where k is permeability (cm^2), n is fracture abundance (cm^{-1}), and d is fracture aperture (cm) (Norton and Knapp, 1977). Measurements of fracture abundance reported here have been made in the units of the expression. It may be seen, however, from the continuously varying value of fracture density for each fracture set (Figure 3.7) that the values of permeability at a constant value of aperture will vary in a similar way. The additional parameter necessary for calculation is that of joint aperture. This dimension, in the altered systems studied here, is not directly determinable and has been estimated elsewhere in only a few reported instances.

Inspection of the character of altered joints reveals that their width, as manifested by their filling and wallrock alteration selvages, is variable, most commonly between millimeters and a few centimeters. Although it is tempting to assign aperture values on the basis of such widths, the textural data indicate that such assessments must be made with caution and in most instances would be in error. Interpretations based on textural evidence and vein habits lead to the belief that the vein apertures were narrow (i.e., millimeters or less, rather than centimeters) in these systems.

Inspection of many altered joint sets from many systems reveals that crack walls are parallel over distances commonly measured in meters. This habit is seen in three dimensions, wherein large blocks appear suspended in the network of parallel-walled, altered fractures. Whereas this characteristic may not establish a rigorous basis for interpretation of the evolution of narrow (millimeter-wide) veins, the case of centimeter-wide veins requires that the vein selvages "grow" (or apparently widen) at a rate sufficiently slowly to restrict rotation or transport of the affected rock masses. Inasmuch as fluid inclusion data indicate cooling during the life of crack formation and filling, it is reasonable to propose that the joints may originate with small apertures (millimeters or less) and maintain some degree of flow porosity and capacity to transmit fluids as a result of continuous thermal contraction. Succinctly stated, wide (millimeters to centimeters) alteration selvages are viewed to be a result of a process of gradual crack widening during cooling and alteration, not the result of the instantaneous development of numerous wide (centimeter) joint openings. The flow porosity is not indicated by the width of vein alteration products. Further, the times at which a crack formed and at which it opened to the passage of fluid may well be different.

Results of the microscopic study of textures of alteration minerals adjoining and within veins reveal that replacement of walls is a dominating character of the joint-altering process. Open-space filling textures, such as inward and interpenetrative growth of quartz toward vein centers, is uncommonly rare. Further, there is rarely any indication of inward growth as might be seen in optical properties of quartz as viewed through the polarizing microscope. The habit of apparent replacement persists across all of the alteration sets studied from these systems; the process results, apparently, in nearly complete filling of the original space available. The microscope commonly reveals that even in wide (centimeters) vein fill, the site of residual porosity remains largely in the central part of the filled-altered joint where, it is interpreted, the last fluids passed. Beyond the evidence from textures, additional observations and inferences from the habit of joints and joint alteration lend support to the interpretation of small values of aperture.

Young vein sets, representing different alteration characteristics from older sets, are imposed on older sets without their visible offset along the older veins (see Figure 3.6). Such nonoffsetting fracturing would be likely to develop only in a rock mass in which the mechanical competence is maintained. From such characteristics and interpretation, Titley *et al.* (1986) proposed that the alteration of time-specific joint sets results in annealing of the rock and restoration of mechanical homogeneity.

SYNTHESIS

High pore pressures in the environment of emplacement of shallow plutons are a result of localized sources of heat acting on pore water. A consequence of this phenomenon is rock failure when its tensile strength is exceeded by the pressures of contained water.

Emplacement of small felsic plutons into shallow portions of the crust is a rapid process in active tectonic regimes where plutons appear to have followed the paths of earlier magmas of volcanic systems. The rapid emplacement results in marked thermal contrasts between cool, shallow crust and magmas; this contrast results in episodic fracturing of porphyry and its wallrocks as cracks form and become annealed by alteration, and magma cooling retreats to progressively greater depths. Reference volumes of wallrock in close proximity to the porphyries undergo heating and then cooling; in the process rocks fail by a process resulting in widespread jointing and, at shallow depths, small bodies of breccia (transported rock fragments). Joints evolve in abundance near the centers of thermal-mechanical energy and diminish in number outward.

Development of joints results in instantaneously imposed permeability characterized by a network of fractures and fluid flow that ultimately results in cooling of the thermal center and synchronous alteration of fracture walls. Evidence of episodic breaking of rocks exists in the super-position of temporally and mineralogically distinct alteration assemblages in their own fracture sets in the same rock volumes; textural evidence permits interpretation that original joint walls continuously separate. Concomitantly, open space is apparently filled, inhibiting transport of blocks but maintaining minimal flow porosity from thermal contraction until the joint space is completely filled by alteration products and deposition of hydrothermal minerals, throttling fluid flow. As the joint space becomes restricted, so does fluid flow, further resulting in episodic increases in pressure above deeper but still cooling magma. This process of hydrothermal flow and reaction brings about constriction of gradually widening vein apertures phenomena necessary to the process of intermittent but continuing rock failure.

ACKNOWLEDGMENTS

This review has been improved by suggestions from D. L. Norton, R. V. Kirkham, and an anonymous reviewer. Some of the research reported here was based on work supported by the National Science Foundation under grant EAR 78-22897. Figures 3.2, 3.4, 3.5A, 3.6, 3.7, and 3.8 are reproduced here by permission, from *Advances in Geology of the Porphyry Copper Deposits: Southwestern North America*, edited by S. R. Titley, University of Arizona Press, Tucson, 1982.

References

Asami, N., and R. M. Britten (1980). The porphyry copper deposits at the Frieda River Prospect, Papua New Guinea, *Society of Mineralogists and Geologists of Japan; Geological Special Issue* 8, 117-139.

Banks, N. G., H. R. Cornwall, M. L. Silberman, S.C. Creasey, and R. G. Marvin (1972). Chronology of intrusion and ore deposition at Ray, Arizona, Part I, K-Ar ages, *Economic Geology* 67, 864-878.

Beane, R. E. (1982). Hydrothermal alteration in silicate rocks, in *Advances in Geology of the Porphyry Copper Deposits, Southwestern North America*, S. R. Titley, ed., University of Arizona Press, Tucson, pp. 117-137.

Beane, R. E., and S. R. Titley (1981). Porphyry copper deposits, Part II, Hydrothermal alteration and mineralization, *Economic Geology, 75th Anniversary Volume*, 235-269.

Brimhall, G. H., Jr. (1979). Lithologic determinations of mass transfer mechanisms of multiple-stage porphyry copper mineralization at Butte, Montana: Vein formation by hypogene leaching and enrichment of potassium-silicate protore, *Economic Geology 74*, 556-589.

Burnham, C. W. (1967). Hydrothermal fluids at the magmatic stage, in *Geochemistry of Hydrothermal Ore Deposits*, H. L. Barnes, ed., Holt, Rinehart and Winston, New York, pp. 34-75.

Burnham, C. W. (1979). Magmas and hydrothermal fluids, in *Geochemistry of Hydrothermal Ore Deposits*, 2nd ed., H. L. Barnes, ed., Wiley-Interscience, New York, pp. 71-136.

Carson, D. J. T., and J. L. Jambor (1974). Mineralogy, zonal relationships and economic significance of hydrothermal alteration at porphyry copper deposits, Babine Lake area, British Columbia, *Canadian Institute of Mining 67*, 1-24.

Creasey, S.C. (1966). Hydrothermal alteration, in *Advances in Geology of the Porphyry Copper Deposits, Southwestern North America*, S. R. Titley, ed., University of Arizona Press, Tucson, pp. 51-74.

Dilles, J. H. (1987). Petrology of the Yerington batholith, Nevada: Evidence for evolution of porphyry copper ore fluids, *Economic Geology 82*, 1750-1789.

Einaudi, M. T. (1982). Skarns associated with porphyry plutons: Description of deposits, southwestern North America. II. General features and origin, in *Advances in Geology of the Porphyry Copper Deposits, Southwestern North America*, S. R. Titley, ed., University of Arizona Press, Tucson, pp. 139-183.

Einaudi, M. T., L. D. Meinert, and R. J. Newberry (1981). Skarn deposits, *Economic Geology, 75th Anniversary Volume*, 317-391.

Gustafson, L. B., and J. P. Hunt (1975). The porphyry copper deposit at El Salvador, Chile, *Economic Geology 70*, 857-912.

Haynes, F. M., and S. R. Titley (1980). The evolution of fracture-related permeability within the Ruby Star granodiorite, Sierrita porphyry copper deposit, Pima County, Arizona, *Economic Geology 75*, 673-683.

Heidrick, T. L., and S. R. Titley (1976). Structural evolution of southwestern North American Laramide porphyry copper deposits and its relationship to the history of plate interactions (abs.), *International Geological Congress, 25th, Sydney, 1976*, 740.

Heidrick, T. L., and S. R. Titley (1982). Fracture and dike patterns in Laramide plutons and their structural and tectonic implications, in *Advances in Geology of the Porphyry Copper Deposits, Southwestern North America*, S. R. Titley, ed., University of Arizona Press, Tucson, pp. 73-91.

Johnson, J. W., and D. L. Norton (1985). Theoretical prediction of hydrothermal conditions and chemical equilibria during skarn formation in porphyry copper systems, *Economic Geology 80*, 1797-1823.

Kanbergs, K. (1980). Fracturing along the margins of a porphyry copper system, Silver Bell district, Pima County, Arizona, Unpublished M.S. thesis, University of Arizona, Tucson, 90 PP.

Kistner, D. J. (1984). Fracture study of a volcanic lithocap, Red Mountain porphyry copper prospect, Unpublished M.S. thesis, University of Arizona, Tucson, 75 pp.

Knapp, R. B., and J. E. Knight (1977). Differential thermal expansion of pore fluids; fracture propagation and microearthquake production in hot pluton environments, *Journal of Geophysical Research 82*, 2515-2522.

Knapp, R. B., and D. Norton (1981). Preliminary numerical analysis of processes related to magma crystallization and stress evolution in cooling pluton environments, *American Journal of Science 281*, 35-68.

Lowell, J. D. (1968). Geology of the Kalamazoo orebody, San Manuel district, Arizona, *Economic Geology 63*, 645-654.

Meyer, C., and J. J. Hemley (1967). Wall rock alteration, in *Geochemistry of Hydrothermal Ore Deposits*, H. L. Barnes, ed., Holt, Rinehart and Winston, New York, pp. 166-235.

Mutch, T. Q., and G. E. McGill (1962). Deformation in host rocks adjacent to an epizonal pluton (the Royal Stock, Montana), *Geological Society of America Bulletin 73*, 1541-1544.

Norris, J. R. (1982). Fracturing, alteration and mineralization in Oxide pit, Silver Bell mine, Pima County, Arizona, Unpublished M.S. thesis, University of Arizona, Tucson, 72 pp.

Norton, D., and R. Knapp (1977). Transport phenomena in hydrothermal systems; the nature of porosity, *American Journal of Science 277*, 913-936.

Norton, D., and J. Knight (1977). Transport phenomena in hydrothermal systems; Cooling plutons, *American Journal of Science 277*, 937-981.

Ohmoto, H. (1986). Stable isotope geochemistry of ore deposits, in *Stable Isotopes in High Temperature Geological Processes*, P. H. Ribbe, ed., Mineralogical Society of America, Reviews in Mineralogy, vol. 16, Washington, D.C., pp. 491-559. Preece, R. K., III, and R. E. Beane (1982). Contrasting evolutions of hydrothermal alteration in quartz monzonite and quartz diorite wall rocks at the Sierrita porphyry copper deposit, Arizona, *Economic Geology 77*, 1621-1641.

Rehrig, W. A., and G. L. Heidrick (1976). Regional tectonic stress during the Laramide and late Tertiary intrusive periods, Basin and Range Province, Arizona, Tucson, *Arizona Geological Society Digest X*, 205-228.

Reynolds, T. J., and R. E. Beane (1985). Evolution of hydrothermal fluid characteristics at the Santa Rim, New Mexico, porphyry copper deposit, *Economic Geology 80*, 1328-1347.

Roedder, E. (1984). *Fluid Inclusions*, Mineralogical Society of America, Reviews in Mineralogy, vol. 12, Washington, D.C., 644 pp.

Rose, A. W., and D. M. Burt (1979). Hydrothermal alteration, in *Geochemistry of Hydrothermal Ore Deposits*, 2nd ed., H. L. Barnes, ed., John Wiley & Sons, New York, pp. 173-235.

Schwartz, G. M. (1947). Hydrothermal alteration in the "porphyry copper" deposit, *Economic Geology 42*, 319-352.

Sheppard, S. M. F., R. L. Nielsen, and H. P. Taylor, Jr. (1971). Hydrogen and oxygen isotope ratios in minerals from porphyry copper deposits, *Economic Geology 66*, 515-542.

Stanton, R. L. (1978). Mineralization in island arcs with particular reference to the south-west Pacific region, *Australian Institute of Mining and Metallurgy Proceedings No. 268*, 9-19.

Titley, S. R. (1978). Geologic history, hypogene features, and processes of secondary sulfide enrichment at the Plesyumi copper prospect, New Britain, Papua New Guinea, *Economic Geology 73*, 768-784.

Titley, S. R. (1982). The style and progress of mineralization and alteration in porphyry copper systems, American southwest, in *Advances in Geology of the Porphyry Copper Deposits, Southwestern North America*, S. R. Titley, ed., University of Arizona Press, Tucson, pp. 93-116.

Titley, S. R., and T. L. Heidrick (1978). Intrusion and fracture styles of some mineralized porphyry systems of the southwestern Pacific and their relationship to plate interactions, *Economic Geology 73*, 891-903.

Titley, S. R., A. W. Fleming, and T. I. Neale (1978). Tectonic evolution of the porphyry copper system at Yandera, Papua New Guinea, *Economic Geology 73*, 810-828.

Titley, S. R., R. C. Thompson, F. M. Haynes, S. L. Manske, L. C. Robison, and J. L. White (1986). Evolution of fractures and alteration in the Sierrita-Esperanza hydrothermal system, Pima County, Arizona, *Economic Geology 81*, 343-370.

Villas, R. N., and D. L. Norton (1977). Irreversible mass transfer between circulating hydrothermal fluids and the Mayflower stock, *Economic Geology 72*, 1471-1504.

Wadsworth, W. B. (1968). The Cornelia pluton, Ajo, Arizona, *Economic Geology 63*, 101-115.

4

Fluid Dynamics During Progressive Regional Metamorphism

JOHN V. WALTHER
Northwestern University

INTRODUCTION

A major part of active tectonic processes is the evolution of continental margins with the progressive burial and metamorphism of lavas and sedimentary strata to deep crustal levels. The nature of fluid components in these rock sequences buried to mid or lower crustal depths (10 to 40 km) is poorly known but of utmost importance. The presence of fluid changes the rheological properties of the rock and thus its response to stress. The fluid also has the potential to carry heat and dissolved mineral components and can therefore dictate the style of metamorphism, the extent of chemical equilibrium, and much about the textural fabric observed in rocks. It has become abundantly clear that the majority of sediments and rocks in active tectonic processes in the upper few kilometers of the Earth's surface have experienced large fluid fluxes. The nature of the flow and its chemical and physical consequences are addressed in other chapters in this volume. This chapter addresses problems regarding the origin, flow dynamics, and chemical consequences of fluid at mid-to-lower crustal depths. Classically, to those who study the mineralogical changes in such rocks, the general process is termed progressive regional metamorphism. All mountain fold belts contain lavas and sediments that have been regionally metamorphosed.

FLUID-ROCK RATIOS AND FLUID FLUX

It is not surprising that large fluid fluxes can be recorded within the upper few kilometers of the Earth's surface where the porosity and/or permeability of many rocks are large and dramatic temperature and topographic gradients can occur. Perhaps more surprising is the evidence for large amounts of fluid flow occurring in deeply buried rocks. Essentially this evidence consists of documenting chemical changes in mineral assemblages during metamorphism and calculating the minimum fluid volume necessary to account for the observed changes, as shown in Figure 4.1.

Perhaps the most straightforward calculation of this type is documentation of the extent of reaction for a simple decarbonation such as the production of wollastonite from calcite and quartz:

$$CaCO_3 + SiO_2 \rightarrow CaSiO_3 + CO_2.$$

At fixed pressure and temperature this reaction fixes the fugacity of CO_2 in the fluid. Thus, the concentration of H_2O or other components in the fluid needed to maintain the fugacity of CO_2 for the extent of production of wollastonite can be calculated (e.g., Rumble *et al.*, 1982).

The calculated minimum fluid necessary to produce the observed chemical changes seen during metamorphism is termed the time-integrated fluid-rock ratio or more commonly the fluid-rock ratio (FRR). FRRs, calculated by a number of approaches from mineral assemblages in basaltic or carbonate units, are as high as 2 to 20 (e.g., Ferry, 1976, 1980; Graham et al., 1983; Tracy et al., 1983). That is, 2 to 20 volumes of fluid have reacted with each volume of rock during progressive metamorphism. Note that these are minimum values since fluids in equilibrium with the mineral assemblage are unrecorded by this technique. Additionally, there is no guarantee that the fluid has reached complete chemical equilibrium with the mineral assemblage, an assumption inherent in the calculation. However, given the high temperatures involved and the large surface area to fluid volume in metamorphic rocks, the equilibrium assumption is probably reasonable (Walther and Wood, 1986). Because the actual grain boundary porosity of metamorphic rock is probably less than 0.1 percent, these high ERRs indicate that fluid within the flow porosity must have been replenished thousands of times if the introduction of fluid does not expand the rock.

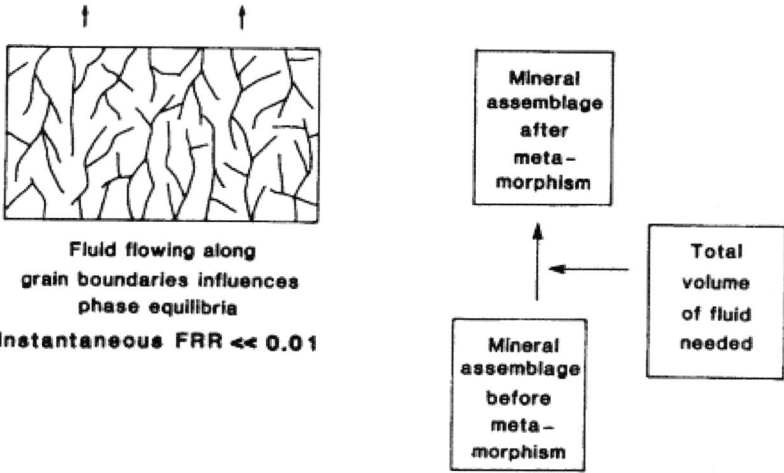

Figure 4.1

A comparison of the amount of fluid within a rock during metamorphism (left) and the fluid-rock ratio (FRR) calculated from mineral equilibria (right) (from Wood and Walther, 1986).

Fluid-rock ratios are often calculated based on changes in the isotopic composition of oxygen in minerals. Unfortunately, these changes are sensitive to fluid-rock interactions at any point in the rock's history, that is, during its burial to peak metamorphic conditions, at the peak of metamorphism, or during its uplift to the Earth's surface. In some cases isotopic exchange indicating substantial fluid-rock interaction may be recording exchange that has occurred near the Earth's surface either during burial or uplift. Thus, they may not be recording ERRs during the peak of metamorphism. Wickham and Taylor (1985; Chapter 6, this volume) have suggested that high-grade metamorphic rocks from the Pyrenees appear to require the introduction of a large volume of seawater deep into the metamorphic pile. Such a flux of fluid to great depths in unlikely (Walther and Orville, 1982; and as discussed below). The data are, however, also consistent with oxygen isotope exchange nearer the Earth's surface.

Fluid-rock ratios are not a measure of the total time integrated fluid flux (IFF) during metamorphism but must be considered in the context of their ability to place limits on the IFF. First, the volume of rock affected must be known, but unfortunately it is often difficult to determine. It is obvious, however, that the same total fluid volume reacting toward equilibrium with a thin rock unit will record a higher ERR than an identical unit that is thicker, because the extent of reaction in the thinner unit is greater per unit rock volume. These problems have been discussed in more detail elsewhere (Wood and Graham, 1986; Wood and Walther, 1986).

Fluid-rock ratios do not record the passage of fluid in equilibrium with a rock. This has important ramifications. Consider a typical divariant (sliding) reaction in a pelitic rock undergoing metamorphism:

$$2HCl + CaCO_3 + 2NaAlSi_3O_8 \rightleftharpoons$$
$$\text{(calcite)} \quad \text{(in plag.)}$$

$$CaAl_2Si_2O_6 + 2NaCl + 4SiO_2 + CO_2 + H_2O.$$
$$\text{(in plag.)} \quad \text{(quartz)}$$

Assuming that calcite and quartz are pure phases, at fixed pressure, temperature, and chloride activity the composition of the plagioclase will dictate the equilibrium ratio of CO_2 and H_2O in the fluid. Imagine a variety of compositional layers, perhaps of sedimentary origin, with varying plagioclase composition. A fluid passing through such layering will react to adjust its CO_2 to H_2O ratio to approach equilibrium with each plagioclase crystal it makes contact with. The extent of reaction will depend then on how far the plagioclase composition is from equilibrium with the fluid. For the same IFF the recorded FRR will be different depending on the variability of plagioclase composition. One may then be drawn to the wrong conclusion that some beds experienced large IFFs while others did not. Thus, it would erroneously appear that some beds acted as metamorphic fluid aquifers while others were aquitards. This could then lead to a model of the major flow occurring along bedding when in fact it may not (Ferry, 1987). However, a number of stable isotopic studies have apparently demonstrated that bed-parallel fluid flow occurs. What will be argued here is that it is not differences in intrinsic permeabilities but differences in the rheological properties of the different layers that control fluid flow.

FLUID FLOW AT MID-TO LOWER-CRUSTAL LEVELS

Fluid-rock ratios at depth are apparently similar to those determined near the Earth's surface, where fluid convection is known to operate. It has, therefore, been proposed that fluid convects to deep crustal levels (e.g., Etheridge *et al.*, 1983; Wickham and Taylor, 1985; Ferry, 1986). Arguments to the contrary have also been presented (Walther and Orville, 1982; Walther and Wood, 1984; Wood and Walther, 1986). Let us assess some of the evidence for the state and transport of fluid at mid or lower crustal depths.

Fluid flow is obviously highly dependent on permeability. Permeabilities of metamorphic rocks are not well known. What is clear is that they increase dramatically in laboratory experiments when fluid pressure equals rock pressure (Brace, 1980). Since most of the experimental studies do not concern themselves with large fractures, such permeability estimates are considered by many to be minimal for the crust as a whole. Use of these permeability values suggests that fluid flow during metamorphism should occur under conditions of fluid pressure significantly below rock pressure. That is to say that for the anticipated fluid flux the permeabilities during metamorphism are great enough to allow fluid pressure to drop below the rock pressure toward a more hydrostatic gradient. A hydrostatic gradient is required to promote convection. By hydrostatic pressure what is meant is the pressure resulting from the density of an overlying column of fluid as opposed to the much greater pressure resulting from the density of an overlying column of rock (lithostatic pressure). Fluid pressure less than lithostatic does not seem to be the case when metamorphic rocks are closely examined (Norris and Henley, 1976; Fyfe *et al.*, 1978). Phase equilibrium studies of devolatilization reactions observed in metamorphic rocks generally seem to require fluid pressure to be near lithostatic pressure. This observation is confirmed by fluid inclusion studies where the trapped fluid has the appropriate density for fluid pressure equal to lithostatic pressure at the temperature of metamorphism.

These two divergent observations could be rectified if the properties of the fluid phase were not those of the bulk fluid. It has been suggested, for example, that the fluid phase present during deep crustal metamorphism is not a discrete fluid phase but rather an absorbed phase on mineral surfaces with presumably greater viscosity and lower fugacity (Elliott, 1973; Rutter, 1976).

Studies at room temperature indicate that multimolecular layers of H_2O are absorbed on mineral surfaces. At metamorphic temperatures and pressures where even structurally bound water becomes unstable in hydrous minerals, it is reasonable to assume that absorption of H_2O is no more than a monolayer in thickness. If we examine the amount of fluid trapped as fluid inclusions along healed microcracks during metamorphism and redistribute it evenly along the entire crack, the width of fluid in the crack is generally about 200 Å or 10 times the thickness of a double monomolecular layer (Walther and Orville, 1982). Because this is the minimum thickness of the fluid film at the time the crack healed, we may reasonably assume that the properties of fluid flowing through such cracks are not significantly modified by absorptive properties of mineral surfaces, and hence the fluid phase can be considered to have the properties of a bulk fluid.

Near the Earth's surface fluid pressure is controlled by the extent of the overlying fluid column (hydrostatic pressure) while the pressure exerted on mineral grains (lithostatic pressure) is considerably greater owing to the much greater density of the minerals. This difference between fluid pressure and rock pressure is maintained by the effective crushing strength of the rock. At some depth the closure of pores and the resultant decrease in permeability causes the fluid pressure gradient to increase dramatically, so that fluid pressure equals lithostatic pressure. Note that fluid pressure equal to rock pressure does not imply that fluid is trapped in a static state but that it is possible that the flow of fluid is balanced by permeability changes near fluid pressure equal to rock pressure.

Figure 4.2 shows fluid pressure as a function of depth determined from well bottom hole fluid pressure measurements from a number of wells in the U.S. Gulf Coast. Note that at a depth just below 3 km fluid pressure begins to depart from hydrostatic, and at 5.5 km or at a lithostatic pressure of 1.5 kbar the fluid pressure is very close to lithostatic. The fluid pressure below 3 km is greater than that exerted by an overlying column of fluid and is there fore considered "geopressured." There are various chemical and physical factors that influence the depth at which geopressuring of the fluid occurs. The decrease of hydraulic conductivity in clay layers is often considered the primary factor in determining the characteristics of the geo-pressured zone (Bredehoeft and Hanshaw, 1968; Chapman, 1972).

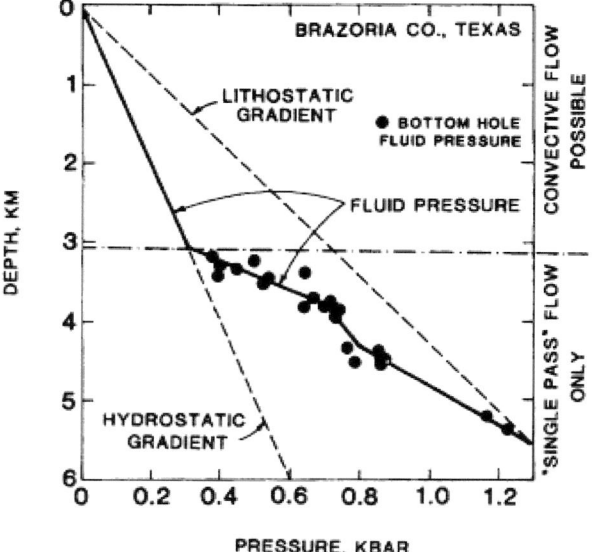

Figure 4.2
Fluid pressure as a function of depth in a sedimentary basin of the U.S. Gulf Coast. Note the crossover from hydrostatic to lithostatic fluid pressure and its implications for fluid convection (from Wood and Walther, 1986).

The depth of onset of geopressuring has been observed from as little as 45 m to depths greater than 8 km, although in most sedimentary basins it is above 6 km. We might imagine that in crystalline rocks in extensional environments hydrostatic gradients in the fluid may be maintained to depths greater than 8 km. However, at a depth of 11 km the difference between hydrostatic pressure and lithostatic pressure is about 2 kbar. Although the crushing strengths of rocks at the temperatures, pressures, and strain rates appropriate for this depth are not well known, it seems reasonable to conclude that hydraulic conductivity would be so greatly reduced by mechanical compaction that fluid below this depth can in general be considered to be close to lithostatic pressure. As mentioned above, this observation is consistent with the evidence from fluid inclusion studies that indicates that fluid pressure equals rock pressure during metamorphism.

It would, therefore, seem that laboratory measurements of the permeabilities of metamorphic rocks do not characterize the permeability of these rocks at the time they are undergoing metamorphism, but overestimate them. It may be that the permeability is reduced by the inevitable recrystallization that occurs when these rocks are subject to high temperatures and pressures for time periods on the order of millions of years during metamorphism. In any event it seems that fluid pressure must be close to lithostatic at mid-to lower-crustal depths. The imposition of a pressure gradient significantly greater than hydrostatic on the fluid during progressive regional metamorphism means that there is no reasonable way to transport the less-dense fluid by flow downward (i.e., no convection of fluid can occur). This means that the large integrated FRRs recorded by some rocks must be recording fluid generated at some greater depth.

FLOW MECHANISM WHERE FLUID PRESSURE EQUALS LITHOSTATIC

Consider a rock initially devoid of fluid undergoing a devolatilization reaction in response to increased temperature during progressive metamorphism at mid-crustal levels. Fluid will be produced by each volatile-containing mineral undergoing destruction by reaction. Depending on the wetting characteristics of the fluid, it will either coat the mineral surfaces or begin to collect in isolated pores at mineral triple junctions (White and White, 1981; Watson and Brenan, 1987), as shown on the left side of Figure 4.3.

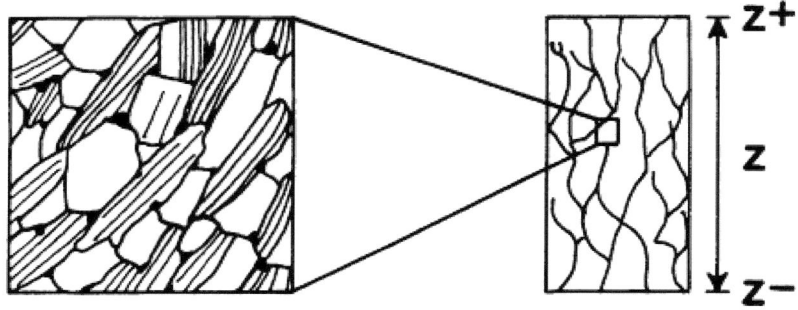

Figure 4.3
Fluid production at reacting volatile containing minerals collecting in isolated pores at mineral triple junctions (left). On a larger scale with increased fluid production, these must interconnect, producing a fluid phase of some vertical extent, z (right).

In any event, continued devolatilization will eventually build an interconnected three-dimensional fluid network of some vertical extent, as shown schematically on the right side of Figure 4.3. A static fluid at lithostatic pressure would rise due to its lower density and therefore to its buoyancy relative to the surrounding rocks if it was not held by the tensile strength of the rock. Because rocks at metamorphic conditions have low tensile strengths, the fluid would become mechanically unstable and hydrofracture its way toward the Earth's surface. Similar arguments have been used to explain the ascent of magmas from depth even though the buoyancy forces are less. As the left side of Figure 4.4 shows, the greater the amount of interconnected fluid space, the greater will be the difference between the pressure on the fluid and surrounding rock and thus the ability of the fluid to hydrofracture the rock. The pressure difference between static fluid and rock is given by

$$\Delta P = gz(\rho_{rock} - \rho_{fluid}) , \qquad (4.1)$$

where g stands for the acceleration due to gravity, z is the vertical distance of interconnectivity of the fluid, and ρ_{rock} and ρ_{fluid} are the rock and fluid densities, respectively. While the tensile strength of a rock acting to prevent propagation of a fracture at a crack tip is poorly known, we might imagine that under metamorphic conditions, with the interplay of deviatoric stresses during active tectonism, it is no more than 10 bars or so, particularly for subcritical crack growth. This means that no interconnected fluid network can exist statically that extends in the vertical direction more than about 60 m. This severely limits the possible vertical extent of fluid convection before the tensile strength is exceeded and fluid migrates upward only. Over this short distance no convection is possible for any reasonable temperature gradient and permeability.

The potential to produce an interconnected fluid network is great. The approximately 2 moles of fluid on average given off per kilogram of pelitic composition rock during medium-to high-grade metamorphism has the potential to hydrofracture the rock many times. This amount of fluid would occupy a volume of approximately 12 percent under metamorphic conditions if it did not escape. While we do not know the volume of fluid necessary to produce an interconnected fluid pathway, it is probably no more than a few tenths of a percent of the volume of the rock. Thus, we might imagine that even on a local scale hydrofracturing may have occurred a large number of times. If we consider the passage of fluid released from minerals lower in the metamorphic pile, it is not surprising that we observe large numbers of healed fluid-filled microcracks in metamorphic rocks.

The predominance of divariant (sliding equilibria) devolatilization reactions during metamorphism suggests that fluid is released from hydrous and carbonate minerals more or less continuously in the metamorphic pile during prograde metamorphism. It is possible that some fractures stay open for extended periods of time fed by a continuous supply of fluid. This situation is shown on the right side of Figure 4.4, where we require a viscous pressure gradient, dP_{vis}, to compensate for the difference in the pressure gradient between the static fluid and the surrounding rock. That is, the loss in hydraulic head by the upwardly flowing fluid due to its viscosity compensates for the difference between the hydrostatic and lithostatic gradients so that fluid pressure in the flowing fluid equals lithostatic pressure along the walls of the fracture, which in turn allows the fracture to remain open:

$$dP_{vis} = g(\rho_{rock} - \rho_{fluid}) . \qquad (4.2)$$

Assuming fluid and rock densities of 0.9 and 2.8 g/cm^3, the viscous pressure gradient required as calculated from Eq. (4.2) is about 2.0×10^3 dyne/cm^3. We can model the fluid flow in a microcrack as steady-state incompressible viscous laminar flow through two parallel plates that extend in all directions to a very much greater extent than d, the distance they are apart. Due to the viscous nature of the fluid, the profile of the flow velocity across the crack is parabolic, zero at the wall, and a maximum in the center. The solution to this fluid mechanical problem is

$$v = \frac{2v_{max}}{3} = \frac{d^2 dP_{vis}}{12v} , \qquad (4.3)$$

where v and v_{max} are the average and maximum fluid velocities, respectively, d is the fracture width, and v is the viscosity of the fluid. Noting that the fluid flux, q, is equal

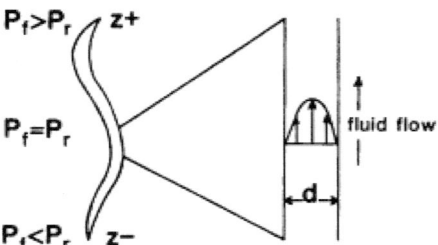

Figure 4.4
The vertical extent of interconnected fluid modeled as static fluid in a fracture must have a lower pressure gradient as a function of z than the surrounding rock. This produces conditions of fluid pressure greater than rock pressure at z^+. The rock hydrofractures and allows fluid to flow in the fracture. Its width is d; velocity profile is also shown.

to the cross-sectional area of the fracture opening times the average velocity, we have

$$q = vdl = \frac{d^3 l\, dP_{vis}}{12\nu}, \qquad (4.4)$$

where l is the length of the fracture opening.

Viscosities of supercritical H_2O-CO_2 mixtures are generally between 0.1 and 0.2 centipoise. For a given flux of fluid we can calculate the length of cracks perpendicular to flow versus their width per square centimeter.

Such calculations have been done (Walther and Orville, 1982; Walther and Wood, 1984) and indicate that the widths of the fractures are in most cases less than 10^5 Å Because of the cubic dependence of fluid flux on crack width, it seems likely that changes in the flux of fluid are accommodated by small changes in fracture width to maintain the viscous pressure gradient at the value necessary to keep the fracture open. Apparently, judging from the extent of fluid inclusions along sealed fractures in metamorphic minerals, if the fracture width falls below about 200 Å the crack will seal. Laboratory investigations indicate that these cracks seal in a matter of days, at least in quartz. It stands to reason that, if the fluid flux through the metamorphic pile is not continuous, many generations of these microcracks will form.

If such a mechanism of fluid flow operates, the usefulness of the concept of intrinsic permeability of a rock is questionable. That is, the permeability is a dynamic function of the fluid flux through the metamorphic rocks. The permeability of the rock is adjusted by the fluid phase to accommodate the flux of fluid, so the fluid pressure is near lithostatic during fluid flow. This is different from the concept of reaction-enhanced permeability due to volume loss of solids because of reaction (Rumble and Spear, 1983). What is argued here is that as a general approximation the permeability adjusts itself, so fluid pressure is always close to rock pressure irrespective of the extent of reaction.

With the deviatoric stresses that operate during metamorphism and the differences in tensile strength of different layers of rock, fracture production and, therefore, fluid channels may develop along preferred layers that are at some angle to the Earth's gravitational field. This would give rise to layers that appear as "metamorphic aquifers" and others that appear as aquitards, at least over limited distances.

CHANNELIZATION OF FLUID FLOW

The interconnectivity, bifurcation, coalition, or distribution of cracks in metamorphic rocks are largely unknown. We do know that some fractures or fracture networks have remained open long enough to experience the passage of a considerable amount of fluid. In pelitic rocks these fractures are often marked by quartz veins. At lower temperatures calcite veins are dominant because calcite has a higher solubility at low temperatures than does quartz. The thickness of the vein represents the accumulated effects of mineral precipitation from the passing fluid. If such major channelways of fluid escape are present, the flow paths of these fluids must to some extent coalesce as fluid is produced at each mineral in the rock that undergoes devolatilization. Thus, there must be some flow of fluid in intimate contact with minerals before fluid enters a major channelway. Obviously, the extent of grain boundary flow versus flow in major channelways controls to a large extent the amount of fluid a particular rock unit may experience during metamorphism and hence IFF. This in turn dictates much about the textural fabric of the rock and the approach to chemical equilibrium that may be expected.

Figure 4.5 shows a quartz segregation/vein in the Bundnerschiefer formation of Switzerland that is considered to have formed during Lepontine metamorphism. If this quartz segregation represents the cross section of quartz deposited along a major conduit for fluid flow, we can calculate the amount of fluid that must have passed to cause the extent of quartz precipitation seen. Because quartz is present in most of the mineral assemblages in the Bundnerschiefer, it is anticipated that fluids responsible for the deposition of quartz along the segregation/vein are at quartz saturation.

Let us calculate the fluid necessary to precipitate a quartz vein 50 cm in diameter, much like the one shown in

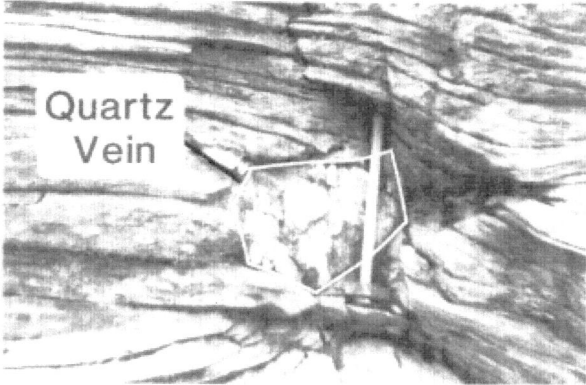

Figure 4.5
Quartz segregation thought to represent the cross section of quartz precipitated during the lifetime of a major fluid conduit during metamorphism. While the width of the fracture was probably less than 10 µm, the large amount of flow precipitated substantial quantities of quartz.

Figure 4.5. For a distance of 1 cm along the vein this amounts to 2000 cm^3 of quartz or about 87 moles. At 500°C and 4 kbar with a geothermal gradient of 20° to 30°C/km, this requires 4 x 10^{8}q moles of H$_2$O or the dehydration of 7.6 x 10^{10} cm^3 of average pelitic rock during metamorphism if 2 moles of H^{2}O are released per kilogram of rock. This corresponds to a cube of pelite 42 m on a side.

Given the large volumes of pelitic and psammitic rock in most metamorphic terrains, the fluid fluxes must be high. For the whole of the metamorphic pile, fluid fluxes of 1 x 10^{-10} to 1 x 10^{-9} g cm^{-2} s^{-1} have been calculated (Walther and Orville, 1982), which means that an average of 3 to 30 kg of fluid must pass through each square centimeter of crust overlying the 400°C isotherm in each million years of progressive metamorphism. The upward flow of fluid may be even greater if significant fluid is released from a subducting slab during metamorphism (perhaps through a magma intermediary) or if mantle fluids (Dawson, 1980) are significant.

Until it is determined to what extent fluid is channeled along structural features such as faults, fold hinges, primary lithological contacts, and layers with low tensile strength or, alternatively, flows at the scale of grain boundaries, the extent of heat and material transport and the very nature of metamorphism will not be understood.

ACKNOWLEDGMENTS

This chapter was written while I was on sabbatical leave at Université Paul Sabatier, Toulouse, France. I would like to thank Jacques Schott for his warm hospitality. Comments by J. M. Ferry and B. J. Wood led to substantial improvement.

References

Brace, W. F. (1980). Permeability of crystalline and argillaceous rocks, *International Journal of Rock Mechanics and Mineral Sciences 17*, 241-251.

Bredehoeft, J. D., and B. B. Hanshaw (1968). On the maintenance of anomalous fluid pressures: I. Thick sedimentary sequences, *Geological Society of America Bulletin 79*, 1097-1106.

Chapman, R. E. (1972). Clays with abnormal interstitial fluid pressures. *American Association of Petroleum Geologists Bulletin 56*, 790-795.

Dawson, J. B. (1980). *Kimberlites and Their Xenoliths*, Springer-Verlag, New York.

Elliott (1973). Diffusion flow laws in metamorphic rocks, *Geological Society of America Bulletin 84*, 2645-2664.

Etheridge, M. A., V. J. Wall, and R. H. Vernon (1983). The role of the fluid phase during regional metamorphism and deformation, *Journal of Metamorphic Petrology 1*, 205-226.

Ferry, J. M. (1976). P, T, fCO$_2$ and fH$_2$O during metamorphism of calcareous sediments in the Waterville-Vassalboro area, south central Maine, *Contributions to Mineralogy and Petrology 57*, 119-143.

Ferry, J. M. (1980). A case study of the amount and distribution of heat and fluid during metamorphism, *Contributions to Mineralogy and Petrology 71*, 373-385.

Ferry, J. M. (1986). Reaction progress: A monitor of fluid-rock interaction during metamorphic and hydrothermal events, in *Fluid-Rock Interactions During Metamorphism*, J. V. Walther and B. J. Wood, eds., Springer-Verlag, New York, pp. 60-88.

Ferry, J. M. (1987). Metamorphic hydrology at 13 kilometers depth and 500-550°C, *American Mineralogist 72*, 39-58.

Fyfe, W. S., N. J. Price, and A. B. Thompson (1978). *Fluids in the Earth's Crust*, Elsevier, Amsterdam, 383 pp.

Graham, C. M., K. M. Greig, S. M. F. Shepherd, and B. Turi (1983). Genesis and mobility of the H$_2$O-CO$_2$ fluid phase during regional greenschist and epidote amphibolite facies metamorphism: A petrological and stable isotope study in the Scottish Dalradian, *Journal of the Geological Society of London 140*, 577-599.

Norris, R. J., and R. W. Henley (1976). Dewatering of a metamorphic pile, *Geology 4*, 333-336.

Rumble, D., and R. S. Spear (1983). Oxygen-isotope equilibration and permeability enhancement during regional metamorphism, *Journal of the Geological Society of London 140*, 619-628.

Rumble, D., J. M. Ferry, T. C. Hoeting, and A. J. Boucot (1982). Fluid flow during metamorphism at the Beaver Brook fossil locality, New Hampshire, *American Journal of Science 282*, 886-919.

Rutter, E. H. (1976). The kinetic of rock deformation by pressure solution, *Philosophical Transactions of the Royal Society of London A283*, 203-219.

Tracy, R. J., D. M. Rye, D. A. Hewitt, and C. M. Schiffties (1983). Petrologic and stable-isotopic studies of fluid-rock interactions, south-central Connecticut. I. The role of infiltration in producing reaction assemblages in impure marbles, *American Journal of Science 283A*, 589-616.

Walther, J. V., and P. M. Orville (1982). Rates of metamorphism and volatile production and transport in regional metamorphism, *Contributions to Mineralogy and Petrology 79*, 252-257.

Walther, J. V., and B. J. Wood (1984). Rate and mechanism in prograde metamorphism, *Contributions to Mineralogy and Petrology 88*, 246-259.

Walther, J. V., and B. J. Wood (1986). Mineral-fluid reaction rates, in *Fluid-Rock Interactions During Metamorphism*, J. V. Walther and B. J. Wood, eds., Springer-Verlag, New York, pp. 194-211.

Watson, E. B., and J. M. Brenan (1987). Fluids in the lithosphere, 1 . Experimentally-determined wetting characteristics of CO$_2$-H$_2$O fluids and their implications for fluid transport, host-rock physical properties, and fluid inclusion formation, *Earth and Planetary Science Letters 85*, 497-515

White, J. C., and S. H. White (1981). On the structure of grain boundaries in tectonites, *Tectonophysics 78*, 613-628.

Wickham, S. M., and H. P. Taylor, Jr. (1985). Stable isotopic evidence for large-scale seawater infiltration in a regional metamorphic terrane; The Trois Seigneurs Massif, Pyrenees, France, *Contributions to Mineralogy and Petrology 91*, 122-137.

Wood, B. J., and C. M. Graham (1986). Infiltration of aqueous fluid and high fluid-rock ratios during greenschist facies metamorphism: A discussion, *Journal of Petrology 27*, 751-761.

Wood, B. J., and J. V. Walther (1986). Fluid flow during metamorphism and its implications for fluid-rock ratios, in *Fluid-Rock Interactions During Metamorphism*, J. V. Walther and B. J. Wood, eds., Springer-Verlag, New York, pp. 89-108.

5

Oxygen and Hydrogen Isotope Constraints On the Deep Circulation of Surface Waters Into Zones of Hydrothermal Metamorphism and Melting

HUGH P. TAYLOR, JR.
California Institute of Technology

INTRODUCTION

The purpose of this paper is to marshal the evidence and try to build a case that (1) shallow (1 to 7 km) circulation of surface waters in the Earth's crust is an extremely widespread and common phenomenon in areas of igneous activity and (2) deep (10 to 15 km) circulation of surface waters can occur in certain favorable geological situations, particularly in rift zones and areas of extensional tectonics. It is shown that very large amounts of water may interact with the rocks in such zones and that this can take place at temperatures high enough for melting and metamorphism to occur. Oxygen and hydrogen isotope studies have proven to be very useful in establishing the characteristics of such deeply circulating hydrothermal systems and in determining the origins of the aqueous fluids involved in producing granitic and rhyolitic magmas in such environments. This is mainly because oxygen-18 and deuterium are constituents of the H_2O molecule itself, and thus stable isotope signatures are by far the best way to characterize hydrothermal fluids of different origins. These conclusions are most clear-cut when low-^{18}O, low-D meteoric waters are involved in the isotopic exchange processes, but ocean waters, sedimentary formation waters, metamorphic dehydration waters, and magmatic waters can also be distinguished from each other in favorable circumstances.

Because of aqueous-fluid interactions, rocks that ultimately undergo partial melting may exhibit isotopic signatures considerably different from those that they started with. Leaving aside effects ascribable to the intrinsic $\delta^{18}O$ and δD values of these different kinds of waters, it is proposed in this paper that stable isotope studies may be used to identify three broad classes of hydrothermal systems, based mainly on the water/rock ratio (w/r), the temperature (T), and the length of time (t) that fluid-rock interaction proceeds.

- Type 1. Epizonal systems with a wide variation in whole-rock $\delta^{18}O$ and extreme $^{18}O/^{16}O$ disequilibrium among coexisting minerals (e.g., quartz and feldspar); these systems typically have $T = 200°$ to $600°C$, $t < 10^6$ yr, and form under hydrostatic pressure conditions.
- Type 2. Deeper-seated and/or longer-lived systems, also with a wide spectrum of whole-rock $\delta^{18}O$, but with equilibrated $^{18}O/^{16}O$ ratios among coexisting minerals ($T = 400°$ to $700°C$, $t > 10^6$ yr, and pressures transitional from hydrostatic to lithostatic.
- Type 3. Thoroughly homogenized and equilibrated systems with relatively uniform $\delta^{18}O$ in all lithologies; these probably form at large w/r, $T = 500°$ to $800°C$, and $t > 5 \times 10^6$ yr.

The most common of these systems, and the type that is

most easily recognized through field and laboratory studies, is Type 1. However, these three categories are not mutually exclusive; prior to melting, many Type 3 systems at an earlier stage may have been subjected to Type 1 or Type 2 conditions. The Type 3 systems very likely pass through an early hydrostatic phase, but at their peak temperatures they almost certainly end up at lithostatic pressures. The best available example of a Type 3 system is probably the Trois Seigneurs area in the Pyrenees; it is described and discussed in detail by Wickham and Taylor (chapter 6, this volume). Also note that even though the whole-rock $\delta^{18}O$ values (and $^{87}Sr/^{86}Sr$ values?) may be radically changed and homogenized in the Type 3 systems, the whole-rock chemical compositions and other isotopic parameters (e.g., end) in many cases may be only slightly modified.

Thus, under certain conditions, fluid-rock interaction can dramatically *increase* the $^{18}O/^{16}O$ heterogeneity of an initially uniform terrane (e.g., a section of volcanic rocks in a Type 1 system), but on the other hand it also can smooth out original heterogeneities (e.g., a Type 3 regional metamorphic system). In either case most natural aqueous fluids usually produce an overall ^{18}O depletion of the rocks (because equilibrium $\delta^{18}O$ mineral-fluid values are smaller at high T than at lower T). However, at least for the relatively lower temperature Type 1 systems, fluids with $\delta^{18}O$ values higher than seawater (i.e., >0) can produce ^{18}O enrichments in the rocks and associated melts (e.g., oceanic plagiogranites).

OXYGEN ISOTOPE KINETICS IN THE QUARTZ-FELDSPAR SYSTEM

Figure 5.1 shows the characteristic patterns of $\delta^{18}O$ feldspar versus $\delta^{18}O$ quartz in some hydrothermally altered granitic rocks from British Columbia (mostly Type 1 systems). The $^{18}O/^{16}O$ envelopes displayed in Figure 5.1 cut across the 45° equilibrium lines at a steep angle as a result of the much faster $^{18}O/^{16}O$ exchange rate of feldspar relative to quartz. Analogous effects are observed for feldspar and pyroxene in hydrothermally altered gabbros (Taylor and Forester, 1979; Gregory and Taylor, 1981). The kinetics of these steep trajectories have recently been quantitatively studied by Criss *et al.* (1987) and Gregory *et al.* (1988). They have shown that if a low-^{18}O aqueous fluid such as meteoric H_2O or ocean H_2O is involved in the exchange process, the slopes of the mineral-pair data points on a graph of $\delta^{18}O$ feldspar versus $\delta^{18}O$ quartz essentially constitute an "oxygen isotope clock," as indicated schematically in Figures 5.2 and 5.3.

Each of the disequilibrium arrays on such diagrams can be regarded as an isochron in which the time increases as the slopes of the isochrons become less steep and approach Although the relative times in Figure 5.2 are well established, where they are normalized to a value of unity for the rate-constant of the slow-exchanging mineral (Criss *et al.*, 1987), the actual times indicated on the "oxygen isotope clock" in Figure 5.3 are obviously dependent on several variables such as temperature, permeability, grain size, etc. However, these effects can all be treated together and described by a single phenomenological parameter, namely the rate constant for hydrothermal isotopic exchange be

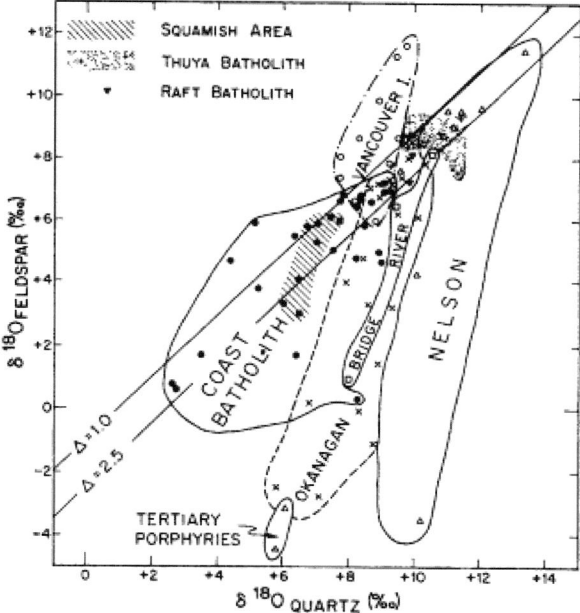

Figure 5.1
A plot of $\delta^{18}O$ feldspar versus $\delta^{18}O$ quartz for various granitic plutons from southern British Columbia. Coast Batholith and Squamish area represent portions of the western part of the Coast Plutonic Complex. In most cases marked isotopic disequilibrium is observed among the coexisting minerals of these hydrothermally altered granitic rocks. The steep trend lines can be extrapolated back to the 45° equilibrium line to obtain the $\delta^{18}O$ values of the original magmas or the unaltered rocks (from Magaritz and Taylor, 1986).45° (Criss *et al.*, 1987). These "times" represent the duration (e.g., in years) of a particular hydrothermal exchange event. These models are not yet perfectly quantified, but for moderate hydrothermal temperatures (300° to 500°C) the actual times indicated in Figure 5.2 are closely constrained by theoretical studies of the "lifetimes" of hydrothermal convective systems (e.g., Norton and Knight, 1977), as well as by available laboratory hydrothermal diffusion measurements (Giletti *et al.*, 1978; Yund and Anderson, 1978; Giletti and Yund, 1984; Giletti, 1986; also see review by Cole and Ohmoto, 1986).

tween the slow-exchanging mineral and the aqueous solution (Gregory et al., 1988). For example, if the rate constant for mineral 2 (k_2) is assumed to be 10^{-14} s^{-1}, which is a very plausible value for quartz, then the values of t on the isochrons in Figure 5.2 (or on the inside of the clock rim in Figure 5.3) should all be multiplied by 10^{14} s (=3.17×10^6 yr) in order to get the actual "lifetimes" of the hydrothermal systems.

A near-vertical slope in Figure 5.2 is characteristic of epizonal hydrothermal systems, which we know typically last for at most only a few hundred thousand years (Norton and Knight, 1977; Norton and Taylor, 1979). The time constraints described above thus *require* that the mineral-pair $\delta^{18}O$ values of all of these kinds of systems must attain near-equilibrium slopes (i.e., 45°) on a time scale of about 4 to 6 million years (m.y.) (Gregory et al., 1988). At higher temperatures (>600°C) the times required for equilibration and attainment of a 45° slope will be much smaller, almost certainly less than 1 to 2 m.y.

In other words, if fluid-rock interactions persist for only about 50,000 to 300,000 yr (the characteristic duration of a modest-sized, epizonal hydrothermal system), we obtain a near-vertical slope in Figures 5.2 and 5.3 (i.e., the clock's hand points downward toward approximately 6:15). This would constitute a Type 1 rock-fluid system as defined above and illustrated in Figure 5.1. If the fluid-rock interactions persist at similar temperatures for more than 5 or 6 m.y., this is sufficient time for isotopic equilibrium among the minerals to be essentially established, resulting in an approximately 45° slope (i.e., the clock's hand will point toward 7:15 or 7:30). This would be termed a Type 2 rock-fluid system. If the fluid-rock interaction continues for even longer times or at higher temperatures and higher fluid-rock ratios, we might envision a Type 3 system in

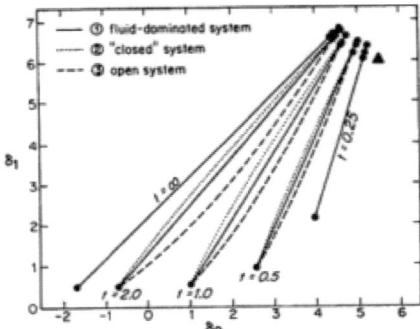

Figure 5.2
Graph of δ_1 versus δ_2, showing sets of isochrons at various normalized times, t = 0.25, 0.5, 1.0, 2.0, and ∞ for two coexisting minerals that undergo isotopic exchange with water in (1) a fluid dominated system, (2) a "closed" system, and (3) a true open system (modified after Criss et al., 1987; Gregory et al., 1988). The triangle indicates the initial values of δ_1 and δ_2. These calculated curves are taken from Figures 6, 7, and 8, and Eq. (12) of Criss et al. (1987), and they all include a uniform ratio of rate constants $k_1/k_2 = 5$. Note that the isochrons for the three different cases are nearly coincident. In fact, for t = 0.25 and ∞ they are essentially identical. The times, t, are normalized to a unit value for k_2. Thus, if in an actual case, $k_2 = 10^{-14}$ S^{-1}, (a plausible value for quartz), the values of t on the isochrons would be a factor of 10^{14} longer than indicated.

Figure 5.3A
schematic "oxygen isotope clock" showing how the slopes of the disequilibrium isochrons (Criss et al., 1987; Gregory et al., 1988) change with time on a plot of the $\delta^{18}O$ of a slow-exchanging mineral like quartz (on the abscissa) versus the $\delta^{18}O$ of a fast-exchanging mineral like feldspar (on the ordinate). As an array of data points (e.g., as in Figure 5.1) change in slope from near vertical to about 45°, with increasing time the array will successively sweep past the positions occupied by the various isochrons during the "lifetime" of a typical epizonal hydrothermal system (one having temperatures of out 250° to 500°C). The times in such a case might change from about 300,000 yr to about 6 m.y., as shown; however, these times are only approximate, as they are arbitrarily predicated on an assignment of 300,000 yr as the "lifetime" of the hydrothermal system appropriate for the 0.09 isochron. The numbers from 0.09 to 1.84 along the inside of the "clock" rim are values of the dimension-less quantity k^2 (see Gregory et al., 1988), where t is the length of time that the hydrothermal system is operating and k_2 is the kinetic exchange rate constant for the slow-exchanging mineral (these may equally well be considered to be normalized values of the "lifetime" of the hydrothermal system; see Criss et al., 1987).

which not only are the mineral $\delta^{18}O$ values equilibrated but the whole-rock samples themselves all attain relatively homogeneous $\delta^{18}O$ values. In an idealized Type 3 situation the 45° array of data points could conceivably shrink down to a single point.

MAJOR FOSSIL METEORIC HYDROTHERMAL SYSTEMS OF WESTERN NORTH AMERICA

Cordilleran Batholiths of Southern British Columbia

Magaritz and Taylor (1986) measured hydrogen and oxygen isotope ratios of 500 samples, mainly from granitic plutons (Figures 5.1 and 5.4) along a 700-km E-W traverse across the "accreted terranes" of southern British Columbia (latitudes 49° to 52°N). Despite the geological complexity and range of intrusive ages (Late Triassic to Tertiary), and even though there are "steps" in the isotopic values at some geologic boundaries (e.g., across the Strait of Georgia between the mainland and Vancouver Island), a clear-cut isotopic pattern was found: the $^{18}O/^{16}O$ and D/H ratios of the waters involved in hydrothermal interactions with the granitic rocks show a regular eastward trend of depletion in D and ^{18}O (Figure 5.5), indicating clearly that these waters were surface (meteoric) in origin (although seawater may have been important in the extreme western-most terrane in Vancouver Island).

Two groups of samples are unique in their high δD values (Figure 5.5). The first group is represented by two geographically isolated batholiths (Guichon and Thuya) that were not affected by the Tertiary meteoric hydrothermal systems and that have preserved a set of Early Jurassic to Triassic K/Ar ages. The second group is represented by the Jurassic plutons of Vancouver Island; there the hydrothermal fluids were both D-rich and ^{18}O-rich ($\delta^{18}O > 0$), as evidenced by the fact that feldspars in the altered granites are enriched in ^{18}O relative to coexisting quartz (see Figure 5.1). Both "anomalies" can be explained if these terranes were located closer to the equator and/or in a maritime environment at the time of intrusive and hydrothermal activity, in agreement with available paleomagnetic data that indicate a considerable northward drift of these terranes prior to their accretion to the western margin of North America (see Magaritz and Taylor, 1986).

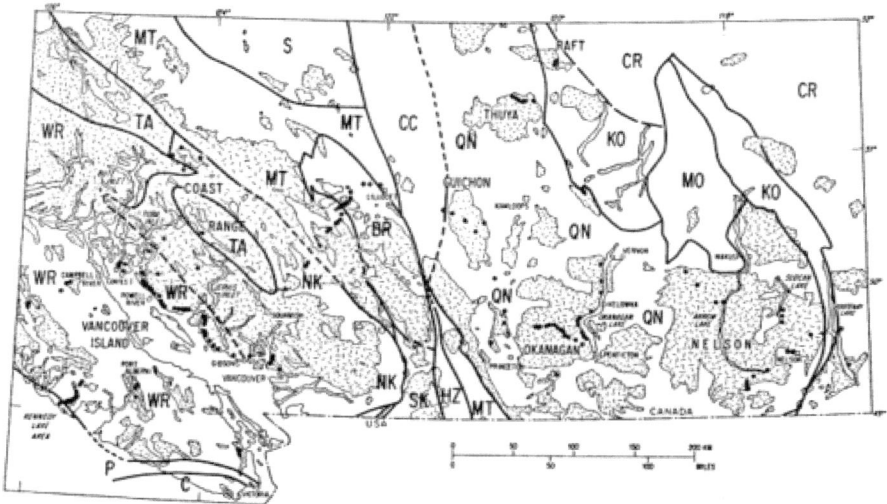

Figure 5.4
Generalized geologic map of the area studied by Magaritz and Taylor (1986) showing the major granitic batholiths, the various tectonostratigraphic terrane boundaries, and the locations of the samples analyzed for $^{18}O/^{16}O$ and D/H. WR, Wrangellia; NK, Nooksack; QN, Quesnel; MT, Methow-Tyaughton; P, Pacific rim; C, Crescent; KO, Kootenay; SK, Skagit; HZ, Hozameen; TA, Tracy Arm; S, Stikine; CC, Cache Creek; CR, Craton.

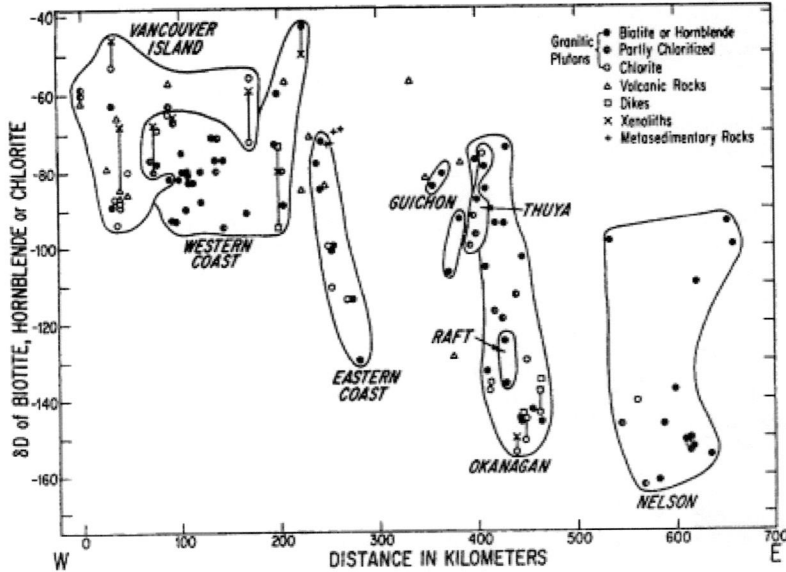

Figure 5.5
Plot of δD of all samples studied by Magaritz and Taylor (1986) versus distance eastward from the Pacific Coast of Vancouver Island.

Excluding these anomalous areas, two distinct ages of meteoric hydrothermal activity can be identified along the 700-km traverse, namely Cretaceous in the west and early to mid-Tertiary in the east. The isotopic trends in the rocks are similar to the present-day patterns of meteoric waters in the region, with one primary difference: the paleowaters are enriched in D by about 20 per mil, compatible with a northward translation of these terranes, a climatic change, or both. The similarities of the patterns suggest a topography similar to that of the present day (a mountain chain along the coast) during the early Tertiary.

Although the freshest, least altered samples in each terrane in British Columbia typically have δD values close to or within the presently accepted "primary magmatic" δD range of -65 to -85, the subset of heavily altered granitic rocks exhibits steadily decreasing δD values from west to east (Figure 5.5). If we confine the discussion just to samples that show hydrothermal $^{18}O/^{16}O$ effects (i.e., those that must have experienced very high water-rock ratios), the measured δD values change systematically, as shown in Table 5.1.

The west-to-east δD changes in Table 5.1 are similar to what is observed in present-day meteoric waters in British Columbia (see Figure 8 in Magaritz and Taylor, 1986). Neither of the two older batholiths that have preserved early Mesozoic K/Ar ages (the Guichon and Thuya batholiths) are listed in Table 5.1 because neither shows any evidence of interaction with such low-δD meteoric waters. All of the Guichon and Thuya samples have essentially "normal" δD and $δ^{18}O$ values (Figures 5.5 and 5.6). The most plausible explanation of this phenomenon is that it is the absence of late Mesozoic and Cenozoic meteoric hydrothermal activity that preserved the "old" K/Ar ages of about 200 m.y. ago (Ma) in these granites. This is very feasible for these two batholiths, because both are relatively small and they crop out at considerable distances from any of the younger batholiths of southern British Columbia (Figure 5.4). Whereas most of the batholiths shown in Figure 5.4 are composite in that they contain plutons with ages ranging from early Mesozoic to mid-Tertiary, the Guichon and Thuya batholiths exclusively exhibit only early Mesozoic ages.

If we examine the geographic variations of both δD and $δ^{18}O$ throughout southern British Columbia, we obtain a series of L-shaped patterns, each one characteristic of a

TABLE 5.1 D/H Ratios of the Most Altered Granites in Various Areas of Southern British Columbia (after Magaritz and Taylor, 1986)

	δD Range, per mil
Vancouver Island (except Kennedy Lake)	-46 to -63
Kennedy Lake area	-80 to -94
Western Coast Batholith	-65 to -95
Central Coast Batholith	-80 to -90
Eastern Coast Batholith	-114 to -130
Princeton area	-107 to -129
Raft Batholith	-125 to -136
Okanagan Batholith	-130 to -151
Nelson Batholith	-147 to -163

specific geographic area. These effects show up nicely when δD is plotted against $\delta^{18}O$ feldspar (Figures 5.6 and 5.7). In both diagrams the horizontal arms of each L represent the samples that have been subjected to the highest water-rock ratios; in each case the horizontal arm displays an approximately constant δD value characteristic of equilibrium with the meteoric hydrothermal waters of that particular geographic area.

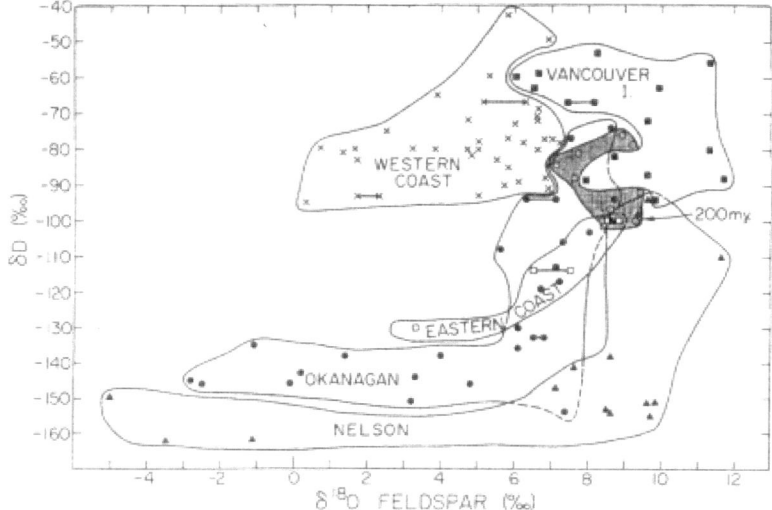

Figure 5.6
Plot of δD versus $\delta^{18}O$ feldspar for samples of the various granitic batholiths studied by Magaritz and Taylor (1986). The stippled area labeled 200 m.y. indicates samples from the relatively old Thuya and Guichon batholiths (see text).

Making certain assumptions about the temperature and other parameters (see Taylor, 1977), we may calculate the δD value of the H_2O that coexisted with the hydrothermal chlorites and biotites in the heavily altered samples, as shown in Figure 5.7. Such a δD value will represent the original value of the surface waters involved in the hydrothermal convective systems at the time of alteration, so by plotting this value on the meteoric water line (Craig, 1961), we can obtain a complete picture of the original isotopic compositions of these waters prior to their entrance into each hydrothermal system; such a calculation was carried out for the various geographical areas delineated in Figure 5.6, with the results as shown in Figure 5.7.

A suite of samples was collected from the Okanagan Batholith along virtually the entire length of Okanagan Lake (Figure 5.4). This suite of samples differs dramatically from those in the main part of the Okanagan Batholith in that every sample is markedly depleted in deuterium, typically -130 to -154. The Okanagan Lake samples are also typically depleted in ^{18}O and intensely chloritized;

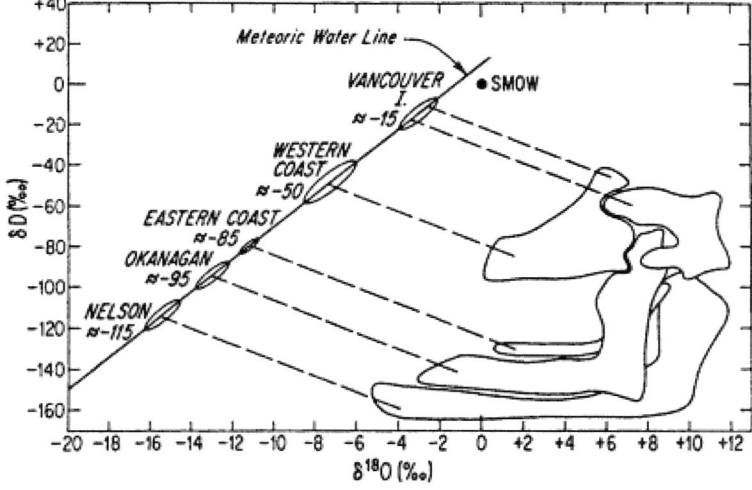

Figure 5.7
Plot of δD versus $\delta^{18}O$ feldspar, showing the various batholith fields from Figure 5.6, the meteoric water line of Craig (1961), and the calculated δD and $\delta^{18}O$ values of the pristine, unexchanged meteoric waters that were the source of the hydrothermal fluids that produced the various "inverted-L" patterns in each batholith in southern British Columbia (see Taylor, 1977, for details of the calculations).

they lie within the horizontal arm of the inverted-L pattern shown for the Okanagan Batholith in Figure 5.6.

Along the Okanagan Lake traverse, the Mesozoic plutons are intruded by several much younger porphyry stocks and dikes (the Eocene Coryell intrusions and associated Princeton volcanics). The dikes and porphyries consistently have lower $\delta^{18}O$ values than the granitic country rocks that they intrude, even though the Mesozoic plutons have locally been thoroughly depleted in ^{18}O; some samples have whole-rock $\delta^{18}O$ as low as +0.2 and +1.1 and feldspar $\delta^{18}O$ as low as -2.8 and -0.2. The dramatically larger ^{18}O depletions observed in the vicinity of Okanagan Lake are almost certainly due in large part to the abundant Tertiary dikes and stocks that occur near this lineament. However, Okanagan Lake also probably occupies a major fracture zone, providing for enhanced permeability that would allow much greater circulation of surface waters to great depths. In fact, this zone of weakness (a rift zone?) may also have provided the access routes followed by the Tertiary magmas that were the immediate cause of the large-scale hydrothermal alteration that is so prominent in the vicinity of this topographic depression.

The Nelson Batholith (Figure 5.4) is also a composite batholith, dominantly Mesozoic in age, whose southwest-em portion is intruded by a group of Tertiary plutons (Coryell intrusions). All of the strikingly ^{18}O-depleted samples from the Nelson Batholith were collected either (1) in close proximity to these Tertiary porphyry intrusions or (2) along the edge of one of the large, narrow, north-trending lakes that are so prominent in southern British Columbia. Analogous to the situation described above for the more extensive sample set from Okanagan Lake, it is probable that both Slocan Lake and Arrow Lake occupy fracture zones that represented major hydrothermal conduits for heated Tertiary meteoric waters as well as access routes for the Coryell intrusions.

Meteoric Hydrothermal Effects of Eocene Magmatism, Southern Idaho Batholith

A series of isotopic studies by Taylor and Magaritz (1978), Criss *et al.* (1982), Criss and Taylor (1983), and Criss and Fleck (1987) showed that widespread meteoric hydrothermal systems formed in the Idaho Batholith about 40 to 45 Ma, associated with the emplacement of several large epizonal Eocene plutons. The Eocene plutons were intruded at rather shallow depths, probably less than 7 km, and some intrude coeval volcanic rocks of the Challis volcanic field. The map in Figure 5.8 illustrates the generalized geology of the southern two-thirds of Idaho Batholith, particularly focusing on the $^{18}O/^{16}O$ effects produced by these hydrothermal systems. The surrounding Mesozoic plutons underwent striking ^{18}O depletions over more than 8000 km^2 (Taylor, 1977; Taylor and Magaritz, 1978; Criss *et al.*, 1982, 1984; Criss and Taylor, 1983). Deuterium depletions in these rocks are observed across an even wider zone, probably at least 25,000 km^2 (practically the entire area of the batholith shown in Figure 5.8 has $\delta D < -120$). The very extensive D/H effects are a result of the fact that only tiny amounts of H_2O are required to "reset" the D/H ratios of the hydroxyl-bearing minerals in a granite (Taylor, 1977).

The hatchured contours in Figure 5.8 indicate the extent of the zones of ^{18}O depletions and disequilibrium quartz-feldspar $^{18}O/^{16}O$ fractionations, analogous to those shown for southern British Columbia in Figure 5.1. So far as can be established with the presently available data, most of the *Eocene* hydrothermal systems in Idaho are Type 1. Only very local development of deeper-seated, hotter, Type 2 hydrothermal systems has been found, and these are only

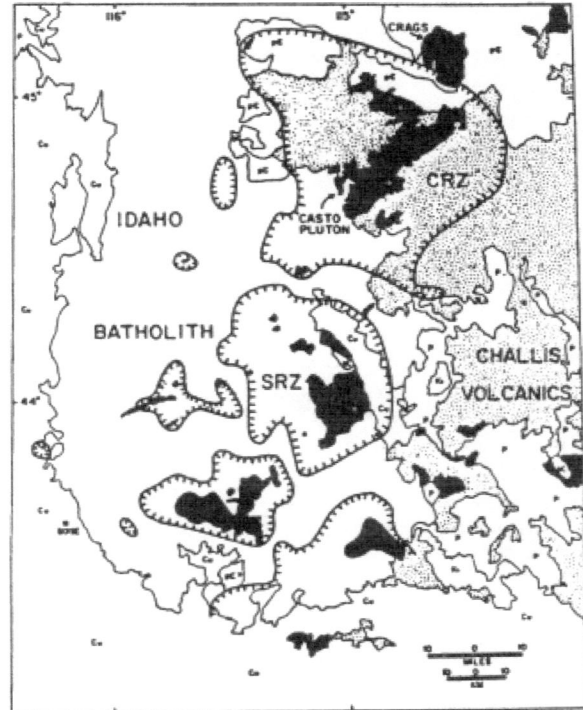

Figure 5.8
Generalized geologic map of south-central Idaho showing the Idaho batholith (blank), Eocene epizonal plutons (solid black), and the Challis Volcanics (stippled). The hachured line denotes the perimeter of identified zones of intense meteoric hydrothermal alteration and ^{18}O depletion, all of which are related to these Eocene plutons. The CRZ is a giant zone of alteration associated with the Casto pluton, and the SRZ is the Sawtooth Ring Zone, both of which are thought to be remnants of giant Eocene calderas (after Criss and Taylor, 1983; Criss et al., 1984).

observed directly adjacent to some of the largest Eocene plutons, where fresh unchloritized biotites exhibit 40 to 45 Ma K/Ar ages (Criss et al., 1982).

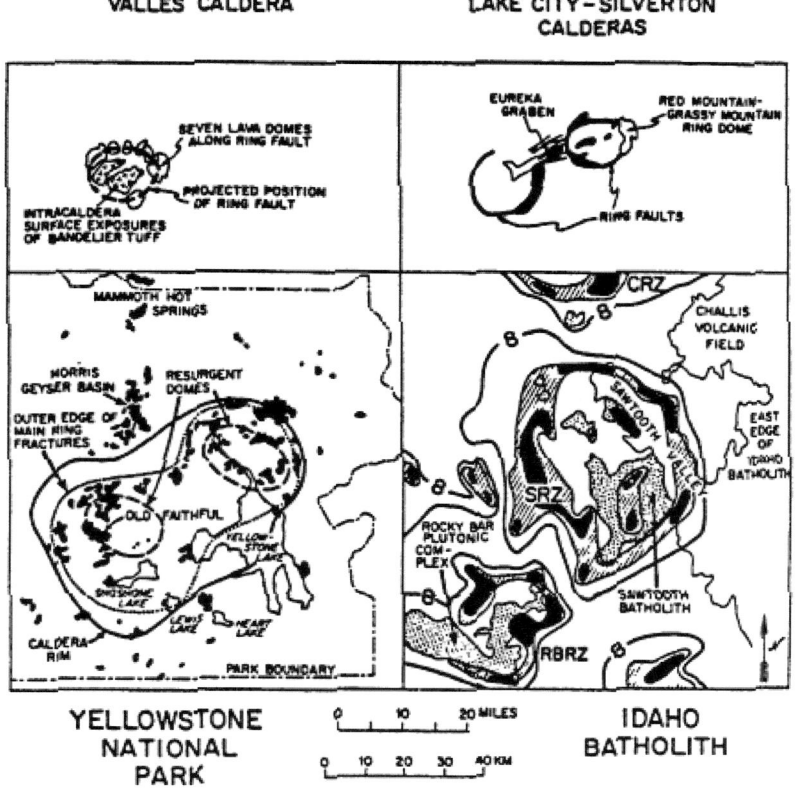

Figure 5.9
Comparison of the sizes and hydrothermal flow patterns of caldera-related hydrothermal systems (modified from Taylor, 1974c; Criss and Taylor, 1983; and Larson and Taylor, 1986c). Black areas shown in the Lake City-Silverton calderas and the Idaho Batholith indicate zones of very strong ^{18}O depletion, typically associated with caldera ring-fracture zones or with central resurgent intrusions within the calderas. Thermal springs in the presently active Yellowstone National Park are also shown as black areas (see Christiansen, 1984). Note the similarity between the present-day distribution of thermal springs at Yellowstone and the Eocene features in the fossil meteoric hydrothermal systems in the SRZ (Sawtooth Ring Zone). The map of the Valles caldera is shown for size comparison [this is another present-day, very active geothermal system that is strikingly similar in size and geologic setting to the Miocene (23 Ma) Lake City caldera studied by Larson and Taylor (1986a,c)].

The largest low-^{18}O zone, the 4500-km^2 Casto Ring Zone (CRZ), is centered on the 700-km^2 Casto pluton and includes the remains of two giant cauldron complexes in the Challis Volcanics (Criss et al., 1984). Another large (2500 km^2) system is the Sawtooth Ring Zone (SRZ), an annular zone of low-$\delta^{18}O$ values that encompasses the Sawtooth batholith (Figure 5.9) and that probably represents the subvolcanic part of the ring-fracture system of an Eocene caldera. Several other zones, mostly a few hundred square kilometers in extent, are associated with hydrothermal metamorphism in the contact zones of smaller igneous bodies. Isotopic material balance w/r ratios of approximately unity characterize most of these anomalous regions, regardless of their size. The principal zones of ^{18}O depletion in the Idaho Batholith are thus associated with either (1) the highly fractured ring zones of caldera complexes or (2) smaller stocks, some of which are resurgent intrusions emplaced into the central portions of deeply eroded calderas.

The Eocene plutons in the northern part of the Idaho batholith are just as numerous and just as large as in the south, but the $^{18}O/^{16}O$ hydrothermal effects there are much diminished compared to those shown in Figure 5.8 (Criss and Fleck, 1987). The sizes of the areas with δD values less than -120 are also very much diminished in the north (Criss and Fleck, 1987). This suggests that the southern area was either more highly fractured or that to the north we may be simply looking deeper into the Eocene continental crust, either of which would imply smaller permeabilities in the north. If it is the latter, this would imply that the Type 1 systems simply die out with depth as the w/r ratios decrease. This in turn supports the idea that Eocene Type 2 systems were rare in Idaho, and in fact such Type 2 systems in general probably encompass much smaller volumes of rock than Type 1 systems; this is because Type 2 systems require either (1) the simultaneous development of *both* high w/r ratios *and* high temperatures or (2) that the hydrothermal circulation persists for an unusually long period of time.

The fossil hydrothermal systems in the Idaho Batholith are among the largest associated with granitic plutons anywhere in the world (Criss and Taylor, 1986), and it is interesting in Figure 5.9 to compare the isotopic relationships observed in Idaho with the relationships found in the Quaternary Yellowstone Volcanic Field (described in more detail below). Despite the massive size of the Eocene hydrothermal systems in Idaho and despite their similarity in size and general character to the Yellowstone caldera systems, in Idaho we have not yet observed any development of the types of low-^{18}O magmas that are so common at Yellowstone (or in the southwestern Nevada caldera complex, in Iceland, or in the Seychelles Islands, as discussed below). Thus, a transition to a Type 2 system may be a requirement for the production of significant volumes of low-^{18}O magmas. On the other hand, the apparent absence of low-^{18}O magmas in Idaho may simply be due to a lack of detailed sampling of the Eocene Challis Volcanics or to the evidence having been eroded away.

Miocene Meteoric Hydrothermal Systems, Western Cascade Range, Oregon

The first ^{18}O/^{16}O study to demonstrate widespread meteoric hydrothermal effects in North America was that of Taylor (1971), who demonstrated that anomalously low δ^{18}O values (-6 to +4) occur around several Miocene diorite and granodiorite plutons that intrude the Tertiary volcanic rocks of the Western Cascade Range. The reason that these were the first fossil meteoric hydrothermal systems to be defined in North America is that the petrographic descriptions in a paper by Buddington and Callaghan (1936) were perceived to be astonishingly similar to features observed in the rocks of the Scottish Hebrides and the Skaergaard intrusion; the latter were the first identified fossil meteoric hydrothermal systems in the world (Taylor, 1968), and it therefore seemed likely that the Oregon localities might also be good candidates for such systems (Taylor, 1971).

The low-^{18}O plutonic and volcanic rocks of the Western Cascades are typically strongly propylitized and are estimated by Taylor (1971) to comprise a total of 1200 km^2 (8 percent) of this volcanic area (Figure 5.10). Rocks collected more than three stock diameters from the intrusive contacts typically have "normal" δ^{18}O values of +5.8 to +8.2. These isotopic data were interpreted in terms of convective circulation of heated groundwaters during the crystallization and cooling of the central intrusions. A 75-km^2 zone of low-^{18}O volcanic rock (average +1.1) occurs in the Bohemia mining district, which is typical of these plutonic centers (Figure 5.10). Two small (<3 km) stocks of augite granodiorite and several dikes and plugs intrude the andesites, tuffs, and breccias in the central part of this concentric low-^{18}O zone, and a 0.3- to 0.6-km-wide contact metamorphic aureole of tourmaline hornfels surrounds the larger stocks and grades into a zone of propylitic alteration at greater distance (Buddington and Callaghan, 1936).

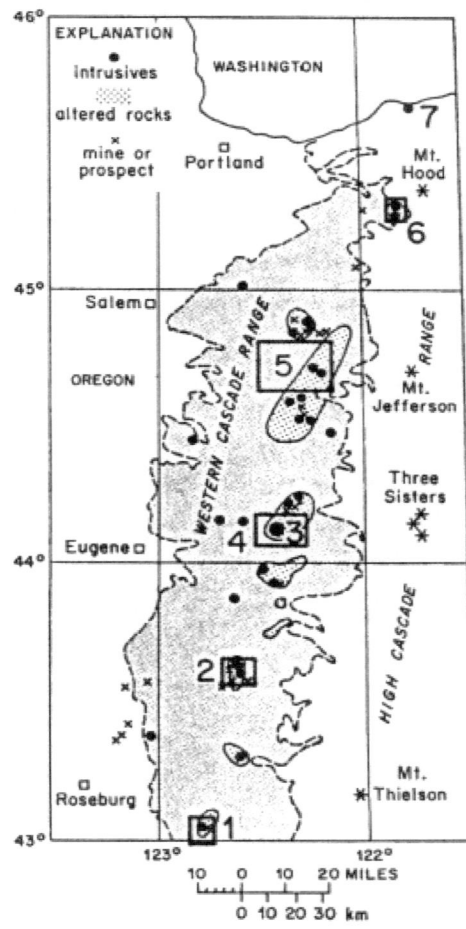

Figure 5.10
Map of part of western Oregon showing the distribution of Tertiary volcanic rocks of the Western Cascade Range (dark stippled pattern); also shown are the occurrences of Tertiary diorites and granodiorites as well as areas of propylitic alteration and mineralization commonly associated with these mediumgrained igneous rocks (after Taylor, 1971). The numbers indicate the various localities studied by Taylor (1971); Area (2) is the Bohemia Mining District referred to in the text.

Other Localities

The above discussion has focused only on the three most widespread regional studies of fossil meteoric hydrothermal systems in the western United States and Canada. A large number of such systems have now been described in the literature, but most are relatively small, isolated systems associated with individual granitic plutons (typically small stocks). Many of these occurrences have been tabulated by Criss and Taylor (1986), and a few of the most notable systems are described below.

Peninsular Ranges Batholith At the present stage of erosion this giant composite batholith in southern and Baja California shows little evidence for interactions with heated surface waters over most of its outcrop area (Taylor and Silver, 1978; Silver *et al.*, 1979; Taylor, 1986). However, all along its western edge in California, which is the shallowest part of this plutonic complex and the only portion where the roof of the batholith is exposed, extremely low $\delta^{18}O$ values and nonequilibrium quartz-feldspar fractionations are observed. Interestingly, just south of the United States-Mexico border, the roof-zone of this batholith changes its $^{18}O/^{16}O$ characteristics, and much higher $\delta^{18}O$ values and "reversed" quartz-feldspar fractionations are observed; Taylor and Silver (1978) attribute these effects to exchange with higher-^{18}O waters, such as marine formation waters or ocean water. Thus, the southern California portion of this batholith was apparently intruded in a subaerial environment, but farther south the surface environment above the plutons must have changed from subaerial to submarine during the time period of emplacement 130 to 105 Ma in the Lower Cretaceous. An analogous situation today might be observed where the Aleutian chain of calc-alkaline volcanoes (submarine) grades onto the Alaskan Peninsula (subaerial).

Lake City and Silverton Calderas, San Juan Mountains, Colorado Most of the mid-Tertiary calderas of the San Juan Mountains appear to have been associated with meteoric hydrothermal systems (Taylor, 1974c). Two of the most heavily investigated examples are adjacent calderas (Silverton and Lake City) that occupy the even older Uncompahgre caldera complex in the Western San Juans; these two calderas are joined by the highly fractured and altered Eureka graben (Taylor, 1974c; Casadevall and Ohmoto, 1977; Forester and Taylor, 1980; Larson and Taylor, 1986a,b,c). As observed in Idaho and Yellowstone (Figure 5.9), the most intense alteration and the most striking ^{18}O depletions in both of these caldera complexes are associated with either (1) the caldera ring-fracture zone or (2) the central resurgent intrusion. The $^{18}O/^{16}O$ effects have been measured over vertical distances of at least 2 km, and they may be readily inferred to have extended to *at least* a depth of 5 km beneath the original land surface.

Boulder Batholith, Montana In western Montana extensive oxygen and hydrogen isotope evidence for meteoric hydrothermal activity was found in the Boulder Batholith by Sheppard and Taylor (1974). They showed that the δD values of various plutons in this composite batholith range from -63 to -155 per mil, with the lowest δD values typically being associated with relatively low $\delta^{18}O$ values, epidote and chlorite alteration, and disequilibrium quartzfeldspar ^{18}O fractionations. The lowest δD and $\delta^{18}O$ values occur in mineralized areas along the western side (roof) of the batholith in the Butte and Wickes mining districts or are associated with small stocks intruded into the Elkhorn Mountains Volcanics. Alterations may have begun in the Late Cretaceous, but the dominant episode was in the early Tertiary; these Paleocene meteoric waters have calculated D values of -100 to -115 and initial $\delta^{18}O$ values of -14 to -16 (compare with Figure 5.7). The mineralization in the main-stage veins in the giant Butte Cu-Zn-Pb-Mn deposit show clear-cut evidence of deposition from meteoric hydrothermal fluids, and these low-^{18}O effects can be observed in the stopes, shafts, and tunnels of the mine itself to extend over a vertical distance of more than 2 km. This Paleocene (~60 Ma) meteoric hydrothermal system formed entirely within the extremely highly fractured Late Cretaceous Butte quartz monzonite pluton; some of the main-stage, sulfide-rich veins are several meters wide, and individual veins can be followed laterally for 4 to 5 km and vertically for 1 to 2 km. The entire meteoric hydrothermal system must have originally extended over a vertical distance of at least 5 km.

Western Nevada Gold-Silver Ore Deposits All of the so-called epithermal, volcanic-hosted, Au-Ag deposits of the Basin and Range Province appear to have formed from heated meteoric groundwaters (Taylor, 1973, 1974a). At least 25 separate localities have been documented by Taylor (1973, 1974a) and O'Neil and Silberman (1974). Some of the largest and most significant examples are Tonopah, Goldfield, Comstock Lode, Bodie, and Aurora. Many of these deposits are associated with the ring fractures of mid-to late-Tertiary rhyolitic caldera complexes.

Central British Columbia Magaritz and Taylor (1976) showed that the entire eastern side of the Coast Plutonic Complex at latitudes 55° to 56°N displayed evidence of meteoric hydrothermal alteration. These effects extend well out into the sedimentary and volcanic country rocks, and they can also be observed in the vicinity of smaller plutons as much as 200 km eastward from the Coast Batholith. The D/H ratios have been lowered to values less than -120 over an area of at least 5000 km^2. The alteration appears to be mainly associated with Eocene plutons, which are abundant throughout the area.

HYDROTHERMAL SYSTEMS ASSOCIATED WITH LAYERED GABBRO BODIES

Layered gabbro plutons that have established meteoric hydrothermal convective systems are the Skaergaard intrusion in East Greenland, Jabal at Tiff in Saudi Arabia, Stony Mountain in Colorado, the Islands of Skye and Mull in Scotland, and Ardnamurchan in Scotland (Taylor and Forester, 1971, 1979; Forester and Taylor, 1976, 1977, 1980; Taylor, 1980, 1983). Analogous examples involving ocean-water hydrothermal systems are observed in all ophiolite complexes, including Cyprus and the Samail ophiolite, Oman (Gregory and Taylor, 1981), and in dredged samples from the Indian Ocean (Ito and Clayton, 1983; Stakes *et al.*, 1983). In all these areas large portions of the layered gabbro complexes display marked $^{18}O/^{16}O$ disequilibrium between coexisting pyroxene and plagioclase, exactly analogous to the effects shown for quartz and feldspar in Figure 5.1.

Mineralogical Alteration Effects in 18O-Depleted Layered Gabbros

Although there is local development of chlorite, epidote, actinolite, talc, sphene, prehnite, and other low-temperature (greenschist facies) minerals in most layered gabbros, particularly in areas of diking, heavy fracturing and veining, or multiple intrusion, it is important to realize that ^{18}O-depleted gabbro bodies in general are astonishingly free of any of the petrographic or mineralogical features that geologists commonly utilize as indicative of hydrothermal alteration. This statement, of course, does not apply where an older gabbro body has been invaded and hydrothermally metamorphosed by a younger intrusion (e.g., the heavily altered Broadford gabbro body on Skye; see Forester and Taylor, 1977), but it definitely applies to *all* layered gabbro bodies involved in a single cycle of meteoric hydrothermal activity (i.e., the convection cells set up by that particular gabbro magma chamber itself).

Because petrographic evidence for hydrothermal alteration is usually very rare in such ^{18}O-depleted gabbros (although it can often be perceived through a careful thin-section study), the $^{18}O/^{16}O$ "reversals" between plagioclase and pyroxene always represent the easiest and most definitive way to characterize such phenomena in gabbros. However, even in the absence of $^{18}O/^{16}O$ studies, fossil hydrothermal systems should be suspected if any of the following fairly subtle petrographic features are observed: (1) clouding or turbidity of some of the plagioclase; (2) development of minor talc-magnetite rims on olivine; (3) coarsening of exsolution lamellae around microfractures in clinopyroxene grains; (4) macroscopic veins that are easily seen on slightly weathered outcrops in glaciated areas or arid regions (on careful examination these veins often prove to contain pyroxene, magnetite, and high-temperature amphiboles such as hornblende); and (5) development of minor amounts of actinolite, biotite, chlorite, and epidote in zones transitional to lower-temperature hydrothermal activity (e.g., late-stage veins). All of the above features are described in the references given above and/or in Norton *et al.* (1984). In addition, Ferry (1985) demonstrated that secondary magnetite (relatively pure Fe_3O_4) is ubiquitous in the Skye gabbros and also that the calculated temperatures of formation of the talc-olivine-orthopyroxene assemblages in these gabbros are in the range of 525° to 545°C. The early veins throughout practically the entire section of layered gabbro in the Skaergaard intrusion contain hydrothermal clinopyroxenes with minimum solvus temperatures of 500° to 750°C (Manning and Bird, 1986).

It is thus an inescapable conclusion that layered gabbro bodies typically undergo meteoric hydrothermal alteration at very high temperatures, in large part in the range 500° to 900°C (Norton and Taylor, 1979; Taylor and Forester, 1979; Taylor, 1987). This conclusion is firm, despite statements to the contrary by Cathles (1983) on this matter. Only at such high temperatures could we simultaneously obtain the clear $^{18}O/^{16}O$ evidence for intense hydrothermal alteration combined with the virtually complete absence of low-temperature hydrous minerals. The temperatures (500° to 900°C) and PH_2O values (200 to 800 bars) are clearly not within the stability fields of chlorite, epidote, serpentine, actinolite, and clay minerals during the bulk of the hydrothermal activity. In fact, much of the mineralogical alteration that does occur in these gabbros takes place at 450° to 550°C, considerably higher than the average temperature of alteration in the granitic plutons described above but still lower than the temperatures at which the bulk of the ^{18}O depletion occurred in these gabbros (Taylor and Forester, 1979; Ferry, 1985).

Why then do we mainly see only the effects of 350° to 400°C H_2O, for example, at the mid-ocean ridge spreading centers and on land in places like the Salton Sea? As Norton and Knight (1977) and Norton (1984) have emphasized, convective systems are strongly controlled by the approximate coincidence of a viscosity minimum, together with a maximum in the isobaric coefficient of thermal expansion in critical to supercritical H_2O (350°C $< T <$ 450°C; 200 $< P <$ 800 bars). These are also the conditions where the density of H_2O undergoes the most rapid change as a function of temperature. Furthermore, the heat capacity of the fluid under these pressure and temperature conditions is quite large and is maximized at the critical point. Thus, in this general pressure-temperature range, the buoyancy and heat transport properties of the fluid are maximized and the drag forces minimized (Norton and Knight, 1977). In fact, it is very likely that these physical properties of H_2O control the observed upper limit of 350° to 400°C observed during surface venting of modern hydrothermal systems, not the "fact" that all rocks hotter than 400°C are impermeable, as proposed by Cathles (1983).

Contrasting Effects in Gabbros and Granites

As indicated above, very distinctive mineralogical features are observed in ^{18}O-depleted layered gabbros compared with those typically observed in ^{18}O-depleted granitic plutons (Taylor, 1987). To recapitulate some of these differences: (1) gabbros with reversed nonequilibrium δ^{18}O plagioclase-pyroxene values but practically no mineralogical alteration are quite common, whereas (2) granodiorite, quartz monzonite, and tonalite plutons with large nonequilibrium δ^{18}O quartz-feldspar values are almost invariably strongly altered, containing abundant chlorite, epidote, sericite, and turbid feldspars, for example. A significant question is: Why does most of the external (hydrostatic) meteoric hydrothermal fluid move through layered gabbro bodies at much higher temperatures than in the case with granitic plutons?

There are several aspects to this problem. Seven of the major differences between gabbros and granites that probably contribute to the observed isotopic contrasts are listed in Table 5.2. It is interesting that each of these seven different igneous rock properties seem to favor a higher temperature of meteoric hydrothermal alteration in the gabbros as opposed to the granites.

The most obvious reason why the meteoric hydrothermal systems of gabbros can exist at much higher temperatures than in granites is that gabbros solidify at much higher temperatures, 1000° to 1050°C. Granitic materials are still liquid at these temperatures and are thus unable to sustain fractures that would allow penetration by hydrothermal fluids. In addition, the latent heats of crystallization of gabbros are higher and the initial percentage of crystals in the gabbroic magmas is typically lower at the time of intrusion. Both features indicate that there is much

TABLE 5.2 Contrasting Properties of Granitic and Gabbroic Plutons (after Taylor, 1987)

	Granitic Plutons	Gabbroic Plutons
Latent heat of fusion	Low (40 cal/g)	High (100 cal/g)
Magma temperature	650° to 900°C	1000° to 1200°C
Initial H$_2$O content of magma	2 to 5 wt.% (second boiling common)	<1 wt.% (second boiling localized in late-stage granophyres)
Initial percent crystallized	20 to 60 percent(reduces latent heat)	<10 percent (little effect on latent heat)
Fracture network	May be very dense within the pluton because of hydraulic fracturing (e.g., porphyry Cu bodies) and caldera collapse; also very large volume change associated with α-β quartz transition	Usually simple contraction cooling and jointing, although occurrence in rift-zone environments is common, suggesting a very deep seated and pervasive extensional fracture network in the country rocks
Presence of magmatic envelope at lithostatic pressure	Very common for H$_2$O-rich magmas at shallow depths in the crust rich in biotite and hornblende	Nonexistent, except perhaps in H$_2$O-rich alkali gabbro magmas
Geometry of crystallization	Homogeneous solidification of the entire body (only local separation of late-stage melt from crystals); forms a single integrated meteoric hydrothermal system or overlapping systems when complicated by multiple intrusions; these systems cannot migrate into the pluton until the magmatic H$_2$O envelope is dissipated	Strong separation of cumulate crystals from silicate melt, with crystallization upward from the floor of the magma chamber; typically forms two decoupled meteoric hydrothermal systems, a lower-T system in the country rocks and upper border zone rocks above the late-crystallizing sheet of magma near the roof of the chamber, and a higher-T system in the layered cumulates below this magma sheet

more energy available in a gabbro for raising external H_2O to a high temperature than there is in a granite.

Possibly of equal importance is the geometry of crystallization of the magma body. The typical granitic magma probably is intruded with a higher percentage of crystals, and, being relatively viscous, both these crystals and any newly formed crystals are probably distributed throughout the magma chamber, with final crystallization taking place in the center and at the lowest levels of the body. This is decidedly not how layered gabbros crystallize. The layered gabbros characteristically solidify upward from the bottom, and the last liquid to crystallize is a sheet-like layer of liquid near the roof of the body. This sheet of late-stage liquid crystallizes very slowly, because crystallization proceeds as the square root of time and the body is at that stage surrounded by a mass of very hot insulating rock. This means that the crystalline fractured gabbro underneath the sheet of late-stage magmatic liquid will remain very hot for a long time. After such material has fractured, it will be penetrated by very hot aqueous fluids that flow inward underneath the magma sheet, which is itself impermeable to the fracture-controlled hydrothermal system. This characteristically seems to produce two decoupled hydrothermal systems in layered gabbros (see Norton and Taylor, 1979; Gregory and Taylor, 1981; and Norton et al., 1984): (1) relatively low temperature, vigorous hydrothermal system above the magma sheet, where water-rock ratios are high and $T = 250°$ to $400°C$, and (2) much higher temperature system below the impermeable magma sheet where $T = 500°$ to $900°C$ and water-rock ratios are much lower (0.1 to 0.3). Both of these de-coupled systems are usually Type 1, as defined above; however, the bottom system can be transitional to Type 2, and Type 3 conditions may occur in local areas within and adjacent to the actual magma body. Only after the final liquid sheet crystallizes can the temperature of the layered gabbro begin to fall rapidly, and at this stage the two decoupled systems become connected and the temperature of the intrusion sharply declines as strong upward fluid flow is now possible out through the top of the intrusion.

Another major difference is the contrasting effects of magmatic H_2O in the two cases. It is well known that granitic magmas typically contain much higher concentrations of H_2O than tholeiitic gabbro magmas. This magmatic water in fact is thought to provide the force that causes abundant fracturing in porphyry copper deposits (Burnham, 1979). Any strong fracturing event will increase the permeability enormously, allowing correspondingly larger amounts of fluid to enter the system on a shorter time scale; the pluton would thus be cooled down to $200°$ to $300°C$ fairly rapidly. These types of effects will not occur in the gabbroic systems. Another contributing aspect is the major volume change that accompanies the α-β quartz transition in granites that is absent in gabbro; this will enhance the microfracture permeability of all quartz-bearing rocks that originally form at temperatures above this transition.

The final feature that may be important is the fact that any magma body that releases significant H_2O under lithostatic pressure can produce a magmatic H_2O envelope that will keep the external meteoric hydrothermal system outside the intrusion until this magmatic H_2O envelope is dissipated. Exactly this effect seems to occur in porphyry copper systems (Sheppard et al., 1971; Taylor, 1974a). Thus, no low-^{18}O effects would be seen in the pluton until the very late low-temperature stages. Because of their low magmatic H_2O contents, these effects would not typically be observed in tholeiitic gabbros. However, they might possibly be seen in volatile-rich alkali gabbros, and this may be the reason for the contrasting isotopic behavior of the Lilloise alkalic hornblende gabbro intrusion as compared to the tholeiitic Skaergaard intrusion. Although both of these intrusions were emplaced into similar country rocks in East Greenland about 55 Ma ago, strong ^{18}O depletions are not observed in the Lilloise body (Sheppard et al., 1977).

LOW-18O BASALTIC AND RHYOLITIC MAGMAS

Muehlenbachs et al. (1974) and Friedman et al. (1974), respectively, made the important discoveries that low-^{18}O basaltic and rhyolitic magmas were produced in large volumes in two of the major late Cenozoic volcanic fields of the world, in Iceland and in the Yellowstone Plateau, Wyoming. There is now a general consensus among all workers who have studied this problem that low-^{18}O meteoric hydrothermal fluids played a major role in the genesis of such magmas. At the present time the principal argument is whether such low-^{18}O magmas are produced by direct influx and exchange with the water (e.g., Hildreth et al., 1984) or whether they were formed by melting and/or assimilation of low-^{18}O hydrothermally altered rocks (e.g., Taylor, 1974a, 1977, 1987).

Low-18O Magmas Produced During Caldera Collapse, Yellowstone Volcanic Field

Hildreth et al. (1984) followed up the original discovery by Friedman et al. (1974) with a detailed chemical and isotopic study of the Quaternary rhyolites of the Yellowstone Plateau. This 17,000-km^2 volcanic field (Figure 5.9) consists of three large, overlapping, rhyolitic calderas, apparently a 115-km-long extension of the axis of the rift system associated with the Snake River Plain in southern Idaho. Brief, caldera-forming, ash-flow eruptions occurred 2.0, 1.3, and 0.6 Ma (Christiansen, 1984), with minimum volumes of 2500, 280, and 1000 km^3, respectively. Very large-scale, Type 1 hydrothermal systems are present all along the ring fractures of the youngest caldera and adjacent to the resurgent domes (Figure 5.9). Given the fact that volcanic and hydrothermal activity has been going on here for more than 2 m.y., and that a large magma body exists at the present time at a depth of only a few kilometers (Christiansen, 1984), it is virtually certain that Type 2 systems also locally occur at depth.

Almost all of the Yellowstone rhyolites are somewhat depleted in ^{18}O relative to the "normal" $\delta^{18}O$ values of +7 to +10 usually observed in silicic volcanic rocks throughout the world (Taylor, 1968, 1974a). However, the postcaldera rhyolites of the first and third caldera cycles include some extraordinarily low ^{18}O eruptive units (Figure 5.11). The combined areal extent of the two first-cycle low-^{18}O flows ($\delta^{18}O = +2.9$ to $+3.6$) is ~66 km^2; their original volume was at least 10 km^3, and they were erupted 30,000 to 350,000 yr after the >2500-km^3 Huckleberry Ridge Tuff. These low-^{18}O lavas cannot be chemically distinguished from the higher-^{18}O Yellowstone rhyolites; they all contain ≈76 wt.% SiO_2 and 15 to 20 percent phenocrysts of quartz, sanidine, plagioclase, clinopyroxene, fayalite, and Fe-Ti oxides.

During the third cycle, enormous ^{18}O depletions were observed (Figure 5.11) in lavas that vented in two separate areas ~45 km apart. These were erupted immediately after the eruption of the >1000-km^3 Lava Creek Tuff, along the compound ring-fracture zone of the Yellowstone caldera. In the northeast part of the caldera, there are three major low-^{18}O units, which together cover ~140 km^2 and represent >40 km^3 of magma, perhaps as much as 70 km^3, all of it having $\delta^{18}O$ between about +0.6 and +1.2. The second area of very low $\delta^{18}O$ third-cycle rhyolites is in the southwestern ring-fracture zone, where scattered exposures in a 25-km^2 area have $\delta^{18}O$ values between -0.1 and +0.6.

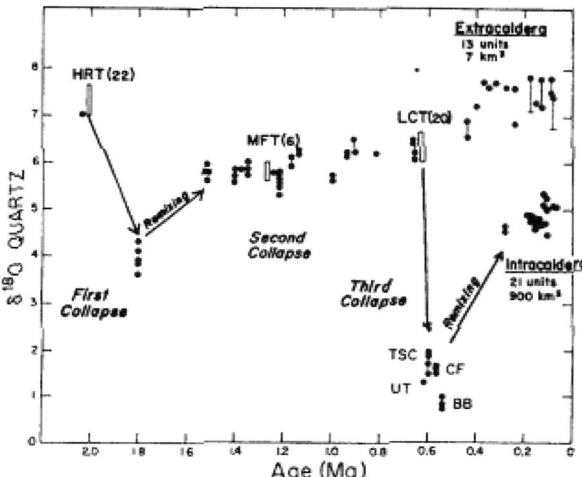

Figure 5.11 Plot of $\delta^{18}O$ quartz versus K/Ar age for the rhyolites of the Yellowstone Plateau volcanic field. HRT, MFT, and LCT are (with numbers of analyzed samples) the major ash-flow sheets referred to in the text (modified after Hildreth et al., 1984).

Hildreth *et al.* (1984) list the following observations that are critical to any theory of origin of the Yellowstone magmas:

1. The narrow range of $\delta^{18}O$ in the three major ash-flow sheets (+5 to +7) contrasts sharply with the wide variation in the postcaldera rhyolitic lavas (Figure 5.11). The biggest ^{18}O depletions directly follow the two largest ash-flow eruptions.
2. The ^{18}O depletions were geologically short-lived events (<300,000 to 500,000 yr) that followed caldera subsidence in some cases by less than 50,000 to 100,000 yr. Successively younger postcaldera lavas show partial recovery of the magma toward precaldera $\delta^{18}O$ values, as a result of mixing with deeper levels of the magma reservoir.
3. The pattern of stepwise ^{18}O depletions is reflected in lavas erupted as far apart as 115 km, indicating that they were part of an integrated magmatic system. More than 100 km^3 of magma was depleted by about 5 per mil and more than 1000 km^3 of magma was depleted by 1 to 2 per mil.
4. Depletion of ^{18}O occurred only in the subcaldera reservoir; contemporaneous rhyolites from outside that caldera have very high $\delta^{18}O$ values.
5. Sr and Pb isotopic ratios of the rhyolites display a zigzag pattern similar to that displayed by $\delta^{18}O$ in Figure 5.11, jumping to more radiogenic values just subsequent to caldera formation (Doe et al., 1982). The caldera-forming events obviously introduced country-rock Pb and Sr into the magma system.

Low-18O Basaltic and Rhyolitic Magmas in Iceland

The data of Muehlenbachs *et al.* (1974), Hattori and Muehlenbachs (1982), Condomines *et al.* (1983), and Gunnarsson *et al.* (1988) can be briefly summarized as follows (see Figure 5.12).

1. The relatively rare alkali olivine basalts on Iceland ($\delta^{18}O = +5.3$ to $+5.7$), which are found only on the periphery of the island at the edges of any deep meteoric hydrothermal circulation systems, have distinctly higher and more uniform (i.e., more "normal") $\delta^{18}O$ values than the much more abundant tholeiites and transitional basalts.
2. The lowest and most heterogeneous $\delta^{18}O$ values in

the Icelandic basalts, +1.8 to +5.4, are all confined to the quartz tholeiites of the eastern rift zone, particularly those from the large Krafla volcano.
3. The olivine tholeiites from the western rift zone have intermediate $\delta^{18}O$ values, +4.0 to +5.7, with the highest (i.e., most "normal") values closest to the coast on the southwest part of Reykjanes Peninsula.
4. In flank-zone volcanoes (e.g., Torfajokull and Hekla), where volcanism is transitional alkalic to alkalic in composition and is superimposed on older crust formed during an earlier period of rifting, the basaltic to andesitic lavas tend to have intermediate $\delta^{18}O$ values; these lavas are consistently only slightly depleted in ^{18}O (by about 0.5 to 1.0 per mil) compared to "normal" basalts, while the spatially related or contemporaneous silicic lavas have similar or much lower $\delta^{18}O$ values (e.g., Torfajokull: +3.6 to +4.4). At Torfajokull, parallel volcanic fissures, 1 to 2.5 km apart, have erupted lavas of variable chemical composition, from basaltic to silicic; these are associated with small, fissure-dependent, and very systematic variations in $\delta^{18}O$ that correlate very well with major and trace elements, suggesting eruption from distinctly different, well-mixed magma chambers beneath the central volcano.
5. On a regional scale there are also some crude correlations between $\delta^{18}O$ and chemical composition; increasing SiO_2 and K_2O tend to be accompanied by decreasing $\delta^{18}O$ for each petrologic class of basalts. In particular, the tholeiite with by far the most extreme $\delta^{18}O$ value (+1.8) also has by far the highest K_2O content.
6. The rhyolites and obsidians all over Iceland tend to have lower $\delta^{18}O$ values than the basalts, and the rhyolites with the lowest and most variable $\delta^{18}O$ values all come from the eastern rift zone.
7. Deep drill holes at several localities show a striking ^{18}O depletion in the lavas as a function of depth, particularly in the eastern rift zone (Figure 5.13).
8. There is a weak correlation between $\delta^{18}O$ and $^3He/\,^4He$, with the intermediate and silicic volcanic glasses typically having $^3He/^4He$ ratios close to the atmospheric value.

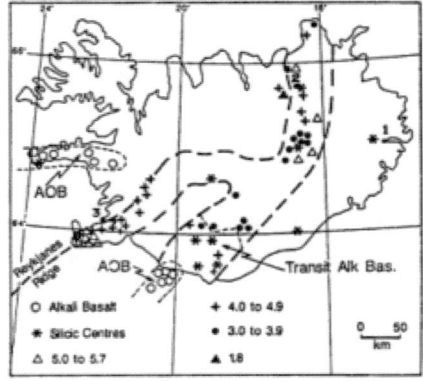

Figure 5.12
Map of Iceland showing the $\delta^{18}O$ values of tholeiitic and transitional basalts within the recently active volcanic zones (modified from Sheppard, 1986). The location of silicic volcanic centers is also shown. Data are from Muehlenbachs et al. (1974) and Condomines et al. (1983). Also shown are the locations of the three drill sites (1 = Reydarfjordur, 2 = Krafla, 3 = Reykjavik) where Hattori and Muehlenbachs (1982) obtained samples for $^{18}O/^{16}O$ analysis (see Figure 5.13).

The isotopic relationships described above strongly support the idea that assimilation and/or partial melting of hydrothermally altered country rocks in the deeper parts of the rift zone is the most likely mode of formation of the low-^{18}O magmas from Iceland. This explanation is compatible with the Pb isotope data of Welke et al. (1968) and the $^{87}Sr/^{86}Sr$ and rare-earth data of O'Nions and Gronvold (1973), and it readily explains why the most contaminated (i.e., most ^{18}O depleted) magmas are either the rhyolites or

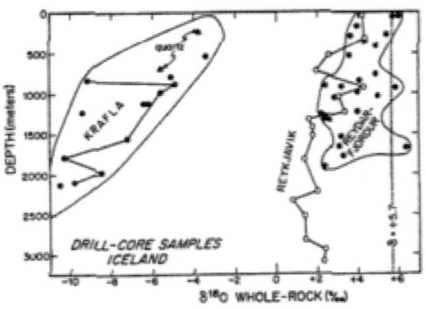

FIGURE 5.13
Plot of depth versus $\delta^{18}O$ (modified after Hattori and Muehlenbachs, 1982) for samples of hydrothermally altered basalt from three drill holes in Iceland (see Figure 5.12 for locations). The vertical dotted line indicates the average $\delta^{18}O$ value of most terrestrial tholeiitic basalts (including mid-ocean ridge samples) and lunar basalts. This figure demonstrates that essentially all of the volcanic rocks in Iceland are depleted in ^{18}O down to at least 3 km, with the ^{18}O depletions becoming more pronounced with depth and with proximity to a major active central volcano in the Eastern Rift Zone (e.g., Krafla). The solid lines connect samples from a single drill core.

those basalts that have probably most strongly interacted with roof rocks or rhyolite melts, namely the K-rich and Fe-rich tholeiites. Direct exchange kith meteoric waters would not be expected to produce these relationships. The overall smaller ^{18}O-depletion effects in the lavas from flank-zone volcanoes compared to those of active rift-zone volcanoes can be attributed to lower temperatures, lower water-rock ratios, and shallower hydrothermal circulation systems within the flank-zone environment (Gunnarson et al., 1988).

It is probably significant that all of the extremely low ^{18}O basaltic and rhyolitic magmas are confined to the eastern rift zone, which has been active only during the past 3 m.y. to 4 m.y. (Saemundsson, 1974). The magmas in this rift zone are penetrating upward through the lower parts of volcanic and plutonic rocks, which presumably were intensively hydrothermally altered in an earlier episode of magmatic activity at the time of their original formation in the western rift zone. Thus, the magmas coming up through the eastern rift zone would be interacting with country rocks that had already suffered heterogeneous ^{18}O depletions and which, through subsidence, have been brought down into a much higher temperature regime (15 to 20 km depth?).

Something on the order of 200 km^3 of low-^{18}O tholeiite has been erupted in the eastern rift zone in the past 12,000 yr (Jacobsson, 1972). It was the scale of this process that most bothered Muehlenbachs et al. (1974) when they rejected the meteoric hydrothermal explanation for the origin of these types of magmas. This mechanism for the origin of the Icelandic volcanic rocks has, however, been strongly favored over the past few years by Taylor (1974a, 1977, 1979), and recently Hattori and Muehlenbachs (1982), Condomines et al. (1983), and Gunnarson et al. (1988) have provided strong new support for this mechanism.

The model of Condomines et al. (1983), based on combined He, O, Sr, and Nd isotopic relationships, is shown in Figure 5.14. It is similar to the models of Taylor (1977, 1979) and Gunnarsson et al. (1988). These workers propose that the primary ^{18}O/^{16}O ratios of mantle-de-rived magmas were changed in a deep magma reservoir by exchange or contamination between the magma and the surrounding meteoric hydrothermally altered basaltic crust. Such processes do not appear to have introduced much water into the magma because Icelandic volcanism, except for subglacial eruptions, is usually not explosive, and hydrous minerals in plutonic ejecta are rare. The rhyolitic magmas are thus assumed to have been produced at relatively shallow crustal levels, either by melting of hydrothermally altered rocks or as the deeper magmas moved upward and underwent further contamination processes. These processes caused introduction of atmospherically derived helium into the magmas.

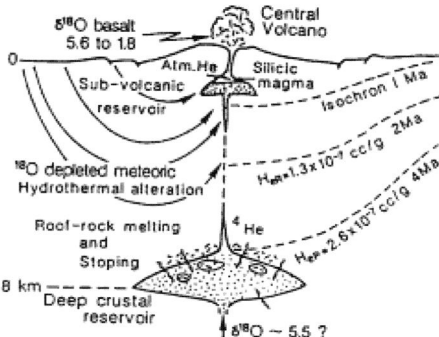

Figure 5.14
Schematic section summarizing current models for the evolution of the Icelandic crust (after Condomines et al., 1983). Radiogenic helium (He$_R$) is generated in the subsided lavas according to their age and U and Th contents, taken to be 0.3 and 1 ppm, respectively. Assimilation, magma mixing, and/ or exchange processes between the magmas and the hydrothermally altered volcanic pile, some of which has now subsided to great depths, are assumed to be the main processes that determine the isotopic and chemical compositions of the magmas.

Seychelles Batholith, Indian Ocean

The S eychelles Islands are unique among oceanic areas in that they are almost wholly composed of granite, a remnant of the breakup of the continent of Gondwanaland left isolated between Africa and India in the middle of the Indian Ocean. The main island of Mahè (Figure 5.15) is made up of a remarkably homogeneous, leucocratic, coarse-grained hornblende microperthite granite (Baker, 1963). This granite is latest Precambrian in age (650 Ma to 670 Ma; Miller and Mudie, 1961; Wasserburg et al., 1964; Michot and Deutsch, 1976). No remnants of older country rocks have been found anywhere on the island, except for a few small xenoliths in the granite. Except for two granite porphyry intrusions on the western side of the island, practically all of Mahè is made up of a single granite pluton that is just as homogeneous in ^{18}O/^{16}O and D/H as it is in terms of texture and mineralogy (Taylor, 1974b, 1977).

The data in Figure 5.15 leave no doubt that the main Mahè pluton crystallized from a very homogeneous, low^{18} granitic magma having a $\delta^{18}O = +3.3 \pm 0.3$. This is proved because (1) the striking ^{18}O/^{16}O homogeneity in the Mahè pluton could not have been produced by later subsolidus hydrothermal exchange and (2) the Δ^{18}O quartz-feldspar values are extremely uniform at 1.25 to 1.50, identical to the equilibrium values in "normal" granites. The existence of a meteoric hydrothermal event prior to emplacement of the pluton is confirmed by the occurrence of amphibolite xenoliths in some of the coastal granite samples that are even lower in ^{18}O than the granite host rock. The extreme ^{18}O depletions and the very large $\Delta^{18}O$ quartz-feldspar values (up to 3.5) in the northern coastal samples (see Figure 5.15) are evidence that the marginal portions of the Seychelles granitic complex underwent subsolidus meteoric hydrothermal alteration in a Type 1 system. We infer that a Type 2 system probably existed at even greater depths and that melting of such rocks produced this large (>300 km^3), remarkably homogeneous mass of low-^{18}O granitic magma that makes up the island of Mahè.

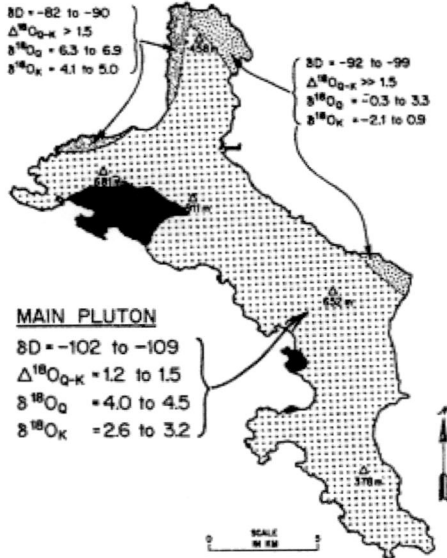

Figure 5.15
Generalized geological map of Mahè, the principal island of the Seychelles Group (after Taylor 1974b, 1977), showing the variation of $\delta^{18}O$ and δD (Q = quartz, K = alkali feldspar). Most of the island is composed of a coarse-grained hornblende microperthite granite with an extremely uniform isotopic composition, as shown in the figure. The only exceptions are some locally distinct facies distributed along the coastal areas: (a) a higher-^{18}O granite border zone, remnants of which locally remain along the northwestern and eastern coasts; (b) a lower-^{18}O facies with very large $\delta^{18}O$ quartz-feldspar values, recognized only along the NE coast. The granite porphyry bodies (shown in solid black), although texturally very distinct, are isotopically similar to the higher-$\delta^{18}O$ granite facies ($\delta^{18}O$ quartz = 5.3 to 6.7; $\delta^{18}O$ feldspar = 3.8 to 4.7; δD = -82 to -95). The dotted contacts shown on the map are only approximate, but they suggest that the present outline of the island probably approximates the original dimensions of the main Mahè pluton.

Meteoric hydrothermal effects are not confined to Mahè A granite sample from Praslin, the second largest island of the Seychelles Group (50 km northeast of Mahè), contains very turbid alkali feldspar with a $\delta^{18}O$ = -1.6 and has a very large nonequilibrium $\Delta^{18}O$ quartz-feldspar (+8.6). All these data clearly document a widespread hydrothermal event in the Seychelles terrane about 670 Ma, involving waters with an initial $\delta^{18}O$ at least as low as -5.

Origin of Low-18O Granitic Magrnas (18O < +6)

One of the most important points to be made about low-^{18}O magmas is simply the immense difficulty of identifying them in the geologic record. This, of course, is not a problem for very young, recently erupted lavas such as those found in Iceland. However, for older volcanic rocks the $\delta^{18}O$ values of even the freshest-appearing samples are notoriously unreliable as a result of incipient weathering and hydrothermal effects (Taylor, 1968). Nevertheless, the $\delta^{18}O$ values of phenocrysts can often be used to calculate the $\delta^{18}O$ of the coexisting magma (see Taylor and Sheppard, 1986). This was in fact the way the Yellowstone $^{18}O/^{16}O$ analyses of Hildreth *et al.* (1984) were carried out (Figure 5.11). However, occasionally the phenocrysts of such volcanic rocks will not be in equilibrium with their coexisting magmas. Also, if the rocks are deeply buried or if they suffer even a low-grade metamorphism or hydrothermal alteration, it may be difficult (or impossible) to determine the $\delta^{18}O$ values of the original magmas.

The above problems are particularly severe in the case of low-^{18}O granitic plutons. Without doing an extremely detailed geological and geochemical study, there is usually no simple way to distinguish a pluton that formed from a low-^{18}O granitic magma from one that started out originally as a normal-^{18}O or high-^{18}O pluton but that later interacted with a Type 2 or Type 3 hydrothermal system. As an example of these ambiguities, look again at the data in Figure 5.1. Disequilibrium $^{18}O/^{16}O$ effects are common in this 700-km traverse from Vancouver Island eastward across southern British Columbia (Magaritz and Taylor, 1986). However, whereas the easternmost batholiths (Okanagan and Nelson) display the characteristic steep slopes of Type 1 systems, to the west in the deeper-seated Coast Plutonic Complex these $\delta^{18}O$ arrays define shallower slopes and swing over closer to the 45° line. In general, the samples that lie near the 45° line in Figure 5.1 probably represent an approach to Type 2 hydrothermal conditions, but this cannot be proved with the available data; many of the samples might represent low-^{18}O mag-mas.

Some of these same phenomena are observed in the Hercynian-age high-grade gneisses, migmatites, and posttectonic granites of the southern part of the Black Forest (Schwarzwald) in West Germany. This classic migmatite area (e.g., Mehnert, 1968) is closely analogous in age and geologic setting to the Hercynian metamorphic systems in the Pyrenees described by Wickham and Taylor (Chapter 6, this volume). One of the main differences is that whereas the waters in the Pyrenees are apparently marine in origin, the Black Forest terrane was subaerial and involved meteoric waters; the latter area locally exhibits remarkably low $\delta^{18}O$ values (0 to +4), and, as Magaritz and Taylor (1981) and Hoefs and Emmerman (1983) have shown, the $^{18}O/^{16}O$ fractionations among coexisting minerals range from nearly equilibrated values (e.g., Type 2) to strongly disequilibrium values (Type 1). These are the lowest-$\delta^{18}O$ values yet found in such deep-seated migmatites, and even though the adjacent host-rock schists are also depleted in ^{18}O, the two kinds of rocks are in general not equilibrated (i.e., Type 3 conditions were either not attained or they were later overprinted by a retrograde Type 1 system). In other areas of the Black Forest the schists and gneisses have $\delta^{18}O$ values that range upward to +7 to + 12. A set of complex Type 1 and Type 2 meteoric hydrothermal systems was clearly produced in this area about 300 Ma, particularly within and along an elongate, down-dropped block of Upper Paleozoic sedimentary and volcanic rocks. The granitic rocks and migmatites developed within this apparent rift zone are strikingly depleted in ^{18}O relative to the rocks on either side (Taylor et al., 1989), and it is virtually certain that at least locally low ^{18}O granitic magmas were developed. More detailed studies are continuing in this fascinating area to try to unravel these problems.

In the light of the above discussion it might appear that the rarity of low-^{18}O magmas in the geologic record is more apparent than real. Nevertheless, a number of workers have been actively searching for more examples of such magma systems for about 20 yr, with only limited success. Two features stand out in this two-decade-long search: (1) such magmas are much less abundant than was originally thought to be the case in the 1970s and (2) no new giant occurrences comparable to Yellowstone or Iceland have been found among Late Cenozoic lavas anywhere in the world. Other than the two examples of low ^{18}O rhyolitic and granitic magmas discussed above, the only other documented major occurrences of low-^{18}O magmas are in Iceland (e.g., Muehlenbachs, 1973; Hattori and Muehlenbachs, 1982; Condomines et al., 1983), in the southwestern Nevada caldera complex (Lipman, 1971; Lipman and Friedman, 1975), and a few rare examples in the Tertiary intrusive complex of Skye (Forester and Taylor, 1977).

Hildreth et al. (1984) explain the development of low-^{18}O magmas by direct influx of large amounts of H^2O into the magma. On the other hand, Taylor (1974a, 1977, 1980, 1983, 1986, 1987) argues that such magmas form either by (1) partial melting of hydrothermally altered rhyolitic country rocks and/or (2) foundering of such ^{18}O-depleted roof rocks into the magma chamber, accompanied by melting, assimilation, or exchange of this foreign material with the magma reservoir. This conclusion is based mainly on the geological factors outlined below and on constraints imposed by the physics and chemistry of H_2O transport through ductile rocks and silicate melts.

The most abundant country rocks above and along the margins of the Yellowstone magma chamber prior to each ash-flow eruption were certainly hydrothermally altered, ^{18}O-depleted, earlier-cycle rhyolites. These young volcanic rocks cannot provide the Pb and Sr isotope signatures that abruptly appear in the postcaldera rhyolites (Doe et al., 1982); therefore, some melting of pre-Tertiary country rocks is required, and, if this occurred, the much more abundant rhyolitic roof rocks would have been very extensively melted. However, the only chemical effect of this process on the magma reservoir would be: (1) lower $\delta^{18}O$ values, (2) lower δD values, and possibly (3) higher H_2O concentrations. This is because these rhyolitic country rocks are chemically and isotopically almost identical to their original parent magmas, except for the subsolidus hydrothermal changes they have undergone, which to a first approximation only involve hydration and depletion in ^{18}O and D.

An important feature of the Hildreth et al. (1984) study is the abrupt interval over which the isotopic changes occur (Figure 5.11). Similar, short time-scale, catastrophic processes involving caldera collapse had been invoked previously by Taylor (1974a, 1977), but up until the Hildreth et al. (1984) study there was no proof that such extreme $\delta^{18}O$ changes do in fact occur on such a rapid time scale. Even though diffusive transport of enormous amounts of H_2O directly into the magma a priori seemed unlikely, there was always the possibility that given enough time it perhaps could occur. The data of Hildreth et al. (1984) are extremely important because they show that $\delta^{18}O$ changes in the magma chamber do not occur by such a long-term process. In fact, the long-term changes are in the opposite direction, toward recovery of the original magmatic $\delta^{18}O$ value. On such a short time scale there is no physical way to separate such a huge volume of pore water from its rock matrix.

Hildreth et al. (1984) raised several objections against the partial melting or bulk assimilation process, among which are the material-balance problem and the fact that there is no evidence of the type of massive cooling or increase in phenocryst content that might be expected in the contaminated magmas, nor is there any obvious change in major or trace element composition. The first objection is invalid because mixing or exchange with large volumes of H_2O would have an even more profound cooling effect. The second argument is also refuted by the fact that the dominant rock types being melted are older-cycle rhy olites with essentially identical chemical compositions to the younger magmas. Such rhyolites are full of hydrous alteration minerals, and their latent heats of fusion are very low (<30 cal/g), both of which make them very susceptible to melting. Hydrothermally altered rocks with $\delta^{18}O$ = -8 to -9 are common at Yellowstone (Hildreth et al., 1984), and at depths of several kilometers their $\delta^{18}O$ values could be expected to be as low as -12 or lower, judging by the data from the deep Krafla drill hole on Iceland (Figure 5.13). Young volcanic rocks are also very porous, so the foundered roof rocks that fall into the magma chamber or that subside into a deep melting zone during such a catastrophic process are not just rock but H_2O-saturated rock, with perhaps 20 to 35 percent pore space. Therefore, the water-saturated rocks could easily have an average $\delta^{18}O$ = -13 to -15, markedly easing the material-balance difficulties.

Why are these low ^{18}O magmas formed in enormous volumes in some volcanic fields, but not others? Why do some eruptive units in the southwest Nevada and Yellowstone volcanic fields show an order of magnitude greater ^{18}O depletion than some other units or any of those from the central Nevada, San Juan, and Superstition volcanic fields? This is the most striking problem raised by the work of Larson and Taylor (1986b), who showed that magmas with remarkably uniform $\delta^{18}O$ values (overall variations of <1 per mil) were generated over time periods of more than 3 to 6 m.y. in two different caldera complexes in Nevada and Colorado. Larson and Taylor (1986b) were able to discern only one *major* factor that separates the low-^{18}O occurrences from the normal-^{18}O occurrences, namely emplacement into a rift-zone tectonic setting.

Emplacement Into an Rift-Zone Tectonic Setting

The only two caldera complexes in North America that are known to have erupted large volumes of low ^{18}O magmas (Yellowstone and Southwest Nevada) are both younger than 15 m.y. The period between 15 Ma and 20 Ma corresponds to the time of transition into the brittle, fracture-dominated extensional tectonic regime of the Basin-Range province (see, for example, Stewart, 1978; Zoback et al., 1981; Eaton, 1982). The Yellowstone caldera, in fact, lies on the eastern end of the currently active Snake River Plain rift system, and the Southwest Nevada caldera complex lies right in the midst of abundant Basin-Range extensional features. Such region-wide extension must produce fractures that penetrate deeply into the crust. These fractures could allow meteoric water to circulate very deeply, as they clearly have in Iceland (Muehlenbachs et al., 1974), which is also a well-defined rift-zone spreading center and where low-^{18}O silicic volcanic rocks are also abundant.

In retrospect, the connection between a rift-zone tectonic setting and low-^{18}O magmas may seem to be a fairly obvious one. After all, where on Earth is there a better chance of bringing into close juxtaposition the two geological materials that are essential to make very high temperature hydrothermal systems and low-^{18}O magmas? Only in rift zones and spreading centers do we find the large-scale extensions and brittle fracturing that are necessary to allow massive amounts of magma to come upward into the crust, as well as providing the greatly increased fracture permeability that allows surface waters to penetrate to depths of at least 10 or 15 km. Such environments represent the best way to attain the required combination of very high temperatures together with large quantities of low-^{18}O, hydrothermally altered rocks and meteoric pore waters.

CONCLUSIONS

The simple fact of the existence of low-^{18}O magmas, together with the gigantic scale at which meteoric hydrothermal and seawater hydrothermal convective systems have been shown to operate on Earth, constitutes the most definitive proof that surface waters can in large amounts locally penetrate to great depths in the Earth's crust. The existence of low-^{18}O magmas also proves that these surface waters can circulate downward into very high temperature environments, essentially into the zone of melting itself. This in turn implies that some of these complex processes take place under lithostatic pressures.

Deep circulation of surface-derived aqueous fluids under hydrostatic conditions appears to be ubiquitous in areas of igneous activity, and in some fossil hydrothermal systems where the deep-seated rocks are exposed by erosion the effects can be shown by direct observation to have extended to at least 8- to 10-km depth. Significant convective circulation and isotopic exchange between rocks and aqueous fluids thus should always occur around igneous intrusions, except in certain cases and under the following conditions.

1. Intrusion into relatively impermeable country rocks— for example, into either (a) limestones and evaporites, which are susceptible to ductile deformation at pressure-temperature conditions where most other rocks undergo brittle fracture, or (b) ordinary silicate rocks at sufficient depth in the crust that the fracture permeability is less than 10^{-15} cm^2 (10^{-4} millidarcy), or where the temperatures and pressures are high enough for recrystallization and ductile

metamorphism to occur, particularly in the absence of extensional tectonics (rifting), rapid strain rates, and brittle deformation (see Norton and Taylor, 1979; Taylor, 1987).

2. In the immediate presence of the silicate melt, which also will be essentially impermeable to aqueous fluids under hydrostatic pressures, as long as the strain rates are low enough and the percentage of melt high enough so that an interconnected fracture network cannot develop. This is very important in layered gabbro plutons because the late-stage magma sheet at the roof of the intrusion can provide a barrier between two decoupled hydrothermal systems: a very hot (>500°C), lower hydrothermal system in the layered cumulates and a cooler system at 250° to 450°C and higher water-rock ratios in the roof rocks.

3. If magmatic aqueous fluids are being exsolved from late-crystallizing portions of a (granitic) pluton, this will produce a magmatic H_2O envelope under lithostatic pressure that fills all available fractures outward from the crystallization front. This can keep the low-^{18}O meteoric waters outside the pluton until the magmatic fluid envelope has dissipated, by which time the temperatures will have fallen into the range of stability (e.g., sericite and chlorite). Thus, two decoupled hydrothermal systems may also commonly occur around H_2O-rich granitic magma chambers (e.g., porphyry Cu deposits; see Taylor, 1974a), but in such cases the two types of aqueous fluids are genetically very different and are under different pressures.

We have shown that layered gabbro cumulates typically undergo hydrothermal interaction with externally derived aqueous fluids (e.g., meteoric water, seawater) at much higher temperatures than in the case of granitic plutons (typically 500° to 900°C versus 200° to 450°C). This is manifested in the absence or rarity of hydrous alteration minerals (e.g., amphibole, chlorite, epidote) in gabbros that show clear-cut $\delta^{18}O$ signatures of having interacted with very large quantities of external H_2O. In such rocks the $\delta^{18}O$ effects may be the only obvious indications of intense hydrothermal exchange. Although a number of characteristics contribute to the higher alteration temperatures of the gabbros, such as higher solidus temperature, greater latent heat of crystallization, higher melt-crystal ratio, and characteristically lower fracture density, the two most important factors appear to be (1) the different geometry of crystallization of a layered gabbro pluton compared to a granite pluton and (2) the higher magmatic H_2O contents of granitic plutons.

Granitic magmas produced in rift zones and areas of extensional tectonics commonly (invariably?) appear to involve anatexis of rocks that were hydrothermally altered at high water-rock ratios. This conclusion also probably applies to local "pull-apart" basins associated with major strike-slip faults. Such extensional zones provide favorable access routes for upward rise of magmas as well as for downward penetration of surface waters; the magma bodies then act as giant "heat engines" that promote convective circulation of the waters. These aqueous fluids can produce large-scale $^{18}O/^{16}O$ changes in the source rocks of certain kinds of granitic magmas, particularly if the water is isotopically distinctive (as when low-^{18}O meteoric groundwaters are involved). The styles of water-rock interaction also change with increasing depth, because with increasing temperature and time of alteration the isotopically heterogeneous, unequilibrated Type 1 systems may locally give way to heterogeneous, equilibrated Type 2 systems, and finally at the highest T, t, and w/r we may in some cases obtain Type 3 systems that are both equilibrated and homogenized.

Prior to or during the formation of these kinds of hy-drothermal-anatectic granitic magmas, there must be a transition from hydrostatic to lithostatic conditions; the nature and timing of this process are not yet well understood. Also, these hydrothermal-anatectic granitic magmas will be readily recognized only if they are abnormally depleted in ^{18}O, which specifically requires deep circulation of enormous volumes of very low ^{18}O groundwaters into the Type 2 and Type 3 hydrothermal regimes; such processes occur only in the most highly fractured, most permeable parts of the Earth's crust. In the absence of other compelling geochemical or geophysical data such as that presented by Wickham and Taylor (1985, 1987; Chapter 6, this volume), Wickham (1987), and Bickle *et al.* (1988), in general *it will not be possible to prove "hydrothermal" origin of such magmas*. The $^{18}O/^{16}O$ signature of the original source of such an aqueous fluid may be completely obscured, particularly if the *w/r* ratios are relatively small. The D/H signature will be preserved but commonly will not be useful, because except in the case of seawaters (very high δD) or high-latitude meteoric waters (very low δD) there will in general not be sufficient isotopic contrast compared to typical magmatic and metamorphic fluids (which have δD = -40 to -90).

Therefore, "disguised" magmas formed by the general type of hydrothermal-anatectic process outlined in this chapter and in Chapter 6 by Wickham and Taylor conceivably could be extremely common but very difficult to recognize. They might even be the dominant silicic magma type in most areas of rift-zone tectonics. The testing of this hypothesis will be an important problem to be addressed in future studies of the petrology and geochemistry of granites. In this connection Wickham *et al.* (1987) discovered a number of isotopic features in northeast Nevada that are similar to those in the Pyrenees, with one major difference: the fluids were low-$\delta^{18}O$ meteoric wa ters rather than high-D marine pore fluids. This comparison is important because it is certain that Nevada was subaerial in the mid-to late-Tertiary and that major crustal extension took place over a broad area at that time.

ACKNOWLEDGMENTS

This work was supported by the National Science Foundation, grant no. EAR83-13106. I am particularly indebted to Robert E. Criss, Robert T. Gregory, Robert I. Hill, Peter B. Larson, Mordekai Magaritz, Denis Norton, Leon T. Silver, G. Cleve Solomon, and Stephen M. Wick-ham for their help and collaboration on the problems discussed in this paper and for numerous discussions over the years on hydrothermal systems and the origin of granites. This is contribution no. 4590, Publications of the Division of Geological and Planetary Sciences, California Institute of Technology.

References

Baker, B. H. (1963). Geology and mineral resources of the Seychelles Archipelago, *Memoir, Geological Survey of Kenya 3*, 1-140.

Bickle, M. J., S. M. Wickham, H. J. Chapman, and H. P. Taylor, Jr. (1988). A strontium, neodynium, and oxygen isotope study of the hydrothermal metamorphism and crustal anatexis in the Trois Seigneurs Massif, Pyrenees, France, *Contributions to Mineralogy and Petrology 100*, 399-417.

Buddington, A. F., and E. Callaghan (1936). Dioritic intrusive rocks and contact metamorphism in the Cascade Range in Oregon, *American Journal of Science 31*, 421-449.

Burnham, C. W. (1979). Magmas and hydrothermal fluids, in *Geochemistry of Hydrothermal Ore Deposits*, 2nd ed., H. L. Barnes, ed., John Wiley and Sons, New York, pp. 71-136.

Casadevall, T., and H. Ohmoto (1977). Sunnyside mine, Eureka mining district, San Juan County, Colorado: Geochemistry of gold and base metal ore deposition in a volcanic environment, *Economic Geology 72*, 1285-1320.

Cathles, L. M. (1983). An analysis of the hydrothermal system responsible for massive sulfide deposition in the Hokuroku Basin of Japan, in *The Kuroko and Related Volcanogenic Massive Sulfide Deposits*, Economic Geology Monograph 5, Society of Economic Geologists, Golden, Colo., pp. 439-487.

Christiansen, R. L. (1984). Yellowstone magmatic evolution: Its bearing on understanding large-volume explosive volcanism, in *Explosive Volcanism*, Geophysics Study Committee, National Academy Press, Washington, D.C., pp. 84-95.

Cole, D. R., and H. Ohmoto (1986). Kinetics of isotopic exchange at elevated temperatures and pressures, in *Stable Isotopes in High Temperature Geological Processes*, J. W. Valley, H. P. Taylor, Jr., and J. R. O'Neil, eds., Reviews in Mineralogy, vol. 16, Mineralogical Society of America, Washington, D.C., pp. 41-90.

Condomines, M., K. Gronvold, P. J. Hooker, K. Muehlenbachs, R. K. O'Nions, N. Oskarsson, and E. R. Oxburgh (1983). Helium, oxygen, strontium and neodymium isotopic relationships in Icelandic volcanics, *Earth and Planetary Science Letters 66*, 125-136.

Craig, H. (1961). Isotopic variations in meteoric waters, *Science 133*, 1702-1703.

Criss, R. E., and R. J. Fleck (1987). Petrogenesis, geochronology, and hydrothermal systems in the northern Idaho batholith and adjacent areas based on $^{18}O/^{16}O$, D/H, $^{87}Sr/^{86}Sr$, K-Ar, and $^{40}Ar/^{39}Ar$ studies , *U.S. Geological Survey Professional Paper 1436*.

Criss, R. E., and H. P. Taylor, Jr. (1983). An $^{18}O/^{16}O$ and D/H study of Tertiary hydrothermal systems in the southern half of the Idaho batholith, *Geological Society of America Bulletin 94*, 640-663.

Criss, R. E., and H. P. Taylor, Jr. (1986). Meteoric-hydrothermal systems, in *Stable Isotopes in High Temperature Geological Processes* , J. W. Valley, H. P. Taylor, Jr., and J. R. O'Neil, eds., Reviews in Mineralogy, vol. 16, Mineralogical Society of America, Washington, D.C., pp. 373-424.

Criss, R. E., M. A. Lanphere, and H. P. Taylor, Jr. (1982). Effects of regional uplift, deformation, and meteoric-hydro-thermal metamorphism on K-Ar ages of biotites in the southern half of the Idaho batholith, *Journal of Geophysical Research 87*, 7029-7046.

Criss, R. E., E. B. Ekren, and R. F. Hardyman (1984). Casto ring zone: A 4500 k_2 fossil hydrothermal system in the Challis volcanic field, central Idaho, *Geology 12*, 331-334.

Criss, R. E., R. T. Gregory, and H. P. Taylor, Jr. (1987). Kinetic theory of oxygen isotope exchange between minerals and water, *Geochimica et Cosmochimica Acta 51*, 952-960.

Doe, B. R., W. P. Leeman, R. L. Christiansen, and C. E. Hedge (1982). Lead and strontium isotopes and related trace elements as genetic tracers in the Upper Cenozoic rhyolite-basalt association of the Yellowstone Plateau volcanic field, *Journal of Geophysical Research 87*, 4785-4806.

Eaton, G. P. (1982). The Basin and Range Province: Origin and tectonic significance, *Annual Reviews of Earth and Planetary Sciences 10*, 409-440.

Ferry, J. M. (1985). Hydrothermal alteration of Tertiary igneous rocks from the Isle of Skye, northwest Scotland, 1. Gabbros, *Contributions to Mineralogy and Petrology 91*, 264-282.

Fleck, R. J., and R. E. Criss (1985). Strontium and oxygen isotopic variations in Mesozoic and Tertiary plutons of central Idaho, *Contributions to Mineralogy and Petrology 90*, 291-308.

Forester, R. W., and H. P. Taylor, Jr. (1976). ^{18}O-depleted igneous rocks from the Tertiary complex of the Isle of Mull, Scotland, *Earth and Planetary Science Letters 32*, 11-17.

Forester, R. W., and H. P. Taylor, Jr. (1977). $^{18}O/^{16}O$, D/H and $^{13}C/^{12}C$ studies of the Tertiary igneous complex of Skye, Scotland, *American Journal of Science 277*, 136-177.

Forester, R. W., and H. P. Taylor, Jr. (1980). Oxygen, hydrogen, and carbon isotope studies of the Stony Mountain complex, western San Juan Mountains, Colorado, *Economic Geology 75*, 363-383.

Friedman, I., P. Lipman, J. D. Obradovich, J. D. Gleason, and R. L. Christiansen (1974). Meteoric water in magmas, *Science 184*, 1069-1072.

Giletti, B. J. (1986). Diffusion effects on oxygen isotope temperatures of slowly cooled igneous and metamorphic rocks, *Earth and Planetary Science Letters 77*, 218-228.

Giletti, B. J., and R. A. Yund (1984). Oxygen diffusion in quartz, *Journal of Geophysical Research 89*, 4039-4046.

Giletti, B. J., M. P. Semet, and R. A. Yund (1978). Studies in diffusion: III. Oxygen in feldspars: An ion microprobe determination, *Geochimica et Cosmochimica Acta 42*, 45-57.

Gregory, R. T., and H. P. Taylor, Jr. (1981). An oxygen isotope profile in a section of Cretaceous oceanic crust, Samail Ophiolite, Oman: Evidence for ^{18}O-buffering of the oceans by deep (>5 km) seawater-hydrothermal circulation at mid-ocean ridges, *Journal of Geophysical Research 86*, 2737-2755.

Gregory, R. T., R. E. Criss, and H. P. Taylor, Jr. (1988). Oxygen isotope exchange kinetics of mineral pairs in closed and open systems: Applications to problems of hydrothermal alteration of igneous rocks and Precambrian iron formations, *Chemical Geology*, in press.

Gunnarsson, B. J., H. P. Taylor, Jr., and B. D. Marsh (1988). Origin of oxygen isotope anomalies in Icelandic lavas: Part I. Flank-zone volcanoes, *Geological Society of America Abstracts with Programs 20*.

Hattori, K., and K. Muehlenbachs (1982). Oxygen isotope ratios of the Icelandic crust, *Journal of Geophysical Research 87*, 6559-6565.

Hildreth, W., R. L. Christiansen, and J. R. O'Neil (1984). Catastrophic isotopic modification of rhyolitic magma at times of caldera subsidence, Yellowstone Plateau volcanic field, *Journal of Geophysical Research 89*, 8339-8369.

Hoefs, J., and R. Emmerman (1983). The oxygen isotopic composition of Hercynian granites and pre-Hercynian gneisses from the Schwarzwald, SW Germany, *Contributions to Mineralogy and Petrology 83*, 320-329.

Ito, E., and R. N. Clayton (1983). Submarine metamorphism of gabbros from the Mid-Cayman Rise: An oxygen isotopic study, *Geochimica et Cosmochimica Acta 47*, 535-546.

Jacobsson, S. P. (1972). Chemistry and distribution pattern of recent basaltic rocks in Iceland, *Lithos 5*, 365-386.

Larson, P. B., and H. P. Taylor, Jr. (1986a). An oxygen isotope study of water/rock interaction in the granite of Cataract Gulch, western San Juan Mountains, Colorado, *Geological Society of America Bulletin 97*, 505-515.

Larson, P. B., and H. P. Taylor, Jr. (1986b). ^{18}O/^{16}O ratios in ash-flow tuffs and lavas erupted from the central Nevada caldera complex and the central San Juan caldera complex, Colorado, *Contributions to Mineralogy and Petrology 92*, 146-156.

Larson, P. B., and H. P. Taylor, Jr. (1986c). An oxygen isotope study of hydrothermal alteration in the Lake City caldera, San Juan Mountains, Colorado, *Journal of Volcanology and Geothermal Research 30*, 47-82.

Lipman, P. W. (1971). Iron-titanium oxide phenocrysts in compositionally zoned ash-flow sheets from southern Nevada, *Journal of Geology 79*, 438-456.

Lipman, P. W., and I. Friedman (1975). Interaction of meteoric water with magma: An oxygen-isotope study of ash-flow sheets from southern Nevada, *Geological Society of America Bulletin 86*, 695-702.

Magaritz, M., and H. P. Taylor, Jr. (1976). ^{18}O/^{16}O and D/H studies along a 500 km traverse across the Coast Range batholith and its country rocks, central British Columbia, *Canadian Journal of Earth Sciences 13* (11), 1514-1536.

Magaritz, M., and H. P. Taylor, Jr. (1981). Low ^{18}O migmatites and schists from the tectonic contact zone between Hercynian (=Variscan) granites and the older gneissic core complex of the Black Forest (Schwarzwald), West Germany, *Geological Society of America Abstracts with Programs 13*, 501.

Magaritz, M., and H. P. Taylor, Jr. (1986). Oxygen-18/oxygen-16 and D/H studies of plutonic granitic and metamorphic rocks across the Cordilleran batholiths of southern British Columbia, *Journal of Geophysical Research 91*, 2193-2217.

Manning, C. E., and D. Bird (1986). Hydrothermal clinopyroxenes from the Skaergaard intrusion, *Contributions to Mineralogy and Petrology 92*, 437-447.

Mehnert, K. R. (1968). *Migmatites*, Elsevier, Amsterdam, 393 pp.

Michot, J., and S. Deutsch (1976). Seychelles, Microcontinent or not? (abs.), *4th European Colloquim of Geochronology, Cosmochronology, and Isotope Geology*, 68.

Miller, J. A., and J. D. Mudie (1961). K-Ar age determinations on granite from the Island of Mahè in the Seychelles Archipelago, *Nature 93*, 1174-1175.

Muehlenbachs, K. (1973). The oxygen isotope geochemistry of siliceous volcanic rocks from Iceland, *Carnegie Institution of Washington Yearbook 72*, 593-597.

Muehlenbachs, K., A. T. Anderson, and G. E. Sigvaldason (1974). Low-^{18}O basalts from Iceland, *Geochimica et Cosmochimica Acta 38*, 577-588.

Norton, D. (1984). Theory of hydrothermal systems, *Annual Reviews of Earth and Planetary Sciences 12*, 55-177.

Norton, D., and J. Knight (1977). Transport phenomena in hydrothermal systems: Cooling plutons, *American Journal of Science 277*, 937-981.

Norton, D., and H. P. Taylor, Jr. (1979). Quantitative simulation of the hydrothermal systems of crystallizing magmas on the basis of transport theory and oxygen isotope data: An analysis of the Skaergaard intrusion, *Journal of Petrology 20*, 421-486.

Norton, D., H. P. Taylor, Jr., and D. K. Bird (1984). The geometry and high-temperature brittle deformation of the Skaergaard intrusion, *Journal of Geophysical Research 89*, 10,178-10,192.

O'Neil, J. R., and M. L. Silberman (1974). Stable isotope relations in epithermal gold-silver deposits, *Economic Geology 69*, 902-909.

O'Nions, R. K., and K. Gronvold (1973). Petrogenetic relationship of acid and basic rocks in Iceland: Sr isotopes and rare earth elements in late post-glacial volcanics, *Earth and Planetary Science Letters 19*, 397-409.

Saemundsson, K. (1974). Evolution of the axial rifting zone in northern Iceland and Tjornes fracture zone, *Geological Society of America Bulletin 85*, 495-504.

Sheppard, S. M. F. (1986). Igneous rocks: III. Isotopic case studies of magmatism in Africa, Eurasia, and oceanic islands, in *Stable Isotopes in High Temperature Geological Processes*, J. W. Valley, H. P. Taylor, Jr., and J. R. O'Neil, eds., Reviews in Mineralogy, vol. 16, Mineralogical Society of America, Washington, D.C., pp. 319-371.

Sheppard, S. M. F., and H. P. Taylor, Jr. (1974). Hydrogen and oxygen isotope evidence for the origins of water in the Boulder Batholith and the Butte ore deposits, Montana, *Economic Geology 69*, 926-964.

Sheppard, S. M. F., R. L. Nielsen, and H. P. Taylor, Jr. (1971). Hydrogen and oxygen isotope ratios in minerals from porphyry copper deposits, *Economic Geology 66*, 515-542.

Sheppard, S. M. F., P. E. Brown, and A. D. Chambers (1977). The Lilloise intrusion, East Greenland: Hydrogen isotope evidence for the efflux of magmatic water into the contact metamorphic aureole, *Contributions to Mineralogy and Petrology 68*, 129-147.

Silver, L. T., H. P. Taylor, Jr., and B. W. Chappell (1979). Some petrological, geochemical, and geochronological observations of the Peninsular Ranges batholith near the international border of the U.S.A. and Mexico, in *Mesozoic Crystalline Rocks*, P. L. Abbott and V. R. Todd, eds., Annual Meeting Guidebook, Geological Society of America, Boulder, Colo., pp. 83-110.

Stakes, D. S., H. P. Taylor, Jr., and R. C. Fisher (1983). Oxygen isotope and geochemical characterization of hydrothermal alteration in ophiolite complexes and modem oceanic crust, in *Ophiolites and Oceanic Lithosphere*, I. D. Gass, S. J. Lippard, and A. W. Shelton, eds., Special Publication of the Geological Society of London, Blackwell Scientific Publishers, London, pp. 199-214.

Stewart, J. H. (1978). Basin-Range structure in western North America, *Geological Society of America Memoir 152*, 1-131.

Taylor, H. P., Jr. (1968). The oxygen isotope geochemistry of igneous rocks, *Contributions to Mineralogy and Petrology 19*, 1-71.

Taylor, H. P., Jr. (1971). Oxygen isotope evidence for large-scale interaction between meteoric ground waters and Tertiary granodiorite intrusions, Western Cascade Range, Oregon, *Journal of Geophysical Research 76*, 7855-7874.

Taylor, H. P., Jr. (1973). $^{18}O/^{16}O$ evidence for meteoric-hydro-thermal alteration and ore deposition in the Tonopah, Comstock Lode, and Goldfield mining districts, Nevada, *Economic Geology 68*, 747-764.

Taylor, H. P., Jr. (1974a). The application of oxygen and hydrogen isotope studies to problems of hydrothermal alteration and ore deposition, *Economic Geology 69*, 843-883.

Taylor, H. P., Jr. (1974b). A low-^{18}O, Late Precambrian granite batholith in the Seychelles Islands, Indian Ocean: Evidence for formation of ^{18}O-depleted magmas and interaction with ancient meteoric ground waters, *Geological Society of America Abstracts with Programs 6*, 981-982.

Taylor, H. P., Jr. (1974c). Oxygen and hydrogen isotope evidence for large-scale circulation and interaction between ground waters and igneous intrusions, with particular reference to the San Juan volcanic field, Colorado, in *Geochemical Transport and Kinetics*, A. W. Hofmann, B. J. Giletti, H. S. Yoder, Jr., and R. A. Yund, eds., Carnegie Institution of Washington, Washington, D.C., pp. 299-324.

Taylor, H. P., Jr. (1977). Water/rock interactions and the origin of H_2O in granitic batholiths, *Journal of the Geological Society of London 133*, 509-558.

Taylor, H. P., Jr. (1980). Stable isotope studies of spreading centers and their bearing on the origin of granophyres and plagiogranites, in *Orogenic Mafic-Ultramafic Association*, C. Allegre and J. Aubouin, eds., Colloques Internationaux du C.N.R.S., No. 272, pp. 149-165.

Taylor, H. P., Jr. (1983). Oxygen and hydrogen isotope studies of hydrothermal interactions at submarine and subaerial spreading centers, in *Hydrothermal Processes at Seafloor Spreading Centers*, P. A. Rona, K. Bostrum, L. Laubier, and K. L. Smith, Jr., eds., NATO Symposium Volume, Plenum Press, New York, pp. 83-104.

Taylor, H. P., Jr. (1986). Igneous rocks: II. Isotopic case studies of circumpacific magmatism, in *Stable Isotopes in High Temperature Geological Processes*, J. W. Valley, H. P. Taylor, Jr., and J. R. O'Neil, eds., Reviews in Mineralogy, vol. 16, Mineralogical Society of America, Washington, D.C., pp. 273-317.

Taylor, H. P., Jr. (1987). Comparison of hydrothermal systems in layered gabbros and granites, and the origin of low-^{18}O magmas, in *Magmatic Processes: Physicochemical Principles*, B. O. Mysen, ed., Special Publication 1, Geochemical Society, Washington, D.C., pp. 337-357.

Taylor, H. P. Jr., and R. W. Forester (1971). Low-^{18}O igneous rocks from the intrusive complexes of Skye, Mull, *Journal of Petrology 12*, 465-497.

Taylor H. P., Jr., and R. W. Forester (1979). An oxygen and hydrogen isotope study of the Skaergaard intrusion and its country rocks: A description of a 55-m.y. old fossil hydrothermal system, *Journal of Petrology 20*, 355-419.

Taylor, H. P., Jr., and M. Magaritz (1978). Oxygen and hydrogen isotope studies of the Cordilleran batholiths of western North America, in *Stable Isotopes in the Earth Sciences*, B. W. Robinson, ed., DSIR Bull. 220, New Zealand Department of Scientific and Industrial Research, Wellington, New Zealand, pp. 151-173

Taylor, H. P., Jr., and S. M. F. Sheppard (1986). Igneous rocks: I. Processes of isotopic fractionation and isotope systematics, in *Stable Isotopes in High Temperature Geological Processes*, J. W. Valley, H. P. Taylor, Jr., and J. R. O'Neil, eds., Reviews in Mineralogy, vol. 16, Mineralogical Society of America, Washington, D.C., pp. 227-271.

Taylor, H. P., Jr., and L. T. Silver (1978). Oxygen isotope relationships in plutonic igneous rocks of the Peninsular Ranges batholith, southern and Baja California, in *Short Papers of 4th International Conference on Geochronology, Cosmochronology, and Isotope Geology*, R. E. Zartman, ed., U.S. Geological Survey Open-File Report 78-701, pp. 423-426.

Taylor, H. P., Jr., M. Magaritz, and S. M. Wickham (1989). Application of stable isotopes in identifying a major Hercynian rift zone and its associated meteoric-hydrothermal activity, Southern Schwarzwald, West Germany, in *Epstein 70th Birthday Symposium Volume*, California Institute of Technology, Pasadena, pp. 86-91.

Wasserburg, G. J., H. Craig, H. W. Menard, A. E. J. Engel, and C. G. Engel (1964). Age and composition of a Bounty Island granite and age of a Seychelles Islands granite, *Journal of Geology 71*, 785-789.

Welke, H., S. Moorbath, G. L. Cumming, and H. Sigurdsson (1968). Lead isotope studies on igneous rocks from Iceland, *Earth and Planetary Science Letters* 4, 221-231.

Wickham, S. M. (1987). Crustal anatexis and granite petrogenesis during low pressure regional metamorphism: The Trois Seigneurs Massif, Pyrenees, France, *Journal of Petrology 28*, 127-169.

Wickham, S. M., and E. R. Oxburgh (1985). Continental rifts as a setting for regional metamorphism, *Nature 318*, 330-333.

Wickham, S. M., and H. P. Taylor, Jr. (1985). Stable isotope evidence for large-scale seawater infiltration in a regional metamorphic terrane: The Trois Seigneurs Massif, Pyrenees, France, *Contributions to Mineralogy and Petrology 91*, 122-137.

Wickham, S. M., and H. P. Taylor, Jr. (1987). Stable isotope constraints on the origin and depth of penetration of hydrothermal fluids associated with Hercynian regional metamorphism and crustal anatexis in the Pyrenees, *Contributions to Mineralogy and Petrology 95*, 255-268.

Wickham, S. M., H. P. Taylor, Jr., and A. W. Snoke (1987). Fluid-rock-melt interaction in metamorphic core complexes— a stable isotope study of the Ruby Mountains-East Humboldt Range, Nevada, *Geological Society of America Abstracts with Programs 19*, 463.

Yund, R. A., and T. F. Anderson (1978). The effect of fluid pressure on oxygen isotope exchange between feldspar and water, *Geochimica et Cosmochimica Acta 42*, 235-239.

Zartman, R. E. (1974). Lead isotopic provinces in the Cordillera of the western United States and their geologic significance, *Economic Geology 69*, 792-434.

Zoback, M. L., R. E. Anderson, and G. A. Thompson (1981). Cenozoic evolution of the state of stress and style of tectonism of the Basin and Range province of the western United States, *Philosophical Transactions of the Royal Society of London, Ser. A 300*, 407-434.

6

Hydrothermal Systems Associated with Regional Metamorphism and Crustal Anatexis: Examples from the Pyrenees, France

STEPHEN M. WICKHAM
University of Chicago
HUGH P. TAYLOR, JR.
California Institute of Technology

INTRODUCTION

Our understanding of the transport of fluids through the deeper parts of the crust, and in particular through rocks undergoing prograde regional metamorphism, is at an embryonic stage. This stems largely from the difficulty of making direct observations of such processes, in contrast to the situation in active shallow hydrothermal systems (which are accessible to study in boreholes). Stable isotope ($^{18}O/^{16}O$, $^{13}C/^{12}C$, and D/H) studies of metamorphic and igneous rocks and minerals may, however, be used to place constraints on the passage of H_2O- and CO_2-rich fluids through the crust. Oxygen, hydrogen, and carbon are major constituents of both rocks and typical crustal fluids, and these three elements show systematic differences in isotopic composition in the different terrestrial reservoirs.

For example, the mantle has a $\delta^{18}O$ value of about +6, distinctly different from the range of values shown by most detrital sedimentary rocks (+9 to +18) and much lower than the values in sedimentary carbonates (+20 to +30). All of these "normal" rock $\delta^{18}O$ values are distinctly higher than the values in seawater ($\delta^{18}O = 0$) or meteoric water ($\delta^{18}O = -25$ to 0) (Taylor, 1974, 1977). Consequently, if aqueous fluids derived from the Earth's surface interact with igneous or metamorphic rocks at elevated temperatures where the equilibrium isotopic fractionations are small, this results in a change in the rock oxygen (and hydrogen) isotopic composition toward that of the water. Stable isotope analyses of such rocks can help to identify the source of infiltrating fluids, and, in the case of oxygen, which is the dominant constituent of H_2O, CO_2, and virtually all crustal rocks, these data can be used to make material-balance calculations constraining the quantity of fluid involved.

In a number of areas it has been recognized that whole-rock $\delta^{18}O$ values tend to decrease with increasing metamorphic grade (e.g., Garlick and Epstein, 1965; Shieh and Taylor, 1969; Shieh and Schwarcz, 1974). For instance, in Idaho the variation is from about +15 in low-grade shales to +11 in sillimanite-grade rocks formed during Cretaceous metamorphism (Fleck and Criss, 1985). There is, however, at present no clear consensus of opinion regarding the origin of this isotopic shift, its timing, or its relationship to fluid transport during metamorphism. For example, the variation of $\delta^{18}O$ from typical sedimentary values to values closer to typical mantle values could involve minimal amounts of fluid and arise simply by exchange and homogenization of oxygen between high-^{18}O metasediments and low-^{18}O mantle-derived igneous lithologies. Such an effect might be expected at increasing depth within the crust, where mantle-derived material becomes progressively more common and sedimentary material more rare. Such exchange conceivably could take place at a relatively low fluid-rock ratio with the fluid merely acting as the agent of isotopic exchange.

On the other hand, the shift in $\delta^{18}O$ to lower values at higher metamorphic grade could reflect exchange of oxygen between higher-grade metamorphic rocks and low-^{18}O fluids at high temperatures, with the fluid itself forming the isotopically light oxygen reservoir. In this case large quantities of fluid would be required to bring about bulk changes in $\delta^{18}O$ by the amounts that are commonly observed (3 to 4 per mil). If the latter process is dominant, then permeabilities must be appreciable and fluid-rich conditions are likely to be common during metamorphism (e.g., Ferry, 1986), with fluid pressure at times equaling or locally exceeding lithostatic pressure. If the former interpretation is more common, a free fluid phase may only be intermittently and/or locally present.

The Hercynian metamorphic terrane exposed in the Pyrenees offers a good opportunity to differentiate between these two alternative scenarios. Here, a characteristic shift to lower-$\delta^{18}O$ values with increasing metamorphic grade is clearly observed, and good exposure, clear-cut geological relationships and access to a wide range of structural levels allow us to place tight constraints on both the nature and scale of fluid-rock interaction during metamorphism (Wickham and Taylor, 1985, 1987; Wickham, 1987a; Bickle *et al.*, 1988).

METAMORPHISM IN THE PYRENEES

Hercynian Basement of the Pyrenees

The Pyrenees are a roughly linear chain of mountains between France and Spain. They were uplifted during the lower Tertiary in response to convergence between Iberia and Western Europe. Uplift has exposed an extensive pre-Mesozoic basement terrane (the Hercynian basement) comprising Paleozoic and Precambrian sediments, metasediments, and gneisses as well as a variety of granitoid plutons (see Figure 6.1). All of these lithologies were metamorphosed or intruded at about 310 to 340 Ma during the Late Carboniferous, Hercynian orogeny (Zwart, 1979; Autran *et al.*, 1980; Bard *et al.*, 1980; Bickle *et al.*, 1988). Subsequent (e.g., Tertiary) deformation and recrystallization of the Hercynian metamorphic and granitic rocks were relatively minor.

The Hercynian basement ranges from virtually unmetamorphosed, fossiliferous Devonian and Carboniferous rocks, some of which were actually being deposited while metamorphism was occurring at depth, through Lower Paleozoic mica schists, marbles, and migmatites, finally to amphibolite and granulite-facies "basal gneisses" (Zwart, 1979). The mineral assemblages in the gneisses formed at pressures of 4 to 7 kbar (Vielzeuf, 1984), compatible with their being the basement upon which the Paleozoic sedimentary sequence was deposited (Vitrac-Michard and Allegre, 1975). The metamorphic sequences are charac

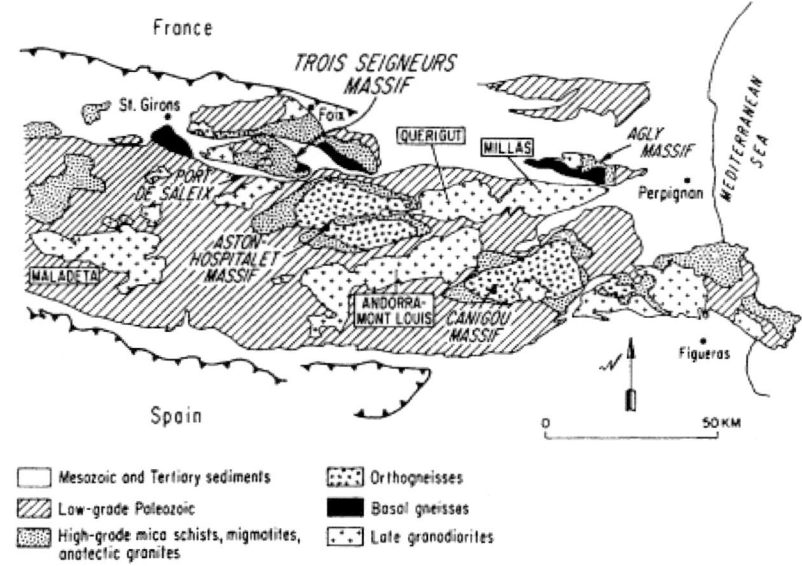

Figure 6.1
Hercynian basement outcrop in the Pyrenees, showing the various localities discussed in the text.

terized by very abrupt transitions from the low- to the high-grade regions and by extensive partial melting of Lower Paleozoic pelitic metasediments at temperatures of about 700°C and depths of only about 10 to 12 km. This is recorded by abundant pelitic migmatites and peraluminous granitoids associated with the high-grade mica schists.

Subdivision of the Hercynian Crust Into Three Structural Levels

Taken together the wide range of structural levels exposed in the different tectonic blocks of the Hercynian basement of the Pyrenees can be interpreted in terms of a composite section through the upper 25 km of the continental crust. Because the oxygen isotope systematics are very distinctive at different structural levels within this section, it is convenient to divide it into three tectonostratigraphic zones: (1) a moderately deformed, fossiliferous sedimentary sequence that was deposited from the Upper Ordovician to the Upper Carboniferous; (2) a metamorphic and migmatitic sequence developed within Cambro-Ordovician (mainly pelitic) metasediments, with equilibration pressures of 2 to 4 kbar, that were locally extensively melted and intruded by peraluminous leucogranites and biotite-cordierite granites; and (3) a region of amphibolite- and granulite-facies orthogneisses and paragneisses equilibrated at 4 to 7 kbar.

Although a continuous section through all of these crustal levels is not exposed in the Pyrenees, rocks from all three levels outcrop within several relatively restricted geographic areas (e.g., in the Agly and St. Barthelemy massifs). Figure 6.2 shows a section through the St. Barthelemy Massif, indicating the relationship between Zones 1, 2, and 3 in that region.

Metamorphism and Anatexis

The very abrupt, albeit progressive and gradational, transition from low to high grade in the metamorphic sequences in the Pyrenees implies the existence of very steep thermal gradients during metamorphism. Typically, a progression in the Zone 2 pelitic rocks (which comprise over 90 percent of the Lower Paleozoic metasediments) from chlorite-sericite phyllites through andalusite and sillimanite schists to migmatites and biotite-cordierite granitoids occurs over horizontal distances of only 3 to 5 km (e.g., in the Trois Seigneurs Massif, Figure 6.3). Thin (10 to 50 m) carbonate-rich beds are interlayered within the Cambro-Ordovician pelites; these have developed various calc-silicate mineralogies consistent with the metamorphic mineral assemblages in the adjacent pelites.

The granitic rocks were mostly derived by partial melting and homogenization of the high-grade pelitic metasediments, and in general the granitic melts have not migrated very far from their region of generation (Wickham, 1987a). Migmatite textures and compositions imply that pelites melted to at least 50 to 60 percent by volume, and the observed rapid increase to these high degrees of melting at relatively low temperatures of 700° to 750°C strongly implies water-rich conditions with aH_2O buffered externally to high values (see Wickham, 1987a). This is also suggested by common pegmatitic textures, widespread muscovitization and tourmalinization, and by calc-silicate mineral assemblages in the marble layers (they commonly contain clinozoisite and sometimes also grossular; see Ferry, 1983).

In contrast to the Zone 2 rocks, much smaller degrees of melting occurred within the amphibolite- and granulite-facies "basal gneisses" (Zone 3). These rocks are lithologically distinct from Zone 2 in that they contain far less

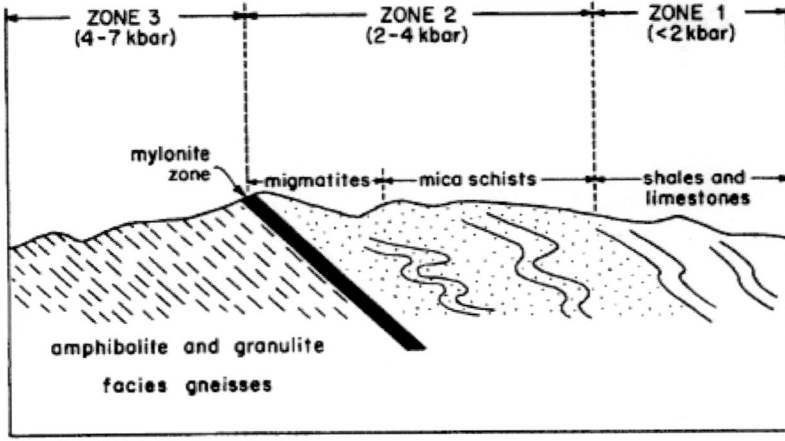

Figure 6.2
Schematic true-scale section through the St. Barthelemy Massif redrawn from Passchier (1984, Figure 3). The line of section is approximately north to south (from low to high grade) and is about 15 km long. The Zone 3 "basal gneisses" have uncertain structural relationship to the other rocks in the massif because they are bounded by a mylonitic shear zone that approximately separates Zone 2 and Zone 3.

pelite and larger amounts of granitic orthogneiss, quartzofeldspathic biotite gneiss, and amphibolite. Zone 3 is locally migmatitic but contains virtually no large (> 100 to 1000 m) bodies of peraluminous granite. Vielzeuf (1984) made a detailed study of the petrology of these rocks, which are commonly orthopyroxene bearing and clearly equilibrated at greater depth (4 to 7 kbar) and under much "dryer" peak metamorphic conditions than those in Zone 2. Although no H_2O activity data are available from this work, some of the granulites probably equilibrated at $aH_2O > 0.5$. This does not in itself preclude the possibility that more water-rich conditions prevailed at an earlier stage during the metamorphism. However, as described below, the stable isotope systematics within the basal gneiss Zone 3 rocks are significantly different from those in the mica schist-migmatite sequences (Zone 2). This almost certainly reflects contrasting fluid-rock interactions during metamorphism at these different levels within the Hercynian crust.

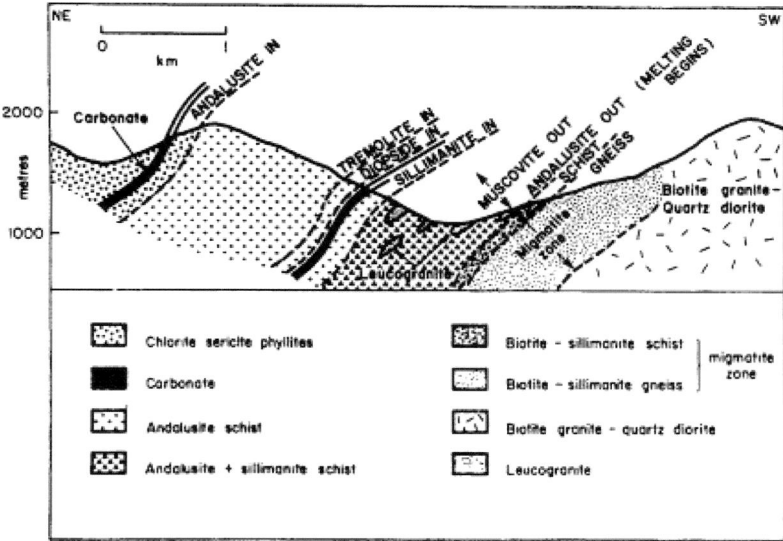

Figure 6.3
Schematic true-scale section through the Trois Seigneurs metamorphic sequence, indicating the progressive gradation from low-grade phyllites through high-grade mica schists and migmatites to the granitoids that were derived mainly through anatexis of the pelitic metasediments. Note the proximity of the low- and high-grade rocks.

OXYGEN ISOTOPE SYSTEMATICS

Zone 1

The Paleozoic rocks from Zone 1 comprise fossiliferous shales and carbonates that have been deformed but only very weakly metamorphosed. Rocks are typically fine grained and may contain chlorite or fine-grained white mica but no higher-grade metamorphic minerals. Because they probably never experienced temperatures in excess of 400°C during the Hercynian metamorphism, it is hardly surprising that these rocks in the main preserve typical sedimentary $\delta^{18}O$ values. However, recent $^{87}Sr/^{86}Sr$ data (Bickle et al., 1988) indicate that even these low-grade phyllites and shales were extensively modified by fluid infiltration during prograde metamorphism. $^{87}Sr/^{86}Sr$ ratios were homogenized (and in some cases lowered) to values of ~0.715 at 310 Ma, a process that is observed in even the lowest-grade Zone 1 shales, which could not have experienced temperatures any higher than 250° to 300°C (see Table 6.1). $^{18}O/^{16}O$ ratios may have been modified at the same time, but we have no way of detecting this because we cannot determine the depositional $\delta^{18}O$ values of these sediments (as we can with $^{87}Sr/^{86}Sr$). The Sr data allow us to infer that significant isotopic modifications resulting from fluid infiltration commenced at the earliest stages of prograde metamorphism and continued up to peak metamorphic temperatures (see Bickle et al., 1988, and below).

The $^{18}O/^{16}O$ data are illustrated in Figure 6.4, where a compilation of $\delta^{18}O$ values for Zone 1 (calcite for the carbonate rocks and whole rock for shale samples) is presented from four different parts of the Pyrenees. In the Trois Seigneurs and Agly massifs these rocks are Ordovician and Silurian shales, whereas in the St. Barthelemy and Arize massifs they are Ordovician to Carboniferous shales and limestones. Calcite from the limestones ranges from +19 to +24, consistent with the typical range of isotopic values shown by many Paleozoic sedimentary carbonates (Baertschi, 1957; Veizer and Hoefs, 1976). Shales range from +13 to +16, with most values between +14 and +15. These kinds of values are also shown by all other Zone 1 Ordovician to Carboniferous sedimentary rocks that we have analyzed from other parts of the Pyrenees.

TABLE 6.1 Oxygen and Strontium Isotope Data (average values) at Various Outcrop Localities of Pelitic Schist, Phyllite, and Shale in the Trois Seigneurs Massif, Arranged into Four Groups as a Function of Grade of Metamorphism

Parameter Studied	Original Sedimentary Rock (model values 450 Ma)[a]	Low Grade (shales and phyllites)	Intermediate Grade (vicinity of the biotite isograd)	High grade (andalusite sillimanite zones)
Mean $\delta^{18}O$ value	?	+14.5 ± 0.7 (8)	+12.1 ± 0.9 (5)	+11.5 ± 0.6 (16)
Range of $\delta^{18}O$	?	+13.3 to +16.0	+10.5 to +13.3	+10.6[b] to +12.7
Mean $^{87}Sr/^{86}Sr$ value	0.7202 ± 0.0053 (5)	0.7130 ± 0.0022 (5)	0.7153 (1)	0.7151 ± 0.0016 (6)
Range of $^{87}Sr/^{86}Sr$	0.7133 to 0.7326	0.7089 to 0.7166	—	0.7129 to 0.7178
Average Sr content	89 ppm (5)	89 ppm (5)	102 ppm (1)	98 ppm (7)
Average Rb content	138 ppm (5)	138 ppm (5)	98 ppm (1)	160 ppm (7)
Average Rb-Sr ratio	1.55	1.55	0.96	1.62

Note: Data are from Bickle *et al.* (1988) and Wickham and Taylor (1985). The tabulated $^{87}Sr/^{86}Sr$, Sr, and Rb data represent the mean values for the various localities, using a single average value for each of these outcrop localities where multiple samples were studied. The ± indicates average deviation from the mean value, with the number of analyzed localities given in parentheses. All $^{87}Sr/^{86}Sr$ values are calculated at 310 Ma except for the original sedimentary rock (model) values, which are calculated for 450 Ma.

[a] The $\delta^{18}O$ values are unknown, but the model strontium isotope values at 450 Ma can be calculated, as discussed in the text.

[b] One sample with an anomalously low-$\delta^{18}O$ value of +8.8 is not included (Wickham and Taylor, 1985).

Zone 2

In contrast to the characteristic sedimentary $\delta^{18}O$ values shown by the shales and carbonates of Zone 1, the Zone 2 mica schists, migmatites, and peraluminous granitoids have lower and much more uniform $\delta^{18}O$ values, mostly in the range +11 to +12 (see Table 6.1 and Figure 6.4). This is particularly clear in the Trois Seigneurs Massif, where the Zone 1 to Zone 2 transition is well exposed and where the Zone 2 carbonates have internally homogeneous $\delta^{18}O$ values that are essentially identical to the $\delta^{18}O$ of the adjacent mica schists (Wickham and Taylor, 1985). A detailed isotopic profile through one of the Trois Seigneurs metacarbonate layers is shown in Figure 6.5. This unit lies between the "sillimanite in" isograd and the migmatite zone and is about 15 m thick, sandwiched between psammitic rocks with $\delta^{18}O$ values of about +12. The calcite has a fairly uniform $\delta^{18}O$ of between +13 and +14 throughout the layer, regardless of the calcite content of the sample or the distance of the sample from the margin of the layer. This indicates an extremely high degree of oxygen isotopic equilibration between this layer and the surrounding metasediments, in contrast to the situation prior to metamorphism when the carbonate layers would have had much higher $\delta^{18}O$ values than the surrounding rocks (as is still the case in Zone 1). This homogeneity of carbonate and pelite oxygen isotopic compositions within the Trois Seigneurs Zone 2 rocks is typical of the oxygen isotope systematics at this structural level throughout the Pyrenees. Preexisting sedimentary heterogeneities in $\delta^{18}O$ have everywhere been smoothed out by pervasive oxygen isotopic exchange over wide regions. The Hercynian age of this process is proved by Rb-Sr isochrons obtained from suites of mica-schist samples (Bickle *et al.*, 1988) from Zone 2, indicating homogenization of $^{87}Sr/^{86}Sr$ at values of ~0.715 at 310 Ma. Interestingly, in the case of Sr the $^{87}Sr/^{86}Sr$ ratios are not smoothed out everywhere; this is probably because the giant reservoir of strontium in the carbonates (containing ~2000 ppm Sr) is much more resistant to change than the pelite reservoir (~200 ppm Sr) and thus tends to retain its original sedimentary value of ~0.708 (Bickle *et al.*, 1988).

In addition to the whole-rock $\delta^{18}O$ values being much more homogeneous in Zone 2 than in Zone 1, there is an obvious, pronounced lowering of $\delta^{18}O$ going from Zone 1 to Zone 2 (Figure 6.4; Table 6.1). This shift reflects a change in the bulk oxygen isotopic composition of the Paleozoic sedimentary pile of 3 to 4 per mil, from about +15 to about +11.5 (similar to that observed in some other regional metamorphic sequences (e.g., Garlick and Epstein, 1967; Rye *et al.*, 1974; Fleck and Criss, 1985). This shift is particularly well characterized in the Hercynian of the Pyrenees, where it takes place fairly abruptly over a stratigraphic distance of 1 or 2 km within the metamorphic sequence.

It is clear that the Zone 1 and Zone 2 pelitic rocks both belong to the same Paleozoic sedimentary succession and that they originally had similar bulk isotopic and chemical compositions. This implies that the observed shift in $\delta^{18}O$ values does not represent any lithological difference between Zone 1 and Zone 2. During metamorphism these rocks were partially dehydrated and decarbonated, con verted to high-grade mica schists and marbles, were isotopically homogenized, and overall had their bulk $\delta^{18}O$ values lowered by about 3 to 4 per mil. The Zone 2 metamorphic rocks clearly must have exchanged oxygen with a large reservoir that had a $\delta^{18}O$ value significantly lighter than +11.

Zone 3

Exposures of Zone 3 rocks in the Pyrenees are limited and principally occur in three regions: the Agly, St. Barthelemy, and Castillon massifs (see Figure 6.1). Additionally, a small area near the village of Lapege in the Trois Seigneurs Massif exposes rocks that probably correlate with the basal gneisses elsewhere because they are lithologically distinct from any other rocks in the Trois Seigneurs area (which clearly belong to Zone 2; see Wickham, 1987a). Zone 3 rocks comprise a lithologically heterogeneous sequence of granitic gneisses, quartzo-feldspathic biotite gneisses, amphibolites, carbonates, and relatively rare kinzigites (pelitic gneisses). Oxygen-isotope data from these regions are summarized in Figures 6.6 and 6.7 (see also Wickham and Taylor, 1987). These data are not as clear cut as those obtained from Zones 1 and 2, mainly because we do not know for certain what the original premetamorphic $\delta^{18}O$ values of the Zone 3 lithologies were; we simply do not have access to the unmetamorphosed equivalents of these rocks (as we do in Zone 2).

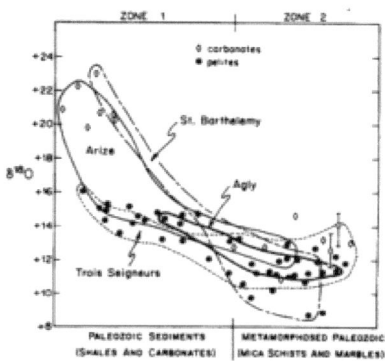

Figure 6.4
Compilation of data for Zones 1 and 2 from the Agly, Arize, St. Barthelemy, and Trois Seigneurs massifs for whole-rock pelite samples (solid circles) and calcite from carbonate-bearing samples (open diamonds). The data are plotted as a function of metamorphic grade with their position on the diagram corresponding roughly to their relative distance from the Zone 1 to Zone 2 boundary in the field. All of the Zone 2 carbonate samples are from Trois Seigneurs, and the two data points with error bars represent the average values (and the $\delta^{18}O$ range) for the two detailed profiles in Figures 6 and 7 of Wick-ham and Taylor (1987). There is a pronounced shift in $\delta^{18}O$ to lower values going from Zone 1 to Zone 2, and the Zone 2 values are much more homogeneous than those in Zone 1 (also see Table 6.1).

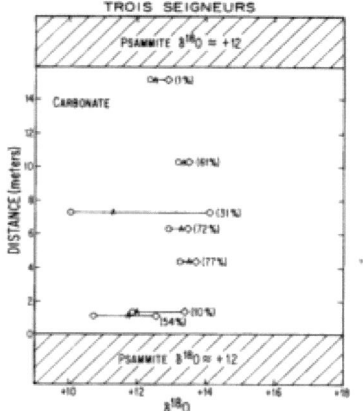

Figure 6.5
Oxygen isotopic profile through a carbonate unit in the Trois Seigneurs Massif (sample SI13, Wickham and Taylor, 1985) that crops out between the "sillimanite in" and "andalusite out" isograds (see Figure 6.3). Note that the $\delta^{18}O$ of calcite is fairly uniform throughout the layer, regardless of either the distance from the margin or the calcite content of the sample. The $\delta^{18}O$ values within the carbonate layer are similar to the psammites to either side, and they are much lower than the original sedimentary values, which must have been at least as high as +22 to +25.

The data in Figures 6.6 and 6.7 show that major lowering of the $\delta^{18}O$ values in carbonates occurred both at St. Barthelemy and at Castillon and that this was also accompanied by some isotopic homogenization. However, the data from Agly and Lapege are rather different. Here the $\delta^{18}O$ values of the carbonates span a wide range from +13 to +22, including both values typical for sedimentary carbonates (as in Zone 1) and values characteristic of the homogenized and ^{18}O-shifted marbles of Zone 2.

In order to investigate these systematics further, we made detailed isotopic profiles across some of the individual carbonate units in the Zone 3 basal gneisses. One such profile from the Agly Massif is shown in Figure 6.8. The calcites in these carbonate layers have essentially retained their original sedimentary $\delta^{18}O$ values, indicating a lack of exchange with the adjacent biotite gneisses (which have $\delta^{18}O$ values of about +12). Steep isotopic gradients of as much as 10 per mil over 50 cm are preserved at the margins of the carbonates. These systematics occur despite the fact that the Agly units are mostly substantially thinner than those at Trois Seigneurs and were metamorphosed at higher temperatures (both of which factors might be expected to favor isotopic homogenization). Similar systematics are observed in the Lapege samples. Again, steep isotopic gradients, internal isotopic heterogeneity, and lack of homogenization with the adjacent lithologies characterize the carbonate unit sampled here (see Wickham and Taylor, 1987; Figure 6.8). This clearly implies a different style of fluid-rock interaction in the Agly and Lapege Zone 3 rocks, as compared with those in Zone 2 in the same areas. In this respect it is important to note that there is no obvious contrast in the present-day permeabilities of Zone 2 and Zone 3 rocks, although major contrasts may have existed between these lithologies at the time of active Hercynian metamorphism.

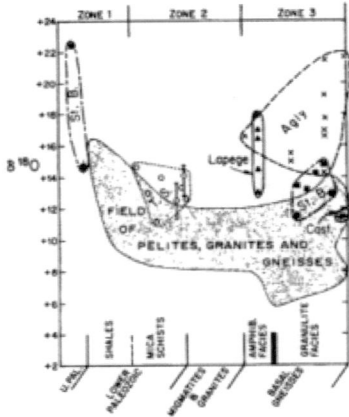

Figure 6.6
Summary of the $^{18}O/^{16}O$ data for metacarbonate lithologies from the Pyrenees (excluding the Arize data shown in Figure 6.4). The calculated whole-rock carbonate $\delta^{18}O$ values (i.e., including the coexisting silicate minerals) are plotted as a function of metamorphic grade, with the relative width of each lithological unit along the abscissa being proportional to the total areal extent of each lithology in the Castillon (Cast.), St. Barthelemy (St. B.), Trois Seigneurs (T.S.), and Agly massifs. Within a given lithological subdivision, the data points are plotted proportional to the actual geographic distance of the sample locality from the boundary of this subdivision in the field. Silicates were not analyzed for two Castillon samples and one lowgrade St. Barthelemy sample (marked with plus signs); these $\delta^{18}O$ values represent calcite rather than whole rock. In the Trois Seigneurs field two of the data points with error bars represent average values (and the $\delta^{18}O$ range) for the two profiles shown in Figures 6 and 7 of Wickham and Taylor (1987). The field of whole-rock $\delta^{18}O$ values of pelites, granites, and gneisses from all regions (Figure 6.7) is shown for comparison. Note the preservation of relatively high (sedimentary) $\delta^{18}O$ values in Zone 1, the ^{18}O-depleted and homogeneous isotopic compositions of both the carbonates and the silicate lithologies in Zone 2, and the wide range of $\delta^{18}O$ values in all rock types in Zone 3.

At St. Barthelemy and Castillon, the oxygen isotope systematics in the Zone 3 basal gneisses have much more in common with the higher-level Zone 2 rocks. Metacarbonates from these regions have $\delta^{18}O$ values similar to those for the Zone 2 rocks from Trois Seigneurs, and the steep isotopic gradients that are a characteristic feature at

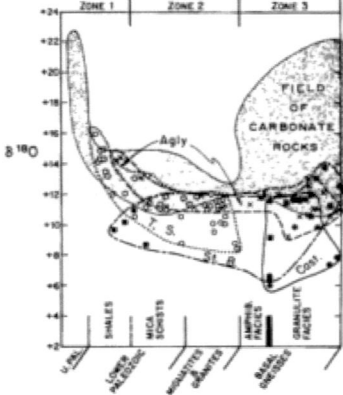

Figure 6.7
Summary of the whole-rock $^{18}O/^{16}O$ data for pelites, granites, and gneisses plotted as a function of metamorphic grade (as described in the caption for Figure 6.6). The data points lying above the heavy black bar on the abscissa represent values for mafic rocks within the basal gneisses. The field for the whole-rock carbonate samples plotted in Figure 6.6 is also repeated in this figure for reference. Note the strong lowering of $\delta^{18}O$ in all lithologies (including the carbonates) going from Zone 1 to Zone 2. The homogeneous $\delta^{18}O$ values in Zone 2 reflect the infiltration of large volumes of aqueous fluid into all lithologies at high temperatures. The $\delta^{18}O$ values are more heterogeneous in Zone 3, indicating that the Zone 3 rocks did not experience such a massive water influx.

Agly and Lapege have not yet been identified. Some isotopic heterogeneity does occur between the rare mafic rocks in these areas and the more abundant biotite gneisses, but it is not as marked as that observed between the carbonates and gneisses at Agly.

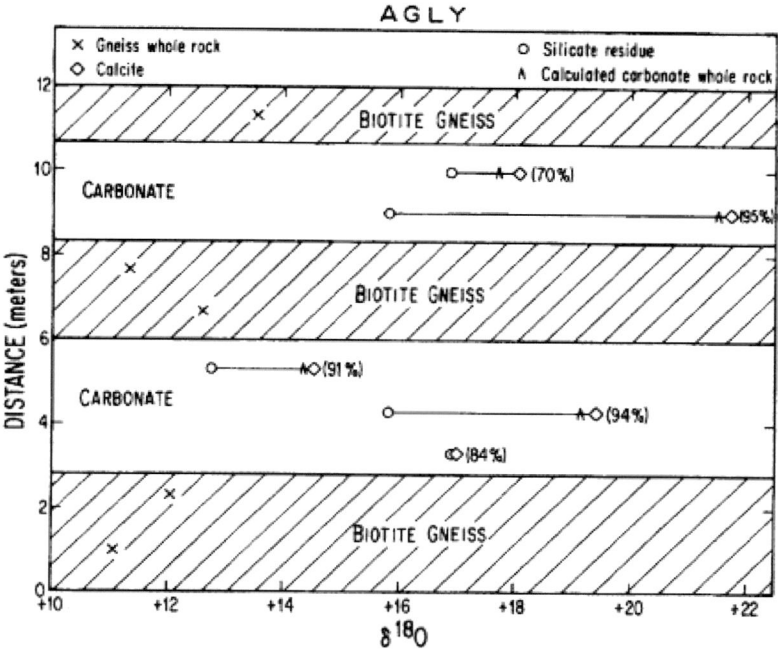

Figure 6.8
Oxygen isotopic profile through two thin carbonate layers within granulite-facies quartzo-feldspathic gneiss in the Agly Massif (see Figure 2 of Wick-ham and Taylor, 1987, for locality). The weight percent calcite in each carbonate sample is shown. Despite the small dimensions of the carbonate layers (<3 m thick), most of the calcites have essentially retained their original sedimentary $\delta^{18}O$ values, indicating a lack of $^{18}O/^{16}O$ exchange with the adjacent gneisses. The original $\delta^{18}O$ values of these gneisses are not known, but they could have been nearly identical to their present metamorphic values (see text).

In general, isotopic heterogeneity within the basal gneisses is more extreme than is observed anywhere in Zone 2, suggesting that the Zone 3 rocks throughout the Pyrenees were not subjected to the same isotopic homogenization process that occurred at higher structural levels. This increase in isotopic heterogeneity with increasing structural depth in the Hercynian crust strongly implies that fluid movement during metamorphism became less important at deeper structural levels.

Interpretation of Zone 1 and Zone 2 Oxygen Isotope Systematics

There are essentially two plausible low-^{18}O reservoirs available to account for the lowering of $\delta^{18}O$ in going from Zone 1 to Zone 2. One is mantle-derived lower crustal rocks, which would probably have had $\delta^{18}O$ values between +6 and +10. The other is a large volume of low-^{18}O aqueous fluid. Geological evidence in the Pyrenees favors the latter reservoir because suitable low-^{18}O rocks are notably absent from the wide range of structural levels exposed. Basal gneisses in Zone 3 contain small amounts of mafic rock with $\delta^{18}O$ values of +6 to +8, but are dominantly composed of metasedimentary biotite gneisses with $\delta^{18}O$ values of +11 to +12. Late granodiorite plutons have values typically in the range +8 to +11, but these were intruded after Hercynian metamorphism and were not available to take part in the earlier isotopic homogenization processes. The synmetamorphic peraluminous granitoids were themselves mainly derived from pelitic metasedimentary material (Wickham, 1987a) and have average $\delta^{18}O$ values of +11 to +12. A suitable low-^{18}O rock reservoir is therefore lacking at any exposed level of the Hercynian crust. Furthermore, the isotopic data from the Zone 3 rocks at Lapege and Agly imply an increase in the degree of isotopic heterogeneity with increasing structural depth. This is opposite to what would be expected if the Paleozoic metasediments were being isotopically homogenized by some large-scale process involving interactions with low-^{18}O materials from the mantle or the lower crust.

If the low-^{18}O reservoir were an aqueous fluid of some type, then simple material-balance calculations could be used to place constraints on the quantity of fluid involved. Clearly, this depends on the original $\delta^{18}O$ value of the fluid and the temperature of the isotopic exchange as well as the magnitude of the isotopic shift observed in the rocks. This is illustrated in Figure 6.9, where water-rock ratio is plotted against $\delta^{18}O$, the initial $\delta^{18}O$ value of infiltrating fluid (see Taylor, 1977, and Wickham and Taylor, 1985, for details), assuming a shift in $\delta^{18}O$ of 3.5 per mil between Zone 1 and Zone 2 for the bulk terrane. Exchange is assumed to occur at a sufficiently high temperature for the final $\delta^{18}O$ values of rock and fluid to be approximately equal. Clearly the lighter the fluid, the smaller is the quantity required to generate the observed change in $\delta^{18}O$.

If the water were magmatic or metamorphic in origin, it is unlikely to have had a $\delta^{18}O$ value of less than +6. From Figure 6.9 it is clear that at this value of $\delta^{18}O_i$, the material-balance, water-rock ratio is about 0.6, equivalent to about 35 percent by weight of the entire isotopically altered rock mass. Such a large quantity of water would need to be derived from plutons many times bigger than the Zone 2 homogenized terrane, and all of the water released would need to be channeled through this region. Metamorphic water from dehydrating minerals would also have to be derived from a very much larger region and effectively focused through the isotopically homogenized region. Either scenario seems implausible and in the latter case implies that large surrounding regions remain isotopically unaffected. We have yet to identify such regions and consider that a magmatic or metamorphic source for the bulk of the water is unlikely, simply because it seems quantitatively inadequate to explain the Zone 1 to Zone 2 isotopic shifts.

The only other plausible water sources are (1) connate formation waters that were trapped in the Paleozoic sediments during deposition and subsequently mobilized during metamorphism or (2) water derived directly from the surface of the Earth. In either case the amounts of water available are far larger and are thus volumetrically easily capable of producing the observed isotopic shifts at all reasonable values of $\delta^{18}O_i$. Formation waters might typically have $\delta^{18}O$ values in the range +2 to +8 (Clayton et al., 1966) and could be drawn in from large regions outside the isotopically homogenized zones. Surface waters represented by seawater ($\delta^{18}O \approx 0$) and meteoric water ($\delta^{18}O > 0$) form an effectively limitless source, but they would have to move from the surface to depths of 10 km or more in order to flow through the Zone 2 rocks during metamorphism.

In summary, the oxygen isotope data for the Zone 1 and Zone 2 rocks imply that pervasive isotopic exchange and homogenization were accompanied by a lowering of $\delta^{18}O$ by several per mil in Zone 2 rocks during metamorphism. The oxygen data do not unambiguously identify the source of this fluid, but simple material-balance constraints imply that huge amounts of H_2O are required and that Paleozoic formation waters or surface waters were the most likely source.

Interpretation of Zone 3 Oxygen Isotope Data

The isotopic data from Zone 3 are more difficult to interpret than the data from Zones 1 and 2, because the oxygen isotope systematics are different in the different localities (see Figures 6.6 through 6.8). At Lapege and Agly there is considerable isotopic heterogeneity between metacarbonates and adjacent biotite gneisses, and the $\delta^{18}O$ values of some of these rocks have been little modified since deposition. At St. Barthelemy, however, the Zone 3 metacarbonates have a range of $\delta^{18}O$ and $\delta^{13}C$ values very similar to those of the Trois Seigneurs Zone 2 rocks. These rocks have lower calcite contents and have therefore been more extensively decarbonated than the rocks at Agly or Lapege. Such decarbonation will tend to lower $\delta^{18}O$ values by a few per mil, but this is unlikely to be the sole explanation for the values at St. Barthelemy and Castillon, which are as much as 10 per mil lighter than typical sedimentary calcites. The $\delta^{18}O$ values are similar to those of the adjacent gneisses and probably reflect isotopic homogenization with these lithologies.

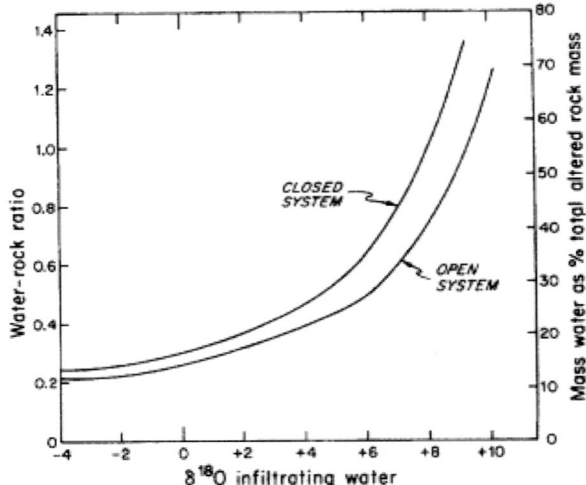

Figure 6.9

$\delta^{18}O$ of the infiltrating water versus calculated material-balance water-rock ratio for the Hercynian Zone 2 metasediments in the Pyrenees, for both closed- and open-system conditions (see Wickham and Taylor, 1985, for details). W/R is also plotted as an equivalent quantity of water, the percent mass of the altered rock. In these calculations the initial and final $\delta^{18}O$ values of the metasediments are taken to be +15.0 and +11.5, respectively, so the Zone 2 rocks are assumed to have been shifted to lower values by 3.5 per mil.

The quartz-feldspar-biotite gneisses that are the dominant lithology in Zone 3 have average $\delta^{18}O$ values of +12 to +10. Unlike the Zone 2 mica schists, these rocks are not true pelites and their sedimentary protolith was probably immature greywacke-type material, which may have had a sedimentary $\delta^{18}O$ value similar to the gneisses. There is therefore no need to postulate a lowering of $\delta^{18}O$ in these gneisses, contrary to the Zone 2 mica schists. The isotopic composition of the gneisses may simply reflect original sedimentary values that have remained essentially unchanged during metamorphism.

Mafic rocks occur sporadically throughout Zone 3 and preserve $\delta^{18}O$ values of typically +6 to +8, not far from their presumed original mantle values. This is true even at Castillon and St. Barthelemy, where the metacarbonates have $\delta^{18}O$ values similar to those of the biotite gneisses. Steep isotopic gradients are preserved at the margins of these mafic layers. Thus, although the nature of the isotopic heterogeneity is different at St. Barthelemy and Castillon in that it does not occur between metacarbonate and adjacent units, it is still more pronounced than anywhere we have sampled in Zone 2. Although we cannot accurately calculate water-rock ratios in Zone 3 because we do not know the original $\delta^{18}O$ values of the rocks prior to Hercynian metamorphism, it seems certain that infiltration has been much less extreme at this deeper level in the crust. This is especially true at Agly and Lapege, where we see extreme oxygen isotope heterogeneity preserved in and adjacent to the metacarbonates.

The $^{18}O/^{16}O$ data strongly support the idea that during
metamorphism water was derived from the overlying Paleozoic supracrustal sedimentary pile or from the surface, because in this case a clear maximum depth of penetration would be expected, possibly due to the expected decrease in permeability with increasing structural depth in the crust. If the water were derived from a deep plutonic source or if the homogenization reflected bulk $\delta^{18}O$ exchange with the deep crust during metamorphism, there should certainly not be an increase in the degree of isotopic heterogeneity at deeper structural levels, as is observed in the Pyrenees. Although the Castillon and St. Barthelemy data require some infiltration of the carbonate lithologies by H_2O- (and CO_2-) bearing fluids during metamorphism, the Zone 3 rocks certainly have not experienced the pervasive flushing at material-balance water-rock ratios in excess of 0.5 that is required by the Zone 2 data at Trois Seigneurs.

HYDROGEN ISOTOPE SYSTEMATICS

Material-balance calculations and other considerations described above imply that the large volumes of aqueous fluid that infiltrated the Zone 2 rocks of the Pyrenees probably were originally derived from the surface of the Earth. To test this hypothesis further, we made D/H analyses on samples of muscovite from the metamorphic and granitic rocks of Zone 2 at Trois Seigneurs and elsewhere in the Pyrenees. The results are summarized in Figure 6.10, where these data are plotted as a function of metamorphic grade.

Most igneous and metamorphic rocks have a relatively restricted range of δD values, typically in the range of -50 to -85 per mil (e.g., Taylor, 1974). This range is distinct from the isotopic composition of present-day seawater ($\delta D = 0$) and many meteoric waters, particularly those found at high latitudes and high altitudes (δD = -100 to -400), so that any interaction at high temperatures between such surface waters and "normal" igneous and metamorphic rocks should result in a modification of rock δD values toward the characteristic meteoric or marine values. Because there is proportionally so much more hydrogen in water than in rocks, rock δD values can be changed significantly even at very low water-rock ratios involving only tiny amounts of infiltrating fluid. Therefore, under favorable conditions, D/H measurements can be very sensitive monitors of the involvement of surface waters with geological systems.

The muscovite data of Figure 6.10 strongly support the notion that marine fluids interacted with the Zone 2

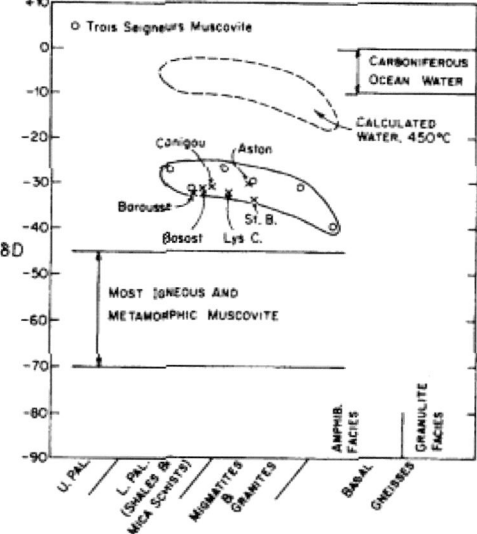

Figure 6.10

δD for muscovite from a variety of Zone 2 lithologies within the metamorphic sequence (leucogranites, quartz-muscovite veins, and boudin-neck fillings) plotted as a function of metamorphic grade. Trois Seigneurs samples are shown as open circles. The other data (shown as crosses) cover a wide area within the Pyrenees, ranging from the Bosost region in the west to the Canigou Massif, about 180 km to the east (see Figure 6.1). The data are very uniform throughout this wide geographic range and have exceptionally high δD values, distinctly higher than the typical range shown by most igneous and metamorphic muscovite. The calculated field of waters with which these muscovites would have been in equilibrium is shown and corresponds closely to the likely δD value of Carboniferous ocean water.

metasediments at high temperatures. All of the δD values are between -40 and -25, which is distinctly heavier than the typical range shown by most muscovites from metamorphic and igneous rocks. The values from the Trois Seigneurs Massif (see also Wickham and Taylor, 1985; Figure 6.11) are particularly heavy (five values lie between -32 and -25 per mil) and uniform over a wide range of structural levels and a variety of lithologies. Interestingly, the sample from the deepest structural level has the lightest (most normal) δD value and may possibly reflect the waning effects of surface water infiltration at deeper structural levels (similar to the downward increase in oxygen isotope inhomogeneity observed in Zone 3 at Agly and Trois Seigneurs).

Figure 6.10 also shows a calculated field of waters that would be in equilibrium with these muscovite samples at plausible metamorphic temperatures (450°C). This field mostly lies between 0 and -10 per mil and overlaps with the hypothetical field of Late Paleozoic (Hercynian) ocean water. In the past, ocean water may have fluctuated between the present δD value (0 per mil) and perhaps -15 per mil when there were no ice caps. Because no major late Carboniferous glaciation has been recognized, seawater was probably closer to this latter value during the Hercynian. The D/H data of Figure 6.10 are thus thoroughly consistent with the proposal that the Trois Seigneurs fluids were ultimately derived from late Paleozoic seawater. Note that by the time this fluid infiltrated Zone 2, it still would have retained its original δD of about -5 to -10, but it almost certainly would not have retained its original $\delta^{18}O$ value of about 0, because it would have undergone extensive exchange with Paleozoic sedimentary rocks enroute. Thus, in this particular case the D/H data are much more informative about the original source of the H_2O than are the $^{18}O/^{16}O$ data, although they of course say almost nothing about the amounts of H_2O involved.

High δD values have also been observed in Hercynian rocks elsewhere in Europe, for example, in Cornwall (Sheppard, 1977), in the Alps (Frey *et al.*, 1976; Negga *et al.*, 1986) and in the Iberian pyrite belt (Munha *et al.*, 1986). These data have generally been interpreted to represent the involvement of connate waters of some type, although not in terms of direct infiltration by marine surface waters. The values obtained from the Pyrenees indicate equilibration of Zone 2 metasediments with marine fluids over a wide region, though they do not tell us if the water was derived from connate fluids within the Paleozoic supracrustal metasediments or whether it had to be derived directly from the surface (see below). Similarly, the D/H data do not in themselves constrain the timing of the infiltration process. However, bearing in mind the homogeneity of δD and $\delta^{18}O$ both between and within individual terranes, it seems likely that both were homogenized during metamorphism more or less simultaneously by the same fluid infiltration event.

STRONTIUM ISOTOPE SYSTEMATICS

To constrain further the timing and scale of the hydrothermal process that have affected the metasediments at Trois Seigneurs and elsewhere in the Pyrenees, Rb-Sr isotope analyses were made on a number of samples from the metamorphic sequence at Trois Seigneurs (Bickle *et al.*, 1988). These data are illustrated in Figure 6.11, where the initial $^{87}Sr/^{86}Sr$ ratios 310 Ma are plotted against $\delta^{18}O$.

The data from both the low-and high-grade metasediments lie on crude Hercynian-age isochrons, indicating that, unlike the $^{18}O/^{16}O$ ratios, $^{87}Sr/^{86}Sr$ was homogenized in both the amphibolite facies mica-schists and in their low-grade protoliths, the Lower Paleozoic shales and phyllites, approximately 310 Ma. Because this $^{87}Sr/^{86}Sr$ homogenization is evident in even the lowest-grade shales (although not so extreme as in the higher-grade samples), it implies that hydrothermal activity commenced at the very lowest grades of regional metamorphism (<200°C?) and continued up to temperatures in excess of 600°C (see Bickle *et al.*, 1988, for detailed discussion).

The Trois Seigneurs pelitic schists occupy a much more homogeneous $(^{87}Sr/^{86}Sr)_{310}$ field (0.713 to 0.718) than they would be expected to occupy if they had evolved as closed systems with their present-day $^{87}Rb/^{86}Sr$ ratios from Cambrian sediments. The model values that they would have had 310 Ma are also plotted on Figure 6.11; these data imply that not only have these values been substantially homogenized, but also that some of the samples have undergone a bulk lowering of $^{87}Sr/^{86}Sr$. This lowering must have occurred in response to mixing with relatively unradiogenic strontium ($^{87}Sr/^{86}Sr$ < 0.715) during prograde Hercynian metamorphism (see Table 6.1).

A possible source for the low $^{87}Sr/^{86}Sr$ might be the metacarbonate layers that are present throughout the Lower Paleozoic sequences and that typically have fixed $^{87}Sr/^{86}Sr$ ratios of 0.708. However, detailed profiles measured through the metacarbonates at Trois Seigneurs (Bickle *et al.*, unpublished data) show that there has been incomplete homogenization of Sr within the central parts of the carbonate layers.

An alternative Sr source is the hydrothermal fluids themselves. For example, the marine formation waters (brines) that may have circulated though these rocks may have contained several hundred parts per million Sr, and, judging by analyses of such waters elsewhere in the world, this Sr would have had a $^{87}Sr/^{86}Sr$ substantially lower than 0.715 (0.708?).

The hydrothermal exchange processes that affected the

regional metamorphism of the Pyrenees have profoundly modified the isotopic composition of both strontium and oxygen and hydrogen throughout huge volumes of the crust. In this respect it would be extremely misleading to take the $\delta^{18}O$ values and the calculated model $^{87}Sr/^{86}Sr$ values of low-grade Cambro-Ordovician shales from the Pyrenees and use them directly as representative protolith end-members for any of the synmetamorphic granitic lithologies magmas produced in this region, despite the fact that modified material of this type undoubtedly formed a major component of some of the Hercynian granitic magmas. The shales and phyllites underwent such profound isotopic modification by hydrothermal processes during prograde metamorphism that in any magma-mixing process this end-member would have very different oxygen and strontium isotopic compositions than if the original sediments had remained perfectly closed systems subsequent to deposition.

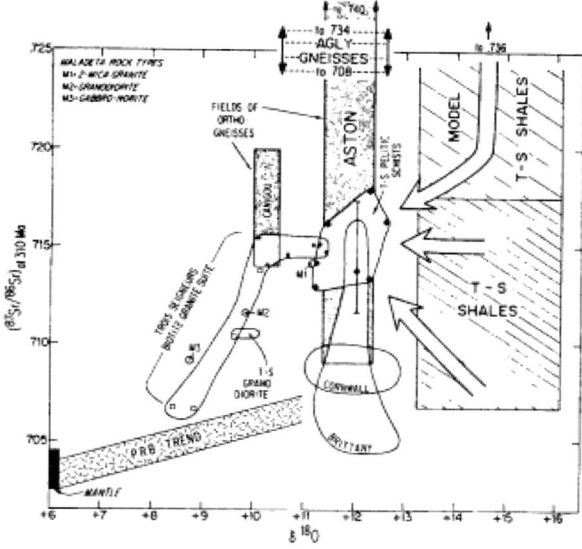

Figure 6.11
Plot of $^{87}Sr/^{86}Sr$ at 310 Ma versus $\delta^{18}O$ (after Bickle *et al.*, 1988), comparing the Trois Seigneurs (T-S) granite and pelite data with data from the Maladeta plutonic complex in the central Pyrenees (M1, M2, and M3; Vitrac-Michard *et al.*, 1980) and with data on some other massifs in the Pyrenees that contain large exposures of Zone 3 granulite-facies basal gneisses (Agly) and of orthogneiss (Canigou and Aston). Also shown are fields for the measured values of Trois Seigneurs shales and phyllites 310 Ma together with the calculated range of model values 310 Ma for these same rocks, assuming that they were originally deposited in the Ordovician with an initial $^{87}Sr/^{86}Sr = 0.707$ (see Table 6.1). The Trois Seigneurs pelitic schists are an obvious high-^{18}O, high-^{87}Sr end-member for the biotite granite site, and they have much more homogeneous $^{87}Sr/^{86}Sr$ values of the shales and phyllites have been changed during prograde Hercynian metamorphism. Fields for the upper mantle, for the main part of the Peninsular Ranges batholith (PRB) from Taylor and Silver (1978) and for the Hercynian granites from Brittany and southwest England (Sheppard, 1986) are also shown. Despite their having major metasedimentary components in their sources, these other Hercynian granitic rocks also have relatively low initial $^{87}Sr/^{86}Sr$ ratios, suggesting that the source rocks for these magmas may also have been affected by the same sort of hydrothermal $^{87}Sr/^{86}Sr$ homogenization process that appears to have affected the Trois Seigneurs pelites.

IMPLICATIONS FOR FLUID TRANSPORT DURING METAMORPHISM

Timing of Infiltration

The massive infiltration of aqueous fluid into the Zone 2 rocks implied by the $^{18}O/^{16}O$ data from the Pyrenees is certainly related in some way to the Hercynian metamorphism. First, the isotopic homogenization is observed only in the (Zone 2) Paleozoic sediments that have been metamorphosed to medium or high grade. Second, this type of large-scale fluid circulation requires some sort of heat engine to drive it, and the intense, localized thermal anomalies generated during metamorphism provide the obvious candidate. Third, the Rb-Sr data (Bickle *et al*., 1988) confirm that there was a major ($^{87}Sr/^{86}Sr$) homogenization event in these rocks and that it occurred ~310 Ma during the Hercynian.

Defining the timing of infiltration relative to the Hercynian metamorphic peak is less straightforward. However, minerals in the granites and mica schists preserve equilibrium $\delta^{18}O$ fractionations (e.g., between coexisting quartz, feldspar, and muscovite in leucogranites; see Wickham and Taylor, 1985). This is contrary to the situation observed in hydrothermal circulation systems associated with cooling plutons (e.g., Criss and Taylor, 1983; Taylor, Chapter 5, this volume) where large disequilibrium effects are commonly observed between quartz and coexisting feldspar. The absence of such effects in the Pyrenean rocks rules out any appreciable postmetamorphic infiltration of the rocks (i.e., following crystallization of the granites).

It is also impossible for all of the infiltration to have occurred contemporarily with the Hercynian metamorphic peak. This is because the Hercynian isograd pattern clearly indicates that very steep thermal gradients (80° to 100°C/km; see Wickham, 1987a) existed at this time, and convective circulation of pore fluids would tend to flatten such gradients. Furthermore, a large proportion of the isotopically homogenized rocks were partially melted and/or undergoing ductile deformation at the metamorphic peak, and this would have reduced their effective permeability, making them resistant to fluid infiltration.

The most plausible scenario is one in which large-scale fluid infiltration of the Paleozoic metasediments occurred over a protracted interval of time during prograde heating of the terrane. This process ceased (or was displaced to higher structural levels) by the time peak metamorphic temperatures were reached. Conceivably, the onset of partial melting in the pelitic lithologies could have sealed them (and any deeper structural levels) to further infiltration, though it remains plausible that some of the water remained available to promote the large-scale melting effects observed in the Zone 2 rocks (see below). This is further supported by strontium isotope data (Bickle et al., 1988) indicating that the $^{87}Sr/^{86}Sr$ ratios were partially homogenized in even the lowest-grade rocks, suggesting that isotopic homogenization commenced at the very earliest stages of regional metamorphism.

Scale of Infiltration

The Hercynian metamorphic terranes in the Pyrenees appear to have been pervasively infiltrated by aqueous fluid over depth ranges of at least 4 km (see Wickham and Taylor, 1985; Figure 6.4). The horizontal extent of the process is much more difficult to estimate due to the fragmentary nature of Hercynian basement exposures. Alpine deformation has broken up the Hercynian crust into a number of discrete tectonic blocks and slices, so it is impossible to know the original regional distribution of the Hercynian isograds with any certainty. However, in the four North Pyrenean Massifs for which data are reported in Figure 6.4, the isotopic homogenization is observed along strike for distances of at least 5 to 10 km. Today the high-grade regions occupy relatively restricted zones commonly cored by bodies of granite or gneiss and separated by geographically more widespread low-grade Paleozoic rocks (Zwart, 1979). It is therefore quite likely that the Hercynian metamorphic and anatectic effects were not continuous regionally at the same level in the crust and that the high-grade terranes formed a pattern of metamorphic hot spots. Models for metamorphism in the Pyrenees have explained such a pattern in terms of diapiric doming of granitoids (Soula, 1982) or localized extensional tectonism and mafic intrusion (Wickham and Oxburgh, 1987).

The origin of the infiltrating marine fluid (whether derived from connate formation waters or directly from the surface) is critically related to the regional Hercynian thermal structure. Material-balance constraints (Figure 6.9) imply that for reasonable $\delta^{18}O$ values of connate formation waters a water-rock ratio of about 0.5 would be necessary to account for the isotopic effects, equivalent to about 25 percent of the isotopically homogenized rock mass (or about 60 percent by volume). Since the maximum reasonable porosity at this depth is perhaps about 2 percent, this implies that large reservoir regions adjacent to the infiltrated terrane would be required to supply this water if it were exclusively connate in origin. Hence, each homogenized region (Zone 2) should be surrounded by a much more extensive terrane that is isotopically unaffected (Zone 1). We have yet to identify such reservoir regions in the Pyrenees and therefore consider it more likely that most of the fluid was derived directly from the surface.

The maximum penetration depth of the fluid is not well constrained. Peak metamorphic mineral assemblages in the Zone 2 rocks suggest pressures of 3 to 4 kbar (10 to 12 km depth) within the Trois Seigneurs migmatite zone (see Wickham, 1987a). Model calculations of Wickham and Oxburgh (1987) suggest that prograde heating of the metasediments probably took only 1 to 2 m.y., so that although fluid infiltration apparently preceded the attainment of maximum metamorphic temperatures, these pressure estimates give us our best indication of the depth of the rocks at the time of infiltration. The petrological and stable isotope data are therefore consistent with the penetration of surface-derived fluid to depths of at least 10 to 12 km below the surface during Hercynian metamorphism.

The data from the Lapege and Agly Zone 3 rocks (basal gneisses) suggest less isotopic homogenization at deeper structural levels (at Agly the mineral assemblages imply Hercynian equilibration pressures of about 5 kbar; Vielzeuf, 1984). However, Figures 6.6 and 6.7 indicate that at St. Barthelemy and Castillon the systematics, particularly in the metacarbonates, are more homogeneous for $\delta^{18}O$. In these latter two localities (which equilibrated at pressures of 4 to 6 kbar), it is possible that some homogenization promoted by fluid infiltration occurred at an early stage in the metamorphic history, although without D/H data we cannot say if this water was marine in origin and therefore surface derived (as we can with the Zone 2 rocks). However, if these terranes are also found to exhibit exceptionally heavy δD values, it would imply penetration of surface waters into some of the deeper-level rocks as well.

Relation of Fluid Transport to Metamorphism and Anatexis

The stable isotope data presented above allow calculation of a plausible material-balance water-rock ratio that could generate the Zone 1 to Zone 2 shift in $\delta^{18}O$. However, consideration of the petrology of migmatite zones such as those exposed in the Trois Seigneurs Massif (Wickham, 1987a) also places constraints on water bud gets during metamorphism and melting. Here there is a rapid increase from 0 percent to >40 percent partial melting of pelitic metasediment over only 300 m of the metamorphic section (representing a temperature increase of no more than 30°C). Such a sharp increase implies that aH_2O is buffered at fixed values close to unity so that melts can be progressively saturated with water as they are generated. Inasmuch as the leucogranitic melt generated from the pelites probably had a water content close to 8 wt.% (see Wickham, 1987b), 50 percent melting would require the partially melted rocks to contain at least 4 wt.% H_2O (disregarding any water that might be contained in hydrous minerals such as biotite). This is two or three times more than the intrinsic water content of high-grade pelite, particularly since much of this water will be held in residual biotite during melting. Hence, an external water source is also required to account for pelite anatexis, even though the size of the source is substantially smaller than that demanded by the oxygen isotope data. Although we cannot at this stage prove that the high-δD marine fluids that flushed the Zone 2 rocks were also responsible for the large-scale melting effects, circumstantial evidence points strongly to a link between the two, implying that the same fluid was available to enter the melting zone during pelite anatexis.

The relationship between the tectonic setting for Hercynian high-thermal gradient metamorphism in the Pyrenees and the large-scale fluid infiltration is still uncertain. However, based on a variety of lines of evidence, including the contemporaneity of marine surface sedimentation with metamorphism and anatexis at depth, the absence of high-pressure metamorphic rocks, the high thermal gradients, and the evidence for deep groundwater circulation, Wickham and Oxburgh (1985, 1986, 1987) proposed that metamorphism occurred in a rift environment and was prompted by intrusion of high-temperature mafic magma into the lower crust. In this scenario localized rift zones (pull aparts?) provided the setting for a focused thermal flux, and deeply penetrating surface water pervasively infiltrated the Paleozoic high-grade metamorphic rocks and promoted melting.

Inasmuch as all active continental rift zones have associated groundwater hydrothermal systems that in many cases penetrate to great but largely unknown depths (see Taylor, Chapter 5, this volume), a rift setting provides the ideal combination of high temperatures at shallow depth with the type of fluid flow system that we infer from the isotope data. In the Pyrenees, where a thick supracrustal pile of Paleozoic sediment (dominated by pelitic material in its lower part) existed prior to metamorphism, rifting would have provided the ideal thermal conditions (i.e., temperatures of 700°C at 10 to 12 km) to cause large-scale melting of the deeper parts of this sedimentary pile.

SUMMARY AND CONCLUSIONS

The Hercynian basement in the Pyrenees is exposed over a wide range of structural levels, and these rocks record the effects of an intense Late Carboniferous metamorphic-anatectic event. $^{18}O/^{16}O$ ratios vary in a systematic fashion across this composite crustal section. At high structural levels the low-grade, moderately deformed Paleozoic sediments (Zone 1) have typical sedimentary $\delta^{18}O$ values (+20 to +25 in carbonates, +14 to +16 in shales). The metamorphosed, strongly deformed equivalents of these rocks (Zone 2) have uniform $\delta^{18}O$ values of about +11.5 in all lithologies. The average $\delta^{18}O$ value of Zone 2 pelitic rocks is about 3 per mil lower than in Zone 1, implying that there has been a bulk change in the oxygen isotopic composition of Zone 2, clearly a result of exchange with some type of low-^{18}O oxygen reservoir. Because low-^{18}O rocks are rare within the Hercynian crust of the Pyrenees at any structural level exposed, it is most likely that this low-^{18}O reservoir was aqueous fluid, which pervasively infiltrated the high-grade rocks, lowering and homogenizing $\delta^{18}O$ values. Mass-balance calculations imply that the quantity of water required to account for the isotopic shift is so large that it can only have been derived from connate formation water or from the Earth's surface. This is supported by exceptionally heavy δD values shown by muscovite from Zone 2 throughout the Pyrenees; such high-δD values are best interpreted as indicating equilibration with marine fluids at high temperatures, because seawater is the major high-D fluid reservoir in the Earth's hydrosphere.

Oxygen-isotope systematics from the deep-level amphibolite and granulite facies gneisses of Zone 3 are in general more heterogeneous than those in Zone 2, although some regions are clearly more isotopically heterogeneous than others. The Zone 3 data cannot be interpreted in such a clear-cut way as for Zones 1 and 2, but they certainly preclude pervasive infiltration of the Zone 3 rocks by large volumes of aqueous fluid. This tends to support the idea that the Zone 2 infiltrating fluids were derived from above rather than below, since in this case a maximum penetration depth would be expected.

At Trois Seigneurs, lower Paleozoic shales and phyllites have $^{87}Sr/^{86}Sr$ values of 0.707 to 0.717 (at 310 Ma), but model values at 310 Ma of 0.709 to 0.736, based on an assumed depositional age of 450 Ma and an initial $^{87}Sr/^{86}Sr = 0.707$. On a regional scale these $^{87}Sr/^{86}Sr$ ratios were homogenized to about 0.713 to 0.717 in both high- and low-grade pelitic schists during the 340- to 310-Ma metamorphic events (see Bickle et al., 1988). Much of this $^{87}Sr/^{86}Sr$ exchange occurred at very low grades (below the biotite isograd), but significant changes also accompanied the $\delta^{18}O$ lowering of the phyllites (+ 13 to +16) during their transformation to andalusite- and sillimanite-grade schists

($\delta^{18}O$ = +11 to +12); all of these effects are attributed to pervasive interactions with hydrothermal fluids.

A scenario for fluid flow in the Pyrenees is depicted in Figure 6.12; this shows a surface water flux penetrating to depths of 10 to 12 km in cool areas, adjacent to a localized thermal anomaly such as would have generated the metamorphic sequences in the Pyrenees. This water migrates through the supracrustal sediments in the heated area during prograde metamorphism, but movement largely ceases once melting is initiated. Such melting may set up lateral gradients in aH_2O that help to draw the deeply penetrating water into the melting zone. This water, however, is largely restricted from entering the Zone 3 gneisses that probably formed the basement to the Paleozoic supracrustal metasediments. This may reflect an intrinsic permeability contrast between the basal gneisses and the supracrustal sediments or, alternatively, the zone of anatexis may have formed a barrier to any deeper penetration of H_2O.

Note that we are not suggesting in Figure 6.12 that there is any multiple cycling of H_2O through the metasedimentary sequence. It is clear from all kinds of considerations that the water that is drawn in laterally toward the thermal anomaly makes only one pass through the terrane before being heated and expelled upward. Even in relatively small hydrothermal convective systems around igneous intrusions, calculations show that there is only time for essentially one pass (see Norton and Taylor, 1979).

Our understanding of fluid transport mechanisms in rocks undergoing high-grade metamorphism is still insufficient to say whether the large-scale fluid infiltration documented in Zone 2 in the Pyrenees is likely to be typical in other metamorphic terranes (for discussion see Ferry, 1986; Wood and Walther, 1986). However, inasmuch as the degree of infiltration appears to be significantly higher in Zone 2 than in Zone 3, this area may provide an opportunity to identify the most important factors controlling effective permeability during metamorphism. These factors obviously must be linked to the characteristics of the rocks at the time of metamorphism, rather than to their present-day physical properties such as permeability. For example, devolatilization during prograde metamorphism may have enhanced permeability in Zone 2 but not in Zone 3, where the rocks had previously been dehydrated and where they already had a high-grade mineralogy by Hercynian times. If we are correct in our interpretation that the infiltrating fluid was derived from the surface or at least from connate fluids within the Paleozoic supracrustal pile, then the Zone 2 to Zone 3 contrast may be depth related, and we may simply be observing the lower, weaker parts of a very deep (hydrostatic?) hydrothermal circulation system.

Some authors (e.g., Wood and Walther, 1986) have suggested that such hydrothermal systems should die out at depths of 3 to 6 km because in some boreholes (e.g., the U.S. Gulf Coast) pore fluid pressures become lithostatic within this depth range. In rift-zone situations like Iceland this is clearly not true, and it is likely that the meteoric hydrothermal convective systems there extended down to at least 10 to 15 km depth (see Taylor, Chapter 5, this volume). However, we are as yet unable to predict where this hydrostatic to lithostatic pressure change will occur in other parts of the crust with different states of stress,

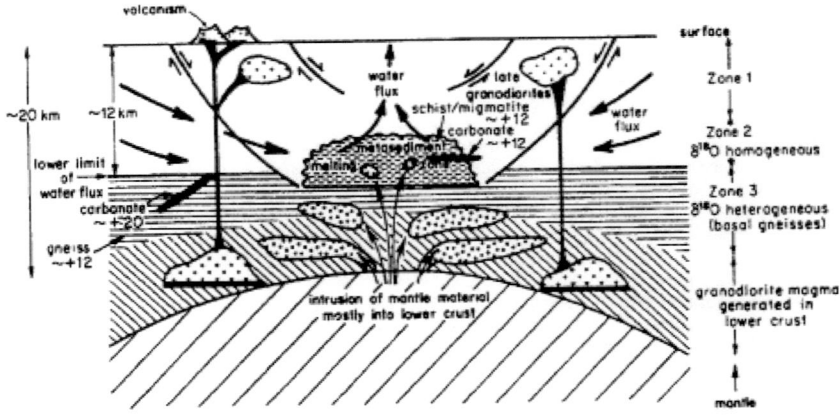

Figure 6.12
An approximately 50-km-wide section through the Hercynian crust of the Pyrenees showing listric normal faults and a schematic rift setting for low-pressure regional metamorphism. Melting occurs at the base of the Paleozoic metasedimentary pile in response to heating by mafic intrusions, which are emplaced predominantly within the lower crust (dot-pattern bodies in center of diagram). Extensive melting at still deeper levels in the crust generates a group of larger, late-stage granodiorite magma bodies (cross-pattern bodies in Zone 1 to the sides of the diagram, fed from "floored" source regions in the lower crust). The flux of surface-derived metamorphic pore fluids (heavy arrows) is very high in the Lower Paleozoic metasedimentary sequences (Zone 2), but either negligible or at least much lower in the underlying basal gneisses (Zone 3), which retain heterogeneous $\delta^{18}O$ values (gneiss = +11, carbonate = +17 to +23), in contrast to the homogenized Zone 2 values (= +12 to +14 in all lithologies). The granulite-facies gneisses thus seem to lie between the two major anatectic levels that affected the Hercynian crust of the Pyrenees.

tectonic regimes, upper crustal composition, and thermal structure. Deep boreholes in continental settings with deep hydrothermal systems (e.g., the Salton Trough, California, and the Taupo Rift Zone, New Zealand) may in the future resolve this important question regarding the maximum penetration depth of groundwater convective systems. The data from the Pyrenees suggest that here at least the surface water reached depths of 10 to 12 km, resulting in profound isotopic modification of the rocks. At depth this presumably early-stage hydrostatic regime (or at least one where $P_{H_2O} < P_{TOTAL}$) must have been transformed to a lithostatic regime with time, probably as wholesale melting began to occur and the style of rock deformation changed from a brittle-fracture mode to a ductile mode. We believe that, at least during the early stages of this transition, the H_2O required to promote pelite anatexis was still the same water as was responsible for isotopic homogenization (and that this water was originally derived from a connate or marine source).

Thus, at some point in the time-temperature history of this terrane (possibly at the onset of anatexis in pelitic lithologies), aqueous fluids that had been moving under a hydrostatic (Darcy's Law) flow regime became (locally?) involved in the lithostatic regime characteristic of a crustal melting process. At present our understanding of these complex three-phase fluid-rock-melt systems is insufficient to predict the nature of this transition in any detail. However, it is possible that aqueous fluids may have still been able to migrate through fractures in refractory layers (e.g., calc-silicates or quartzites) embedded in the volumetrically more extensive, partially molten pelitic layers, thereby providing a water supply to the anatectic zones and maintaining water-rich conditions throughout the melting event. The lithostatic-hydrostatic boundary probably migrated upward and outward from the zone of enhanced heat flow, at least up until the time that the thermal anomaly began to decay. Understanding the details of this complex lithostatic-hydrostatic transition and the effects of intercalated brittle and ductile layers in terms of space, time, temperature, and permeability will be difficult. It will necessarily involve a wide variety of geological, geochemical, and geophysical efforts and approaches.

ACKNOWLEDGMENTS

Financial support for this work was provided by NSF grant no. EAR-8313106. We are grateful to M. J. Bickle, H. J. Chapman, and Ron Oxburgh for their collaboration and help on these problems. Discussions with Denis Norton and Steve Sparks were also very helpful. This is contribution no. 4442, Division of Geological and Planetary Sciences, California Institute of Technology.

References

Autran, A., B. Barriere, B. Bonin, J. Didier, P. Fluck, S. Fourcade, P. Giraud, J. Jonin, J. Lameyre, J-B. Orsini, and G. Vivier (1980). Les granitoides de France, in Évolutions Géologiques de la France, *Bulletin de Recherche Géologique et Minéralogique, Mémoire 107*, Colloque C7, 51-97.

Baertschi, V. P. (1957). Messung und Deutung relativer Häufigkeitsvariationen von ^{18}O and ^{13}C in karbonatgesteinen und mineralien, *Schweizerische Mineralogische und Petrographische Mitteilungen 37*, 73-151.

Bard, J-P., B. Briand, J-M. Cantagrel, G. Guitard, J-R. Kienast, J. Kornrobst, B. Lasnier, C. LeCorre, and D. Santallier (1980). Le metamorphisme en France, in Évolutions Géologiques de la France, *Bulletin de Recherche Géologique et Minéralogique, Mémoire 107*, Colloque C7, 161-189.

Bickle, M. J., S. M. Wickham, H. J. Chapman, and H. P. Taylor, Jr. (1988). A strontium, neodynium, and oxygen isotope study of the hydrothermal metamorphism and crustal anatexis in the Trois Seigneurs Massif, Pyrenees, France, *Contributions to Mineralogy and Petrology 100*, 399-417.

Clayton, R. N., I. Friedman, D. L. Graf, T. K. Mayeda, W. F. Meerts, and N. F. Shimp (1966). The origin of saline formation waters: I. Isotopic composition, *Journal of Geophysical Research 71*, 3869-3882.

Criss, R. E., and H. P. Taylor, Jr. (1983). An $^{18}O/^{16}O$ and D/H study of Tertiary hydrothermal systems in the southern half of the Idaho batholith, *Geological Society of America Bulletin 94*, 640-663.

Ferry, J. M. (1983). Regional metamorphism of the Vassalboro formation, South-Central Maine, U.S.A.: A case study of the role of fluid in metamorphic petrogenesis, *Journal of the Geological Society of London 140*, 551-76.

Ferry, J. M. (1986). Infiltration of aqueous fluid and high fluid-rock ratios during greenschist facies metamorphism: A reply, *Journal of Petrology 27*, 695-714.

Fleck, R. J., and R. E. Criss (1985). Strontium and oxygen isotopic variations in Mesozoic and Tertiary plutons of central Idaho, *Contributions to Mineralogy and Petrology 90*, 291-308.

Frey, M., J. C. Hunziker, J. R. O'Neil, and H. W. Schwander (1976). Equilibrium-disequilibrium relations in the Monte Rosa granite, Western Alps: Petrological, Rb-Sr and stable isotope data, *Contributions to Mineralogy and Petrology 55*, 147-179.

Garlick, G. D., and S. Epstein (1965). Oxygen isotope ratios in coexisting minerals of regionally metamorphosed rocks, *Geochimica et Cosmochimica Acta 31*, 181-214.

Munha, J., F. J. A. S. Barriga, and R. Kerrich (1986). High-^{18}O ore-forming fluids in volcanic-hosted base metal massive sulphide deposits: Geologic, $^{18}O/^{16}O$ and D/H evidence from the Iberian pyrite belt; Crandon, Wisconsin; and Blue Hill, Maine, *Economic Geology 81*, 530-552.

Negga, H. S., S. M. F. Sheppard, J. M. Rosenbaum, and M. Cuney (1986). Late Hercynian U-vein mineralization in the Alps: Fluid inclusion and C, O, H isotopic evidence for mixing between two externally derived fluids, *Contributions to Mineralogy and Petrology 93*, 179-186.

Norton, D., and H. P. Taylor, Jr. (1979). Quantitative simulation of the hydrothermal systems of crystallizing magmas on the basis of transport theory and oxygen isotope data, *Journal of Petrology 20*, 421-486.

Passchier, C. W. (1984). Mylonite-dominated footwall geometry in a shear zone, central Pyrenees, *Geological Magazine 121*, 429-436.

Rye, R. O., R. D. Schuiling, D. M. Rye, and J. B. H. Jansen (1976). Carbon, hydrogen and oxygen isotope studies of the regional metamorphic complex at Naxos, Greece, *Geochimica et Cosmochimica Acta 40*, 1031-1049.

Sheppard, S. M. F. (1977). The Cornubian batholith, SW England: D/H and $^{18}O/^{16}O$ studies of kaolinite and other alteration minerals, *Journal of the Geological Society of London 133*, 573-591.

Sheppard, S. M. F. (1986). Igneous rocks: III. Isotopic case studies of magmatism in Africa, Eurasia, and oceanic islands, in *Stable Isotopes in High Temperature Geological Processes*, J. W. Valley, H. P. Taylor, Jr., and J. R. O'Neil, eds., Reviews in Mineralogy, vol. 16, Mineralogical Society of America, Washington, D.C., pp. 319-371.

Shieh, Y. N., and H. P. Schwarcz (1974). Oxygen isotope studies of granite and migmatite, Grenville Province of Ontario, Canada, *Geochimica et Cosmochimica Acta 38*, 21-45.

Shieh, Y. N., and H. P. Taylor, Jr. (1969). Oxygen and hydrogen isotope studies of contact metamorphism in the Santa Rosa range, Nevada, and other areas, *Contributions to Mineralogy and Petrology 20*, 306-356.

Soula, J-C. (1982). Characteristics and mode of emplacement of gneiss domes and plutonic domes in the central-eastern Pyrenees, *Journal of Structural Geology 4*, 313-342.

Taylor, H. P., Jr. (1974). The application of oxygen and hydrogen isotope studies to problems of hydrothermal alteration and ore deposition, *Economic Geology 69*, 843-883.

Taylor, H. P., Jr. (1977). Water/rock interactions and the origin of H_2O in granitic batholiths, *Journal of the Geological Society of London 133*, 509-558.

Taylor, H. P., Jr., and L. T. Silver (1978). Oxygen isotope relationships in plutonic igneous rocks of the Peninsular Ranges batholith, southern and Baja California, in *Short Papers of 4th International Conference on Geochronology, Cosmochronology, and Isotope Geology*, R. E. Zartman, ed., U.S. Geological Survey Open-File Report 78-701, pp. 423-426.

Veizer, J., and J. Hoefs (1976). The nature of $^{18}O/^{16}O$ and $^{13}C/^{12}C$ secular trends in sedimentary carbonate rocks, *Geochimica et Cosmochimica Acta 47*, 697-706.

Vielzeuf, D. (1984). Relations de phases dans le facies granulite et implications geodynamiques. L'exemple des granulites des Pyrénées, *Annales Scientifiques Université Clermont H 79*, 189 pp.

Vitrac-Michard, A., and C. J. Allegre (1975). A study of the formation and history of a piece of continental crust by $^{87}Sr/^{86}Sr$ method: The case of the French oriental Pyrenees, *Contributions to Mineralogy and Petrology 50*, 257-285.

Vitrac-Michard, A., F. Albarede, C. Dupuis, and H. P. Taylor, Jr. (1980). The genesis of Variscan (Hercynian) plutonic rocks: Inferences from Sr, Rb, and O studies of the Maladeta igneous complex, Central Pyrenees, Spain, *Contributions to Mineralogy and Petrology 72*, 57-72.

Wickham, S. M. (1987a). Crustal anatexis and granite petrogenesis during low pressure regional metamorphism; the Trois Seigneurs Massif, Pyrenees, France, *Journal of Petrology 28*, 127-169.

Wickham, S. M. (1987b). The segregation and emplacement of granitic magmas, *Journal of the Geological Society of London 144*, 281-297.

Wickham, S. M., and E. R. Oxburgh (1985). Continental rifts as a setting for regional metamorphism, *Nature 318*, 330-333.

Wickham, S. M., and E. R. Oxburgh (1986). A rifted tectonic setting for Hercynian high thermal gradient metamorphism in the Pyrenees, in *The Geological Evolution of the Pyrenees*, E. Banda and S. M. Wickham, eds., *Tectonophysics 129*, 53-69.

Wickham, S. M., and E. R. Oxburgh (1987). Low pressure regional metamorphism in the Pyrenees and its implications for the thermal evolution of rifted continental crust, *Philosophical Transactions of the Royal Society of London A321*, 219-243.

Wickham, S. M., and H. P. Taylor, Jr. (1985). Stable isotopic evidence for large-scale seawater infiltration in a regional metamorphic terrane: The Trois Seigneurs Massif, Pyrenees, France, *Contributions to Mineralogy and Petrology 91*, 122-137.

Wickham, S. M., and H. P. Taylor, Jr. (1987). Stable isotope constraints on the origin and depth of penetration of hydrothermal fluids associated with Hercynian regional metamorphism and crustal anatexis in the Pyrenees, *Contributions to Mineralogy and Petrology 95*, 255-268.

Wood, B. J., and J. V. Walther (1986). Fluid flow and its implication for fluid-rock ratios, in *Fluid-Rock Interactions During Metamorphism*, J. V. Walther and B. J. Wood, eds., Springer-Verlag, New York, pp. 89-108.

Zwart, H. J. (1979). The geology of the Central Pyrenees, *Leidse Geol. Med. 50.1*, 1-74.

7

Time-Dependent Hydraulics of the Earth's Crust

AMOS NUR
Stanford University
JOSEPH WALDER
University of Washington

ABSTRACT

The deceivingly simple question posed here is how deep free water extends in the Earth's crust. This question is found to be inseparable from crustal porosity, permeability, and pore pressure and most significantly their dependence on time at depth. If we assume that porosity and permeability are time independent, it follows from *hydrological evidence* that ambient pore pressure must be hydrostatic. However, *geological evidence* suggest that rocks in situ rapidly seal hydraulically. To reconcile this conflict, the hypothesis is explored that crustal porosity, permeability, and hence pore pressure are in general time-dependent due to the gradual closure of crustal pore space via healing, sealing, and inelastic deformation. It is found that when the hydraulic conductivity of the system is large, so that the ratio of porosity reduction rate to permeability k, $/k$, is small, the initially porous water-saturated crustal rock mass will gradually lose its porosity and fluid, until it becomes essentially dry. Pore pressure, P_p, throughout this process will remain around hydrostatic. If on the other hand, system permeability is small and $/k$ is large, the pore fluid cannot escape fast enough and pore pressure will build-up. It is envisioned that when the pore pressure reaches the level of the least compressive stress in the crust, natural hydraulic fracturing takes place, leading to some fluid release, pore pressure drop, and resealing of the system. With time P_p builds up again, leading to another cycle of hydrofracturing.

Analysis shows that tens to hundreds of such P_p cycles are possible and that the duration of one cycle may be 10^3 to 10^5 yr; the duration of the entire dewatering process can be 10^6 to 10^7 yr. If subducted lithosphere supplies additional waters, the process may last as long as 10^8 yr.

The conclusion that the crust is most likely undergoing repeated (1) cyclic episodes of pore pressure buildup to lithostatic and (2) water expulsion events associated with natural hydraulic fracturing may explain such observations as the episodic nature of some types of ore deposits and veins, the recurrence time between large earthquakes, the emplacement of foreland thrust systems, and the seismic and mechanical nature of crustal detachment zones.

INTRODUCTION

One of the simplest yet profound questions that can be asked about the state of the Earth's crust is the depth to which free water extends. The top of the free water zone in the crust, in the form of groundwater table, is present almost everywhere. The depth to the water table typically varies between 0 and 1000 m or so and is the subject of extensive hydrological exploration, especially in arid and semiarid regions. A much more difficult question is the depth of the bottom of the free water or groundwater in the crust (see Table 7.1): Is groundwater limited to the top 1 to 2 km, with the crust below being dry, or does free water extend to much greater depth, for example, the depth to which the crust is brittle (~10 to 15 km), or even deeper? And if free water is present at the greater depth, how much is present or, equivalently, what is the porosity in the crust? This question in turn is found to be inseparable from what the water's ambient, or steady-state, pore pressure P_p must be (Figure 7.1). Furthermore, as discussed in this chapter, the pore pressure question is itself inseparably linked to the question of permeability at depth.

In this chapter we consider the following three interlinked questions: What is the depth in the crust of free water and its pressure? What are typical crustal porosity and permeability? Are or can these parameters be time invariant? These simple questions are actually very important because their answers are keys to understanding a surprisingly wide range of geological and geophysical crustal phenomena. For example, the mechanisms by which crustal rocks deform tectonically are strongly influenced by the presence or absence of water as well as by the level of pore pressure (e.g., Carter, 1976; Brace and Kohlstedt, 1980). Circulation of crustal water has important effects on heat flow (e.g., Sleep and Wolery, 1978; Lachenbruch and Sass, 1980; Smith and Chapman, 1983), on the distribution of oxygen and hydrogen isotopes (e.g., Taylor, 1977; O'Neil and Hanks, 1980), and on the formation of

TABLE 7.1 Possible Indicators for the Presence of Free Water in the Earth's Crust and Estimated Depth Values

Indicator	Depth range	References
Water table	0 to 2 km	—
Deep wells	to 12 km	—
Reservoir induced seismicity	to 12 km	Bell and Nur (1978)
Crustal low velocity zones	7 to 12 km	Berry and Mair (1977)
		Feng and McEvilly (1983)
		Jones and Nur (1984)
Crustal electrical conductivity zones	10 to 20 km	Nekut et al. (1977)
		Shankland and Ander (1983)
Oxygen isotopes	to 20 km	Taylor (1977)
Metamorphism	>20 km	Fyfe et al. (1978)
		Etheridge et al. (1984)
Crack healing and sealing	?	Richter and Simmons (1977a,b)
		Sprunt and Nur (1979)
		Ramsay (1980)
		Smith and Evans (1984)
Formation of hydrothermal ore deposits	>5 km	Norton and Knight (1977)
		Cathles and Smith (1983)
Crustal seismic attenuation zones	7 to 15 km	Hermann and Mitchell (1975)
		Winkler and Nur (1982)
Low stress on faults	0 to 10 km?	Raleigh and Evernden (1982)

hydrothermal ore deposits (e.g., Norton and Knight, 1977). The role of pore pressure is also central in understanding the processes of the earthquake failure process (Byerlee, 1967), earthquake prediction (Nur, 1972), induced seismicity (Bell and Nur, 1978) (Figure 7.2), the mechanical processes in and below the accretionary wedge in subduction zones (Zhao et al., 1986), the rate and depth of magmatic melting and volcanism associated with subduction zones (McGeary et al., 1985), the nature of deep crustal seismic reflectors (Jones and Nur, 1984), and the state of stress in the crust (Zoback et al., 1987).

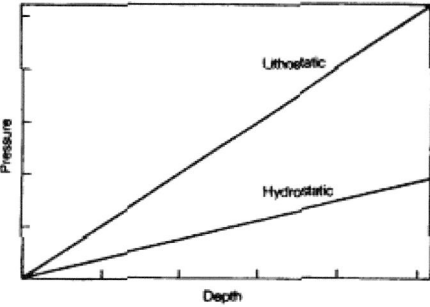

Figure 7.1
Rock and water pressure in porous rock in the Earth's crust. The gradients of the two lines are proportional to rock density and water density, respectively. When pore pressure is equal to hydrostatic, the difference between the two pressures gives rise to the tendency of the pores to inelastically close with time.

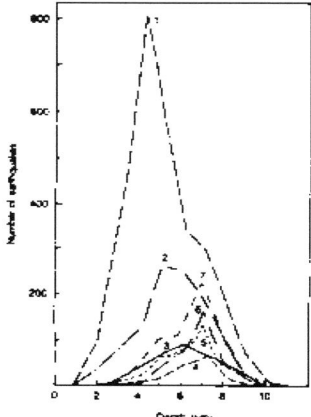

Figure 7.2
Example of the depth and time migration of earthquakes induced by reservoir impounding (from Chung-Kang et al., 1974). As shown by Bell and Nur (1978), this induced seismicity requires the presence of free water at the depth of these induced earthquakes prior to the formation of the reservoir.

Answers to the three questions posed must at present be based largely on indirect evidence. For example, in situ measurements of crustal hydrologic properties typically reach to depths of only 2 to 3 km (Brace, 1980). However, the presence of free water to much greater depths in the

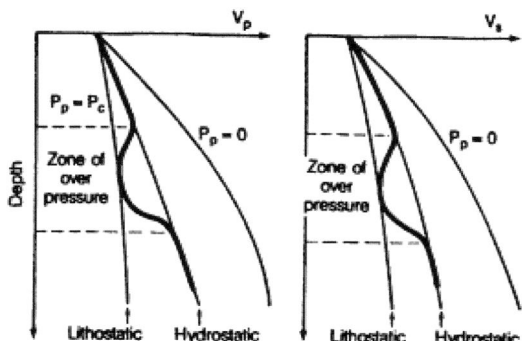

Figure 7.3
Schematic illustration of the development of a low compressional and shear wave velocity zone due to anomalously high pore pressure, based on extensive laboratory measurements (Nur and Simmons, 1969) and some field observations.

crust is suggested by isotopic studies of batholithic rocks (e.g., Taylor, 1977; Norton and Taylor, 1979), studies that indicate that meteoric water may circulate to depths of up to 20 km. Deep crustal electromagnetic soundings have revealed zones of relatively low electrical resistivity, which suggests the presence of a continuous water phase (e.g., Nekut et al., 1977; Thompson et al., 1983), as inferred from laboratory studies of the electrical properties of rocks (Olhoeft, 1981; Shankland and Ander, 1983).

Seismology has also contributed to ideas about the hydrologic character of the crust. For example, Berry and Mair (1977) argue that crustal low-velocity zones could be due to P_f locally in excess of hydrostatic. This argument is based on experimental results such as those of Nur and Simmons (1969) (Figure 7.3), who showed that even in very low porosity saturated rocks, compressional velocity drops markedly as P_f approaches the confining pressure P_c Feng and McEvilly (1983) discovered a prominent low-velocity zone near the near San Andreas Fault in California, which has been interpreted to result from high pore pressure (Raleigh and Evernden, 1982; Figure 7.4). Similarly, crustal low Q zones, such as observed by Hermann and Mitchell (1973; Figure 7.5), are also consistent with the presence of zones with pore pressure close to lithostatic (Winkler and Nur, 1982). Using a similar line of reasoning, Jones and Nur (1984) suggested that reflections from deep crustal fault zones may be associated with elevated P_f within or below these zones. The hypothesis of elevated P_f has also been suggested by Raleigh and Evernden (1982) to explain the low deviatoric stresses thought to exist along plate boundaries such as the San Andreas Fault.

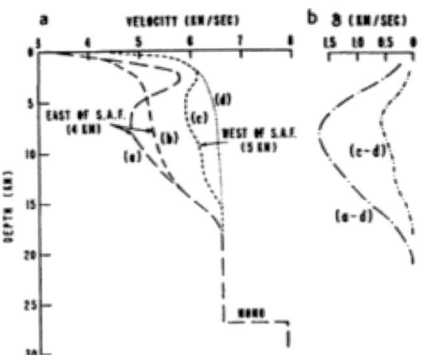

Figure 7.4
Compressional low-velocity zones near the San Andreas Fault in central California. The figure is taken from Raleigh and Evernden (1982), who based it on extensive crustal reflection measurements by Feng and McEvilly (1983). Raleigh and Evernden (1982) suggest that the (a) low-velocity zone and (b) velocity anomaly δ are due to high pore pressure in the depth range of 5 to 10 km or so.

Other inferences about crustal hydrology are derived from geological evidence. For example, Fyfe *et al.* (1978) and Etheridge *et al.* (1984) have reviewed geological indicators for free water, with P_f often exceeding hydrostatic, being widespread during low-to medium-grade regional metamorphism. Two principal lines of evidence follow. (1) The ubiquity of mineralized fractures whose microstructure and orientation indicate that they formed in extension (Ramsay, 1980). On the basis of commonly accepted criteria for brittle failure, this requires that P_f exceed the minimum principal confining stress at the time of fracture formation. (2) Experimentally determined phase equilibria (with P_f equal to confining pressure P_c) are consistent with natural distributions of metamorphic mineral assemblages.

Together the above arguments suggests that free water must be fairly common at upper and mid-crustal levels. Furthermore, it appears that elevated P_f directly implies that the permeability of crustal rocks must be very low. Bredehoeft and Hanshaw (1968) and Hanshaw and Bredehoeft (1968), for example, studied simple models of crustal P_f development and concluded that, in general, maintenance of elevated P_f for geologically significant periods of time requires the presence of some crustal horizons with very low permeability, down to 10^{-21} m² (1 ndarcy) and lower.

The evidence and arguments favoring the presence of water at great crustal depth at high P_f and the consequent implication that crustal permeability must be low are in remarkable disagreement with the conclusion of Brace (1980), who after reviewing direct and indirect estimates of crustal permeability (Figure 7.6) argued that zones with

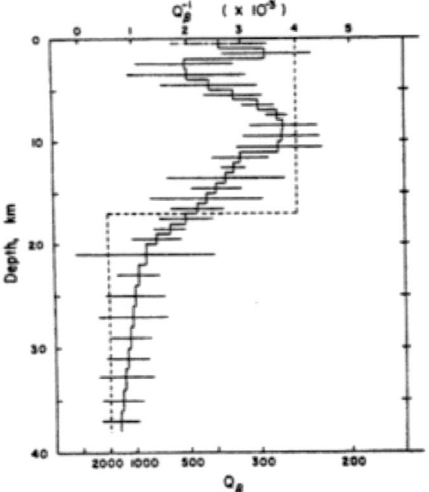

Figure 7.5
Estimated shear wave specific attenuation versus depth obtained from surface wave measurements in the stable North American continent (from Hermann and Mitchell, 1975).

permeability of about 10^{-15} m^2 (1 mdarcy) or higher must exist down to at least 10 km depth. Brace therefore concluded that P_f is very unlikely to exceed hydrostatic pressure in regions where crystalline rocks extend to the surface and that P_f above hydrostatic in crystalline rocks could be maintained only if a cover of very low permeability (e.g., argillaceous) rocks were present. Also, Jones (1983) examined the hypothesized relationship between seismically reflective zones in the crust and elevated P_f by combining simple models of P_f development with synthetic seismograms. Jones concluded that, although elevated pore pressure can significantly affect the existence and amplitudes of reflected waves, such effects persist for geologically significant periods of time "only for a permeability lower than that generally observed in laboratory measurements on crustal rocks."

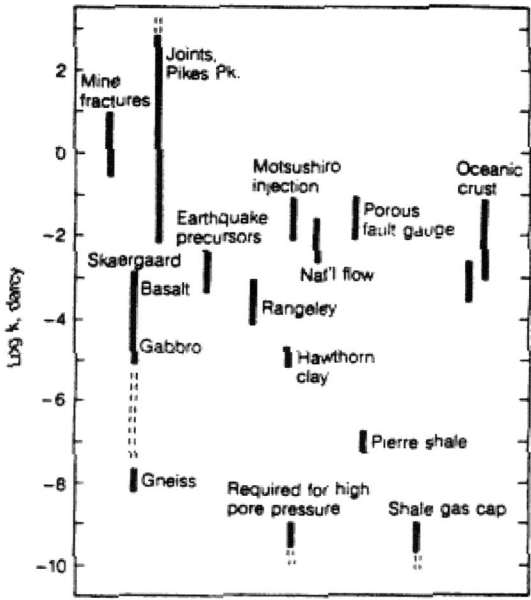

Figure 7.6
In situ permeability inferred from various large-scale phenomena. Numbers in parentheses refer to accompanying notes, which explain the calculation of k (from Brace, 1980).

Clearly, a model is needed that can reconcile the two conflicting lines of evidence regarding crustal hydrology: evidence for ubiquitous high pore pressure and hence low permeability on the one hand, and evidence for relatively fast flow and fast pore pressure dissipation on the other. We will outline such a model by considering two general cases of crustal hydraulics: (1) time-invariant crustal porosity (and hence permeability) and (2) time-dependent porosity, permeability, and pore pressure.

TIME DEPENDENCE OF POROSITY AND PERMEABILITY

The simplest model one can envision for the hydraulics of the crust is based on the assumption that porosity and, consequently, permeability remain unchanged with time or are time invariant. In that case hydraulic permeability or diffusivity inferred from in situ phenomena (Brace, 1980) directly provides an estimate of the typical steadystate or ambient crustal permeability. In fact, Brace's compilation of numerous such estimates (Figure 7.5), made for a wide variety of rock types and geological settings, results in typical values of crustal permeabilities ranging from as much as 10 to 100 darcy for shallow fractured rock masses to 10^{-1} to 10^{-5} darcy for most rocks down to 10^{-6} to 10^{-7} darcy for shales and some gneisses.

As pointed out by Brace (1980), pore pressure in a crust with these rock permeabilities will generally have to be close to hydrostatic, with the water at depth thus sufficiently connected to the free surface of the crust, so that the pressure at any depth is simply the weight per unit area of a column of water reaching the Earth's surface. Episodes or regions of overpressure will thus be special cases, transient in time and localized in space.

The apparent contradiction between the indirect evidence for elevated crustal P_f, on the one hand, and the inferred relatively high permeability throughout the crust, on the other hand, cannot be reconciled with this model. Suppose instead that porosity, permeability, and consequently pore pressure can vary significantly with time rather than remain static. Such variations have been inferred by D. Norton (University of Arizona) and co-workers in communication with fluid flow and mineralization associated with magmatic intrusion in the crust. Norton (Chapter 2, this volume) suggests that the pore fluid pressure induced by such intrusions sufficiently exceeds lithostatic pressure to cause natural hydraulic fracturing. Ramsay (1980) described indirect evidence for repeated episodic P_f pulses in the form of composite crack filing veins, presumably episodically precipitated from solution. In this chapter we explore the feasibility of this kind of episodic P_p buildup not as limited to magmatic intrusions but as a general behavior of the crust. Such general time dependence might allow for periods of fast fluid flow, for example, during episodes of P_f equal to lithostatic, bracketing periods of lower pore pressure, and no or little flow. Brace's geological estimates may thus represent only the periods or episodes of fast flow and not the crust in general. The immediate question that arises is: How must porosity deep in the crust decrease with time, and can this decrease be rapid enough (e.g., due to inelastic processes in porous rocks)? If so, what are the conditions under which permeability changes through time to maintain P_f well in excess of hydrostatic? Three most obvious mechanisms potentially responsible for time-dependent changes in porosity and pore space configuration are (1) inelastic pore deformation, leading to pore closure; (2) dissolution, including pressure solution, and redeposition of solutes in the pores; and (3) the creation of fractures, with their subsequent healing and sealing. As porosity changes due to these processes, so will permeability.

Possible relationships between porosity reduction and strain can be indirectly inferred from rocks that were once at considerable depths in the crust. Porosity reduction processes in rocks such as indicated by detailed studies by optical and scanning electron microscopy (e.g., Richter and Simmons, 1977a; Sprunt and Nur, 1979; Padovani et al., 1982) clearly demonstrate that crack healing and sealing are quite ubiquitous in a wide variety of crustal rock types, particularly in crystalline rocks. We use the term *healing* for former cracks in which the mineral filling is the same as the host grain and the term *sealing* for cases in which the crack filling is mineralogically different from the host grain. The material source for healing is likely to be local (i.e., nearby grains), whereas sealing may require an external or more remote source of crack-filling material transported via the pore fluid.

Although there has been little laboratory work on porosity reduction in crystalline rocks, experimental studies provide some indications that porosity reduction could be relatively fast. For example, Sprunt and Nut (1976, 1977; Figure 7.7) measured appreciable porosity loss, presumably due to local pressure solution, in sandstone samples subjected to elevated temperatures (to 250°c), pressures, and macroscopic shear stress (to ~500 bars) for 2 weeks. Their results indicate that porosity reduction rates in rocks subjected to tectonically induced deviatoric stresses can be fast, suggesting that porosity reduction rates in situ may also be geologically fast.

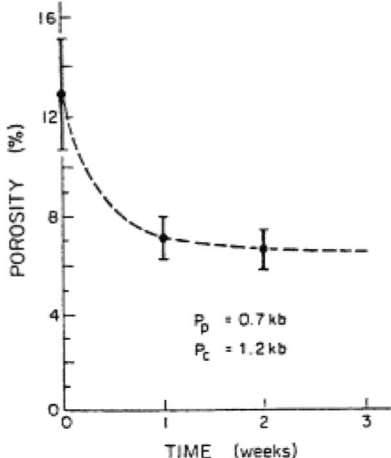

Figure 7.7
Experimentally induced inelastic porosity reduction in a sandstone sample, subject to an overburden pressure of 1.2 kbar and pore pressure of 0.7 kbar for a period of 2 weeks at 250°C. Note the rapid porosity decrease in a period of a week or so at this temperature (from Sprunt and Nut, 1976).

Smith and Evans (1984) examined healing of cracks in synthetic quartz under elevated pressure ($P_f = P_c = 200$ MPa) and temperature (200° to 600°C). Morphologically, healed cracks were strikingly similar to fluid inclusions and "microtubes" commonly seen in thin sections (e.g., Richter and Simmons, 1977b). Smith and Evans found that thin cracks healed extremely rapidly (in less than 1 hr) at 400°C. It is likely, therefore, that healing and sealing are quite rapid on the geologic time scale under mid-crustal conditions.

Evans and co-workers also found that the sealing rate falls off very rapidly with increasing crack aperture, so thick cracks (>1 mm?) may seal or heal very slowly. As pointed out by Brace (Massachusetts Institute of Technology, private communication, 1988), we do not know the crack aperture versus depth in the crust. However, the ubiquitous occurrence of fully sealed cracks in exposed or exhumed mid-crustal rocks and the evidence for sealing by repeated episodes (Ramsay, 1980) suggest that, in general, sealing is widespread and fast enough to leave only bubble chains behind.

POROSITY REDUCTION WITH TIME

The discussion above implies that porosity must generally tend to decrease with time. This reduction leads to two important interrelated effects (Figure 7.8): (1) the gradual expulsion of water out of the crustal pore space (if permeability is high enough) and (2) the gradual buildup of pore pressure within the pore space of crustal rocks if permeability is sufficiently low. Accordingly, two cases (Figure 7.9) for the development of crustal porosity with time, permeability, and pore pressure can be envisioned.

Case 1

To illustrate this case, consider an element of porous crust at depth, subject to overburden stress, as shown in Figure 7.8. Due to porosity reduction processes, the pore space gradually decreases with time. If a permeable path exists between the subsurface rock element and the Earth's free surface, the fluid in the pores will gradually be squeezed out. The pore fluid pressure during this process will obviously depend on the rate of porosity reduction on the one hand and the resistance to flow through the permeable path to the surface. If the resistance is low or, equivalently, if the permeability is sufficiently high, P_p, throughout the duration of the porosity reduction process will be only slightly higher than hydrostatic. Under these conditions a gradual, continuous process of porosity loss accompanied by gradual water loss will take place, ultimately leading to a dry, pore-free crust. For this condition to be satisfied, the initial permeability of the rock element must be high and the rate of permeability decrease with porosity $dk/d\phi$ must be small enough so that drainage can occur.

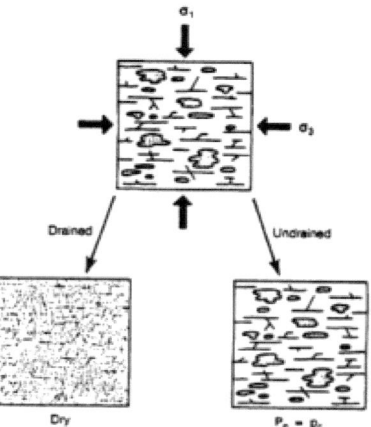

Figure 7.8
Cartoon illustrating the possible development of a porous fluid-saturated rock, initially saturated and subject to stress and undergoing irreversible porosity reduction. If fluid escape is sufficiently fast relative to the pore pressure buildup rate (due to porosity reduction), the rock will gradually lose its porosity and hence the pore fluid. If fluid escape is slow relative to pore pressure buildup, P_p, will reach lithostatic pressure and cause natural hydraulic fractures to occur.

A process that may lead to this sort of crustal porosity and permeability reduction without major P_p build-up is crack healing, which has been investigated in the laboratory (e.g., Smith and Evans, 1984). In this process progressive elimination of pore space in the form of cracks in crystals due to mechanical closure and the healing of the bonds across the crack surfaces takes place. Under a fairly wide range of circumstances little trapped fluid remains behind and little connected pore space remains in the form of fluid inclusions.

Case 2

A drastically different development of pore pressure will take place when permeability decreases more rapidly with time than porosity due to sealing and healing (Bernabe et al., 1982; Walder and Nur, 1984), so that $\dot{k}/k > \dot{\phi}$. Consider again an element of porous crustal rock at depth, again subject to overburden stress as shown in Figure 7.8. Again we expect the porosity to decrease inelastically with time. If a permeable path does not exist between the element and the free surface of the crust or the permeability of such a path is sufficiently low, the fluid in the pores will not be squeezed out fast enough, and the pore pressure of the trapped fluid will thus rise with time. If the porosity reduction rate $\dot{\phi}$ is sufficiently large, permeability k sufficiently small, and their rate of change with time or with each other are such that $\dot{k}/k > \dot{\phi}$, then pore pressure must increase with time. The rate of such pore pressure buildup will depend on the rate of porosity reduction $\dot{\phi}$ and the permeability k. Eventually P_p, will reach its upper possible limit—the least normal stress acting one the element plus its cohesive strength. When P_p reaches that stress level, natural hydrofracturing will occur, involving a rapid episode of fluid release together with a sudden reduction in pore pressure. As soon as the pore pressure drops, the flow path through the hydrofracture will close again. Because the process of porosity reduction continues, P_p will again build up toward lithostatic, leading to another cycle of hydrofracturing, fluid release, and sealing, etc.

In this case, dominated by low permeability, pore pres

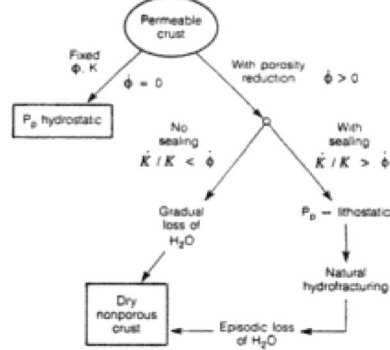

Figure 7.9
Flow diagram illustrating the possible paths that initially permeable, fluid-saturated crustal rocks can follow, depending on permeability, porosity, and their rates of change.

sure oscillates episodically between lithostatic and less than lithostatic, with each oscillation involving a pulse of fluid release and flow and a short episode of fracturing, followed by a longer period of little or no flow, pore pressure recovery, and sealing.

RATES AND MAGNITUDES OF PORE VOLUME STRAIN AND PORE PRESSURE BUILD-UP

To determine whether the transient processes described above—the draining, drying, and porosity elimination of crustal rocks—are actually geologically important and whether they are gradual or episodic, we need to estimate four parameters: (1) the magnitude of the porosity reduction rate $\dot{\phi}$ required for P_p buildup, (2) the permeability k and its dependence on porosity to ensure P_p buildup, (3) the time required for pore pressure buildup in relevant crustal rock masses and correspondingly the duration of the P_p cycles, and (4) the time required or duration of these processes before they cease due to the elimination of connected porosity and fluid removal in situ.

Walder (1984) and Walder and Nur (1984) investigated the conditions under which lithostatic pore pressure will develop and estimated the length of time required for its buildup due to porosity reduction. By dimensional analysis they show that the controlling factor is the dimension-less grouping

$$\frac{\mu \dot{\phi} H}{\Delta \rho k g} = \dot{F}, \quad (7.1)$$

where H is depth in the crust, Dr is the difference in density between rock and fluid, μ is fluid viscosity, $\dot{\phi}$ is porosity reduction, k is hydraulic permeability, and $\dot{F}$ is the buildup index. When $\dot{F} < 1$, pore pressure development is largely unaffected by porosity reduction. For $F > 1$, fluid pressure is strongly affected by porosity reduction. Furthermore, because $\dot{F}$ increases as the depth H to which porosity reduction occurs increases, high pore pressure is more likely at greater depth. This may be further enhanced by the lower permeability k that is more likely at greater depth.

As an example, Walder and Nur (1984) considered a 10-km-thick section of granitic rock undergoing porosity reduction. For the decrease of permeability as decreases, they assumed the following relationship:

$$k = k_0 \frac{\phi^n - \phi_c^n}{\phi_0^n - \phi_c^n}, \quad (7.2)$$

where k_0 is the initial value of permeability, $_0$ is the initial value of porosity, $_c$ is the percolation threshold porosity for throughflow, and n is the exponent. Also assume that $n = 2$ and $_c = 2 \times 10^{-4}$ which is well below porosities typically measured in crystalline rocks ($\phi \geq 10^{-3}$). Assuming the initial permeability of 5×10^{-20} m^2 (50 ndarcy) throughout the section (Brace et al., 1968), water viscosity $\mu = 2 \times 10^{-4}$ Pa, $\Delta_p = 1.7 \times 10^3$ kg/m^3, Walder and Nur found that $\dot{F} = 1$ when $\dot{\phi} = 4 \times 10^{-16}$/s (Figure 7.10). In other words, if porosity reduction were to proceed throughout the 10-km-thick section at a rate higher than 4×10^{-16} /s or 1 percent per million years, excess pore pressure would be generated and maintained.

The next question we need to consider is whether the time required for pore pressure to build up to lithostatic is fast enough to be geologically important. This question is difficult to answer because few details are known about the processes involved and only little relevant experimental evidence is available. However, much of this evidence indicates that porosity and permeability reduction can be very rapid, geologically speaking, if temperatures are sufficiently high. Smith and Evans (1984) found that crack healing rates, based on a model by Evans and Charles (1977), are most likely governed by processes with activation energies around 50 to 100 kJ/mol, which in turn can be used to estimate in situ rates. Smith and Evans (1984) also found in the laboratory that crack healing in quartz requires several hours at 600°C and several days at 400°C. At the crustal depth of interest here (5 to 12 km or so) with temperatures typically ranging from 100° to 300°C, the

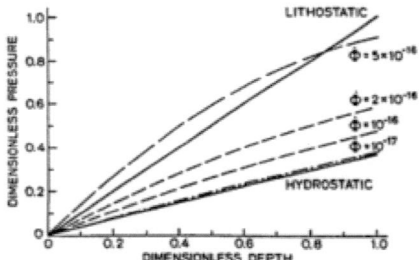

Figure 7.10
Fluid pressure as a function of depth for a 10-km-thick section undergoing uniform porosity reduction for several values of porosity reduction rate $\dot{\phi}$ (in s^{-1}). Initial permeability is 5×10^{-20} m^2 (50 ndarcy). Solid lines show hydrostatic and lithostatic pressure gradients. Dashed curves show fluid pressure profiles that would develop after porosity reduction for 2500 yr at indicated rates. Note that the fluid pressure would exceed lithostatic for the largest value, which slightly exceeds the "critical" value for this geometry and permeability (from Walder and Nut, 1984).

corresponding rates increase to a few hundred years. Sprunt and Nur (1977) found that porosity can be reduced significantly (by 30 to 40 percent) in laboratory experiments in sandstones subject to moderate shear stress at 250°C over a period of 2 weeks. The corresponding period of porosity reduction at 150°C would be, using the Smith and Evans activation energy, 102 to 103 longer, yielding at least several tens of years.

Walder (1984) and Walder and Nur (1984) used a different approach to estimate the minimum time for pore pressure to reach lithostatic pressure by considering the extreme case of a totally sealed rock mass. In this case the rate of pore pressure increase with time, $\Delta P/\Delta t$, is simply proportional to the rate of porosity change with time $\dot{\phi}$

$$\left.\frac{\Delta P}{\Delta T}\right|_{no\,flow} = \gamma \dot{\phi}, \qquad (7.3)$$

where the proportionality constant γ is given by

$$\gamma = \frac{2}{\beta}\left(\frac{\beta_s}{\beta}\right)\left(\frac{1-2\nu}{1-\nu}\right), \qquad (7.4)$$

where β is the compressibility of the rock, β_s is grain compressibility, and ν is the rock's Poisson's ratio, which, by using the parameters listed earlier, reduces to

$$\Delta P = (2 \times 10^{10}\,\text{Pa}) \times \dot{\phi} \times \Delta t. \qquad (7.5)$$

Thus, for a given porosity reduction rate $\dot{\phi}$, it is possible to estimate from Eq. (7.1) the critical permeability required to cause pore pressure to build-up and from Eq. (7.5) to estimate the time for this buildup to take place. As mentioned above, it is the decrease of pore volume or porosity with time that drives the pore pressure up, when permeability is sufficiently small. It is reasonable, for example, on the basis of the results of Sprunt and Nur (1977), to expect that porosity reduction would be especially likely in porous rocks undergoing tectonic deformation (and hence subject to shear stresses). Strain rates $\dot{\varepsilon}$ associated with active crustal deformation such as in orogenesis or accretionary wedges are generally estimated at 10^{-14}/s to 10^{-11}/s (Price, 1970; Heard and Raleigh, 1972; Rutter, 1974), whereas strain rates at or near faults are much higher (Pfiffner and Ramsay, 1982; Woitel and Mitra, 1986). If we assume that the pore volume strain rate is only a small fraction of the total strain rate (e.g., $\dot{\phi}/\dot{\varepsilon} = 0.01$), we obtain $\dot{\phi}$ in the range of 10^{-16} to 10^{-13} s^{-1}, and for $\dot{\phi}/\dot{\varepsilon} = 0.10$, $\dot{\phi}$ is in the range of 10^{-15} to 10^{-12} s^{-1}. Clearly, even these conservative estimates of porosity reduction rates as fractions of total strain rates are more sufficient to drive the pore pressure toward lithostatic while porosity lasts.

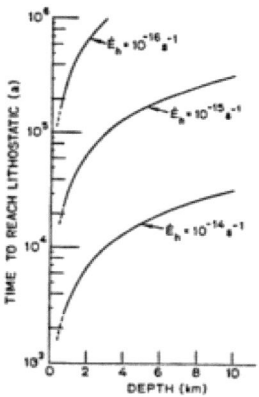

Figure 7.11
Time for initially hydrostatic pore pressure in a "sealed" rock mass to reach lithostatic as a result of lateral strain-induced overpressuring, as a function of depth to the rock mass, for three values of strain rate. For strain rates typical of convergent plate margins, pore pressure may reach lithostatic within only a few thousand or tens of thousands of years (from Walder, 1984).

With the above estimated ranges of $\dot{\phi}$, it is now possible to estimate the time τ_l for P_p to reach lithostatic pressure. Figure 7.11, from Walder (1984), provides relations between time to reach lithostatic pressure τ_l, the depth, and the strain rate. Clearly, τ_l can be surprisingly short at strain rates of 10^{-13} to 10^{-15} s^{-1}, with $\tau_l \approx 10^3$ and 10^5 yr, respectively. Thus, if in a given crustal section subject to tectonic strain there is a sufficient thickness of porous low permeability units, lithostatic pore pressure can develop in a few thousands to tens of thousands of years.

Next we estimate how much of the total porosity must be lost to sustain these porosity reduction rates over a period τ_l. At a rate of pore volume reduction of 10^{-15} s^{-1}, porosity will decrease by 3×10^{-4} (or 0.03 percent) in 10^4 yr, by 3×10^{-3} (0.3 percent) in 10^5 yr, and by 3×10^{-2} (3 percent) in 10^6 yr. When the rate $\dot{\phi}$ is higher, the pressure buildup is faster and thus the duration τ_l is less. From this simple analysis we can conclude that the total porosity reduction associated with one cycle of pore pressure buildup and natural hydraulic fracturing is only a small fraction of the porosity of typical crustal rocks. Consequently, maybe the porosity reduction process could involve many such cycles.

GEOLOGICAL DURATION

Next we need to determine how many cycles of P_p release can take place at a given location. For this purpose we need to estimate the duration τ_2 of the crustal drying process and the porosity elimination. One estimate can be made by assuming that the cyclic buildup and release of P^p will proceed as long as porosity remains or $\tau_2 = \dot\phi^{-1}$. Thus, for $\dot\phi = 10^{-16}$ s^{-1}, $\tau_2 = 300 \times 10^6$, $\tau_2 = 300 \times 10^6$ yr, and for $\dot\phi = 10^{-15}$ s^{-1}, $\tau_2 = 30 \times 10^6$ yr. These values are probably high estimates, first because porosity may begin to lose its connectivity at some finite value, thus reducing the amount available for the reduction process considered here. Furthermore, the rate of porosity reduction may be slower with lowered porosity. But even if we allow only one-tenth of the total porosity reduction to influence pore pressure increase, we still obtain periods of 30×10^6 yr (for $\dot\phi = 10^{-16}$ s^{-1}) and 3×10^6 yr (for $\dot\phi = 10^{-15}$ s^{-1}). These shorter values for τ_2 over τ_2 yield the number of natural hydraulic fracturing episodes n

$$n = \frac{\tau_2}{\tau_1} \qquad (7.6)$$

with $\tau_1 = 10^4$ to 10^6 yr, and $\tau_2 = 10^7$ to 10^8 yr, we obtain $n = 10$ to 10^4 events.

A different estimate of the number of P_f cycles n can be made by considering the amount of porosity reduction needed to raise P_p to lithostatic pressure per P_p cycle. Assuming that the mass of fluid in the decreasing pore space is conserved during the P_p buildup phase of each cycle and is being reduced only during the expulsion phase, we can write

$$\frac{\Delta V_f}{V_f} = \beta_f \Delta p, \qquad (7.7)$$

where V_f is the pore fluid volume, ΔV_f is the change of pore fluid volume due to pore pressure ΔP, and β_f is the fluid's compressibility. The quantity Δp represents the magnitude of the induced P_p fluctuation during a cycle. If we ignore, as first approximation, the change of fluid density during the porosity reduction cycle, we can write

$$\frac{\Delta \phi}{\phi} = \frac{\Delta V_f}{V_f}, \qquad (7.8)$$

where Df is the porosity change during a cycle. Combining Eqs. (7.7) and (7.8) we obtain

$$\frac{\Delta \phi}{\phi} = \beta_f \Delta p \qquad (7.9)$$

taking $\beta_f = 3 \times 10^{-5}$ per bar for water, Eq. (7.9) indicates that pressure rises to lithostatic of $\Delta p = 100$, 300, and 1000 bars required 1/3, 1, and 3 percent reduction of the initial porosity, respectively. Accordingly, if these Δp values represent realistic P_p fluctuation, somewhere between 30 and 300 episodes are possible. This value is in reasonable agreement with the estimate of n based on the duration ratio of Eq. (7.6) and is consistent with crack seal layers documented by Ramsay (1980).

WATER REPLENISHED BY SUBDUCTION

The duration τ_2 of the time over which the crust dewaters could be significantly prolonged if additional waters are supplied to the crust from the lower crust or mantle below (Figure 7.12). The most obvious possible source of such waters is associated with subducted oceanic slabs. It is generally thought that the oceanic slab is rich in water and hydrous minerals and that this water plays major roles in controlling the onset, amount, and rate of melting in and above the downgoing slab. A recent study by McGeary et al. (1985) showed, for example, that most of the prominent gaps in the circum Pacific active volcanic chains are associated with the subduction of anomalous oceanic crust, usually in the form of thick-rooted oceanic rises. One explanation for these gaps is that the rises somehow reduce the supply of water to the melt zones, thus raising the

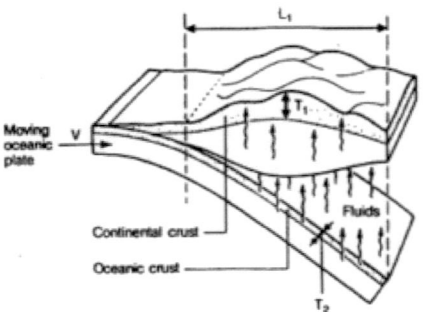

Figure 7.12
Cartoon illustrating the supply of subducted water into the crust overlying a subduction zone and some of the parameters that determine the magnitude of this supply and hence its effects on prolonging the episodic hydraulic behavior of the crust.

TIME-DEPENDENT HYDRAULICS OF THE EARTH'S CRUST

melting temperature and causing a temporal cessation or reduction of volcanism. This model suggests that the subduction of normal oceanic crust does provide for a continuous supply of water into crustal regions below which subduction is taking place. Most probably this water is initially stored in the heavily fractured oceanic lithosphere, some of which migrates upon subduction into the overlying lithosphere and crust.

TABLE 7.2 The Main Time Constants for Time-Dependent Hydraulic Behavior of the Crust

Duration of a P_p cycle	10^3 to 10^5 yr
Time for crust to dry up	10^6 to 10^7 yr
Duration of subduction	10^7 to 10^8 yr

Constraints on the amount of water that might be added this way to the crust overlying a slab may be obtained from a comparison between the pore fluid volume in the crust and the pore fluid volume that passes underneath while subduction lasts. The volume of fluid in the continental crust V_c per unit length of subduction zone (Figure 7.12) is roughly $V_c = T_1 D$, where T_1 is thickness, D is the width of the crustal region overlying the subducted slab and tectonically affected by the subduction process, and   is the average porosity over the thickness T.

In the continental crust we consider the pore fluid to be present in the top $T_1 = 12$ km, with the average porosity   = 1 percent. With D = 300 km as a representative length of the crust overlying the subducting slab and which is tectonically affected by the subduction process, the volume of water present per unit length of the crust is on the order of 30 km^2. The total volume of fluid that is sub-ducted with the oceanic crust per unit subduction zone over the duration of subduction is $V_0 = v \times \tau \times T_2 \times$ $_0$, where v is subduction rate, τ_3 is the subduction duration (and v $\times \tau_3 = L_2$ is the total length of slab subducted), T_2 is the thickness of the porous oceanic crust, and $_0$ is its average porosity. For the subducted oceanic crust we take crustal thickness $T_0 = 6$ km, average porosity also 1 percent, duration of subduction of 100 Ma, and subduction rate of 5 cm/yr, which yields a volume of subducted free water per unit width of slab on the order of 300 km^2. This value of V^0 is about 10 times V_c, which suggests that the waters released from the slab at depth can replenish dissipated crustal water, adding as much as several times the volume of initial fluid to the continental crust. As a result, the duration of the cycles of pore pressure buildup, natural hydrofracturing, and sealing may continue for longer periods of time up to the duration of the subduction process, on the order of 10^3 yr (see Table 7.2).

DISCUSSION AND CONCLUSIONS

The original question posed in this chapter was whether free water is generally deep in the crust. As it turns out this question is inseparable from the question of the nature of porosity, permeability, and pore pressure in the crust and, most significantly, their dependence on time, as illustrated in Figure 7.9. If we *assume* that porosity and, consequently, permeability and pore pressure are time invariant, and consider the hydrological evidence for easy fluid flow in the crust, it follows that crustal pore pressure must be generally hydrostatic to a depth of 10, 15, or even 20 km. This conclusion is in conflict with geological and laboratory evidence that implies that permeable paths in rock tend to clog very rapidly by healing, sealing, and inelastic deformation, all of which become very effective in crustal rocks even at moderately elevated temperatures.

One way to reconcile the conflict between rapid flow on the one hand and rapid clogging on the other is to consider crustal hydrological behavior as varying with time. Although direct data are very sparse, the analysis in this chapter suggests that the most likely state of crustal porosity and water content is transient, with both porosity and water content as well as permeability decreasing with time. It is quite unlikely that these quantities are constant over geological time because inelastic deformation below a depth of a few kilometers in the crust must tend to cause pore closure. Such a tendency will be accompanied by squeezing of water out of the crust due to induced pore pressure. If crustal permeabilities are high enough and the rate of permeability decreases with time or porosity is sufficiently small, the dewatering of the crust will be a gradual, monotonic process. However, if permeability is low and its rate of reduction with porosity is relatively fast, the pore pressure of the trapped fluid will rapidly rise to overcome the least principal stress, leading to natural spontaneous hydrofracturing accompanied by the pulsed release of water and the loss of a little porosity (Figure 7.13). This is followed by a drop in pore pressure, a prolonged buildup period, another hydrofracturing episode, etc. It is especially intriguing that the episodic release process is most probable when tectonic deformation, even at fairly low strain rates, is taking place.

Estimates of the time required for P_b to reach lithostatic and the amount of porosity reduction, especially in tectonically active areas, suggest that this kind of episodic hydrological and mechanical behavior of the crust is quite probable. The number of cycles expected at a given site depends on strain rate, permeability, depth, and other rock parameters, but simple analysis suggests that tens to hundreds of cycles may be expected. The duration of each cycle is estimated at 10^3 to 10^5 yr, whereas the duration of the crustal dewatering process is estimated at 10^6 to 10^7 yr if no additional waters are supplied to the crust. Such supply may be common where the ocean lithosphere was or is being subducted underneath the system. In such situations the geologic duration of the cyclic pore pressure behavior may last as long as subduction does, on the order of 10^8 yr.

The likelihood that the Earth's crust experiences cyclical pore pressure buildup as suggested by Norton and Knight (1977) in conjunction with igneous intrusions, with P_p magnitude oscillating somewhere above hydrostatic to lithostatic, has profound implications for crustal processes and our understanding of these processes. One such process is the formation of hydrothermal ore deposits and mineralized veins (Cathles and Smith, 1983; Vrolijk, 1987). Extensive geological evidence indicates that several types of ore deposits are formed by periodic precipitation from brines. Similarly many types of veins in rocks have been shown to have formed from repeated episodic precipitation from fluids in fractures (Ramsay, 1980). The cyclic pore pressure behavior described in this chapter provides a very simple and compelling mechanism for the formation of these bodies. The cyclic fluctuation in crustal pore pressure may also play an intriguing role in the control and initiation of large earthquakes, which tend to be cyclic in time. In the past it has often been assumed that the repeat time of large earthquakes is basically controlled by the rate at which tectonic strain is accumulating, with fault strength essentially constant. More recently it has become apparent that some sort of time-dependent fault strength is required, if only to reconcile laboratory rock failure results with in situ fault rupture. The increase of P_p toward lithostatic during a pore pressure cycle as described above provides a simple mechanism for cyclic fault weakening, which in turn leads to cyclic failure. Unlike pore pressure changes associated with elastic and elastic dilatant deformation (e.g., Nur and Booker, 1971; Nur, 1972; Rice, 1975, 1979; Rudnicki, 1977) in which P_p buildup is very sensitive to geometrical details, pore pressure buildup due to the inelastic process considered here is fairly insensitive to the details of the stress field buildup around the impending failure zone. Rather it is a robust process of sealing and trapping of the pore fluid in and around the fault zone and the gradual buildup of P_p due to porosity decrease until fault rupture begins. A related effect of this process has been suggested by Sibson et al. (1975) in which inelastic deformation and dilatancy work together to induce seismic pumping, which enhances the instability during failure.

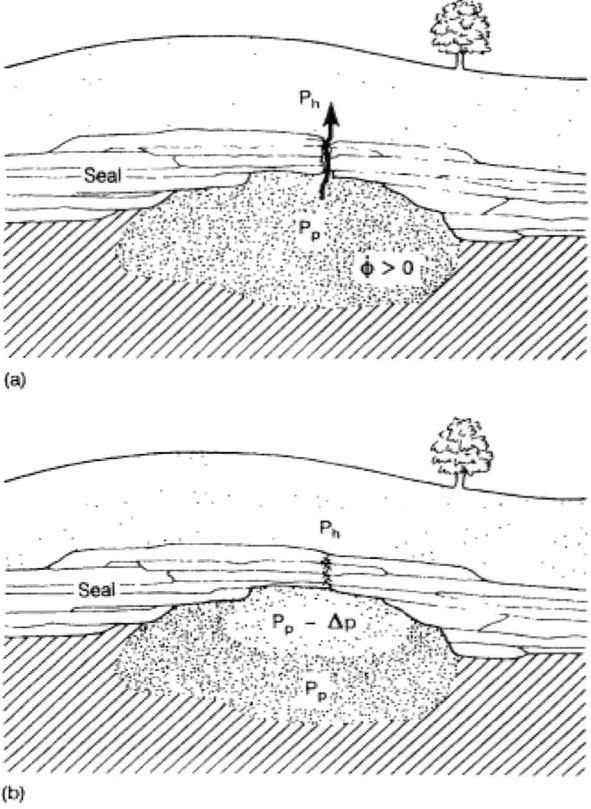

Figure 7.13
Cartoon illustrating the pore pressure buildup and release envisioned in this chapter. (a) Pore pressure in the region of porosity reduction reached the minimum compressive stress and caused natural fractures to occur in the sealed rock surrounding this region. Some fluid escapes through these fractures into the overlying lower pore pressure (P_h) region, causing a drop ΔP in pore pressure (dotted region) before (b) the induced fractures close again.

A third manifestation of the presence of high P_p may be deep crustal reflectors. As discussed by Eaton (1980), laboratory results (Nur and Simmons, 1969; Todd and Simmons, 1972) and field observations (Berry and Mair, 1977) together have suggested that high pore pressure zones in the crust would show up as seismic low-velocity zones. Eaton further suggested that such zones may come about when crustal waters are trapped under a permeability barrier or seal. Jones and Nur (1984; Figure 7.14) and Walder and Nur (1984) showed that such a seal can be effective but only when its permeability is maintained at very low values. Because such a seal is most likely to be broken repeatedly in areas subject to earthquakes, it can be effective only if healing and sealing processes are continuously active. It is quite possible that the process we outlined in this chapter is therefore responsible for dynamically trapping high pore pressure zones in mid-crustal depth. These zones will last as pronounced seismic reflec tors as long as enough water remains in the pore space. The possibility that crustal seismic reflectors may be high pore pressure zones and hence mechanically weak is especially intriguing in view of the growing evidence that these reflectors may represent subhorizontal crustal detachment zones. Such detachments are mechanically very difficult to explain unless high pore pressure is actually involved. Finally, Oliver (1986; Chapter 8, this volume) suggested that deep crustal fluids can migrate horizontally over very large distances away from consumption or collision zones. Our proposed processes of pore pressure development, and its consequent episodic behavior, especially in continental crust overlying subducted slabs, provide a mechanism that can drive fluids horizontally, the pore pressure needed to sustain such flow, and the low-permeability barrier needed to prevent the escape of pressurized brines upward.

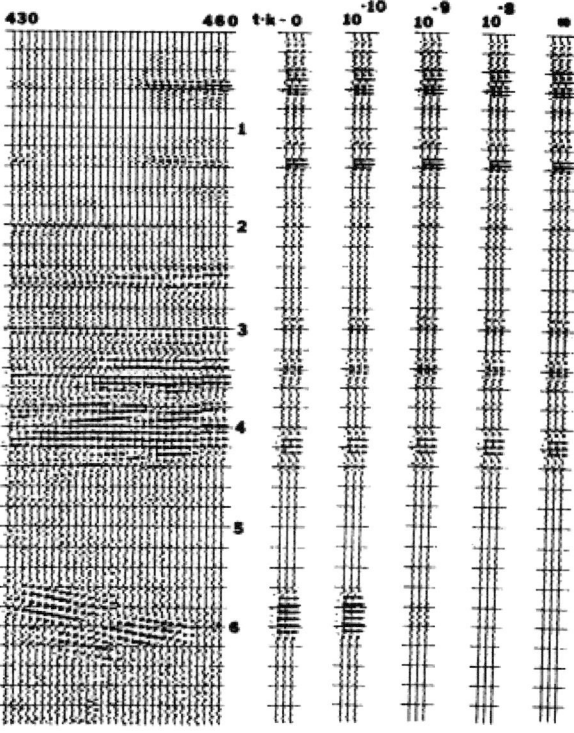

Figure 7.14
The possible role of high pore pressure zones on crustal reflections (from Jones and Nur, 1984). Portion of the Wind River line 1 on the left, showing reflections from the Pacific Creek thrust at 6 s. Synthetics computed for a series of zones 110 m in thickness, alternating ones that have an initial lithostatic pore pressure. Several cases show the change in reflectivity as the pore pressure diffuses out as the product tk increases.

ACKNOWLEDGMENTS

The ideas and speculations presented in this chapter are the result of collaboration and discussions with students and colleagues for over a decade. Many of the ideas here can be traced back to work and papers by Eve Sprunt, Terry Jones, Lee Bell, Ken Winkler, Sue McGeary, Mark Zoback, Bill Brace, Gene Simmons, Brian Evans, Jim Byerlee, Mike Etheridge, Denis Norton, and Rick Sibson. The work by these investigators and others has made it possible to understand as much as we do at present, but they should not be held responsible for any errors or mistakes that we may have made in this chapter. We also wish to thank Steve Kirby and Bill Brace for their insightful comments, which helped clarify several important points. Much of the funding for the research reported here was provided by the Division of Mathematical and Physical Sciences of the U.S. Department of Energy.

References

Bell, M. L., and A. Nur (1978). Changes due to reservoir-induced pore pressure and stresses and application to Lake Oroville, *Journal of Geophysical Research* 83(B9), 4469-4483.

Bernabe, Y., B. Evans, and W. F. Brace (1982). Permeability, porosity and pore geometry of hot-pressed calcite, *Mechanics of Materials 1*, 173-183.

Berry, M. J., and J. A. Mair (1977). The nature of the Earth's crust in Canada, in *The Earth's Crust: Its Nature and Physical Properties*, J. G. Heacock, ed., Geophysical Monograph Seties, American Geophysical Union, Washington, D.C., pp. 319-348.

Brace, W. F. (1980). Permeability of crystalline and argillaceous rocks: Status and problems, *International Journal of Rock Mechanics in Mineral Science and Geomechanical Abstracts 17*, 876-893.

Brace, W. F., and D. L. Kohlstedt (1980). Limits on lithospheric stress imposed by laboratory experiments, *Journal of Geophysical Research 85*, 6248-6252.

Brace, W. F., J. B. Walsh, and W. T. Frangos (1968). Permeability of granite under high pressure, *Journal of Geophysical Research 73*, 2225-2236.

Bredehoeft, J. D., and B. B. Hanshaw (1968). On the maintenance of anomalous fluid pressures. I. Thick sedimentary sequences, *Geological Society of America Bulletin 79*, 1097-1106.

Byerlee, J. D. (1967). Frictional characteristics of granite under high confining pressure, *Journal of Geophysical Research 72*, 3639.

Carter, N. J. (1976). Steady state flow of rocks, *Reviews of Geophysics and Space Physics 14*, 301-360.

Cathles, L. M., and A. T. Smith (1983). Thermal constraints on the formation of Mississippi Valley-type lead-zinc deposits and their implications for episodic dewatering and deposit genesis, *Economic Geology 78*, 983-1002.

Chung-Kang, S., Hou-Chun, C. Chu-Han, H. Li-Sheng, L. Tsu-Chiang, Y. Cheng-Yung, W. Ta-Chun, and L. Hsueh-Hai (1974). Earthquakes induced by reservoir impounding and their effect on the Hsinfengkiang Dam, *Scientifica Sinica 17*(2), 239-272.

Eaton, G. P. (1980). Geophysical and geological characteristics of the crust of the Basin and Range Province, in *Continental Tectonics*, Geophysics Study Committee, National Academy of Sciences, Washington, D.C., pp. 96-114.

Etheridge, M. A., S. F. Cox, V. J. Wall, and R. H. Vernon (1984). High fluid pressures during regional metamorphism and deformation: Implications for mass transport and deformation mechanisms, *Journal of Geophysical Research 89*, 4344-4358.

Evans, A. G., and E. A. Charles (1977). Strength recovery by diffusive crack healing, *Acta Metallurgica 25*, 919-927.

Feng, R., and T. V. McEvilly (1983). Interpretation of seismic reflection profiling data for the structure of the San Andreas Fault zone, *Seismological Society of America Bulletin 73*(6), 1701-1720.

Fyfe, W. S., N. J. Price, and A. B. Thompson (1978). *Fluids in the Earth's Crust*, Elsevier, Amsterdam.

Hanshaw, B. B., and J. D. Bredehoeft (1968). On the maintenance of anomalous fluid pressure. II. Source layer at depth, *Geological Society of America Bulletin 79*, 1107-1122.

Heard, H. C., and C. B. Raleigh (1972). Steady-state flow in marble at 500 to 800°C, *Geological Society of America Bulletin 83*, 935-956.

Hermann, R. B., and B. J. Mitchell (1975). Statistical analysis and interpretation of surface-wave anelastic attenuation data for the stable interior of North America, *Seismological Society of America Bulletin 65*(5), 1115-1128.

Jones, T. D. (1983). Wave propagation in porous rock and models for crustal structure, Ph.D. thesis, Stanford University.

Jones, T. D., and A. Nur (1984). The nature of seismic reflections from deep crustal fault zones, *Journal of Geophysical Research* 89(B5), 3153-3171.

Lachenbruch, A. H., and J. H. Sass (1980). Heat flow and energetics of the San Andreas Fault zone, *Journal of Geophysical Research 85*, 6185-6222.

McGeary, S., A. Nur, and Z. Ben-Avraham (1985). Spatial gaps in arc volcanism: The effect of collision or subduction of oceanic plateaus, *Tectonophysics 119*, 195-221.

Nekut, A., J. E. P. Connerney, and A. F. Kuckes (1977). Deep crustal electrical conductivity: Evidence for water in the lower crust, *Geophysical Research Letters 4*, 239-242.

Norton, D., and J. Knight (1977). Transport phenomena in hydrothermal systems: Cooling plutons, *American Journal of Science 277*, 937-981.

Norton, D., and H. P. Taylor, Jr. (1979). Quantitative simulation of the hydrothermal systems of crystallizing magmas on the basis of transport theory and oxygen isotope data: An analysis of the Skaergaard intrusion, *Journal of Petrology 20*, 421-486.

Nur, A. (1972). Dilatancy, pore fluids, and premonitory variations of t_s/t_p travel times, *Seismological Society of America Bulletin 62*(5), 1217-1222.

Nur, A., and J. R. Booker (1971). Aftershocks caused by pore fluid flow? *Science 175*, 885-877.

Nur, A., and G. Simmons (1969). The effect of saturation on velocity in low porosity rocks, *Earth and Planetary Science Letters 7*, 183-193.

Olhoeft, G. R. (1981). Electrical properties of granite with implications for the lower crust, *Journal of Geophysical Research 86*, 931-936.

Oliver, J. (1986). Fluids expelled tectonically from orogenic belts: Their role in hydrocarbon migration and other geologic phenomena, *Geology 14*, 99-102.

O'Neil, J. R., and T. C. Hanks (1980). Geochemical evidence for water-rock interaction along the San Andreas and Garlock fault zones of California, *Journal of Geophysical Research 85*, 6286-6292.

Padovani, E. R., S. B. Shirey, and G. Simmons (1982). Characteristics of microcracks in amphibolite and granulite facies grade rocks from southeastern Pennsylvania, *Journal of Geophysical Research 87*, 8605-8630.

Pfiffner, O. A., and J. G. Ramsay (1982). Constraints on geological strain rates: Arguments from finite strains of naturally deformed rocks, *Journal of Geophysical Research 87*, 311-321.

Price, N. J. (1970). Laws of rock behavior in the Earth's crust, in *Symposium on Rock Mechanics*, 11th ed., H. Somerton, ed., American Institute of Mining and Metallurgical Engineers, Berkeley, California, pp. 3-23.

Raleigh, C. B., and J. Evernden (1982). Case for low deviatoric stress in the lithosphere, in *Mechanical Behavior of Crustal Rocks*, N. L. Carter, M. Friedman, J. M. Logan, and D. W. Steams, eds., Geophysical Monograph Series, American Geophysical Union, Washington, D.C., pp. 173-186.

Ramsay, J. G. (1980). The crack-seal mechanism of rock deformation, *Nature 284*, 135-139.

Rice, J. R. (1975). On the stability of dilatant hardening for saturated rock masses, *Journal of Geophysical Research 80*(11), 1531.

Rice, J. R. (1979). Earthquake precursory effects due to pore fluid stabilization of a weakening fault zone, *Journal of Geophysical Research 84*(B5), 2177.

Richter, D., and G. Simmons (1977a). Microcracks in crustal igneous rocks: Microscopy, in *The Earth's Crust: Its Nature and Physical Properties*, J. G. Heacock, ed., Geophysical Monograph Series, American Geophysical Union, Washington, D.C., pp. 149-180.

Richter, D., and G. Simmons (1977b). Microscopic tubes in igneous rocks, *Earth and Planetary Science Letters 34*, 1-12.

Rudnicki, J. W. (1977). The inception of faulting in a rock mass with a weakened zone, *Journal of Geophysical Research 82*(5), 844.

Rutter, E. H. (1974). The influence of temperature, strain rate and interstitial water in the experimental deformation of calcite rocks, *Tectonophysics 22*, 311-334.

Shankland, T. J., and M. E. Ander (1983). Electrical conductivity, temperature, and fluids in the lower crust, *Journal of Geophysical Research 89*, 9475-9484.

Sibson, R. H., J. M. Moore, and A. H. Rankin (1975). Seismic pumping--a hydrothermal fluid transport mechanism, *Journal of the Geological Society of London 131*, 653-659.

Sleep, N. H., and T. J. Wolery (1978). Egress of water from mid-ocean ridge hydrothermal systems: Some thermal constraints, *Journal of Geophysical Research 83*, 5913-5922.

Smith, D. L., and B. Evans (1984). Diffusional crack healing in qu artz, *Journal of Geophysical Research 89*, 4125-4135.

Smith, L., and D. S. Chapman (1983). On the thermal effects of groundwater flow. 1. Regional scale systems, *Journal of Geophysical Research 88*, 593-608.

Sprunt, E. S., and A. Nur (1976). Reduction of porosity by pressure solution: Experimental verification, *Geology 4*, 463-466.

Sprunt, E. S., and A. Nut (1977). Destruction of porosity through pressure solution, *Geophysics 42*, 726-741.

Sprunt, E. S., and A. Nur (1979). Microcracking and healing in granites: New evidence from cathodoluminescence, *Science 205*, 495-497.

Taylor, H. P., Jr. (1977). Water/rock interaction and the origin of H20 in granitic batholiths, *Journal of the Geological Society of London 133*, 509-588.

Thompson, B. G., A. Nekut, and A. F. Kuckes (1983). A deep crustal electromagnetic sounding in the Georgia Piedmont, *Journal of Geophysical Research 88*, 9461-9473.

Todd, T., and G. Simmons (1972). Effect of pore pressure on the velocity of compressional waves in low-porosity rocks, *Journal of Geophysical Research 77*, 3731-3743.

Vrolijk, P. (1987). Tectonically driven fluid flow in the Kodiak accretionary complex, Alaska, *Geology 15*, 466-469.

Walder, J. S. (1984). Coupling between fluid flow and deformation in porous crustal rocks, Ph.D. thesis, Stanford University.

Walder, J., and A. Nur (1984). Porosity reduction and crustal pore pressure development, *Journal of Geophysical Research 89*(B13), 11,539-11,548.

Winkler, K. W., and A. Nur (1982). Seismic attenuation: Effects of pore fluids and frictional sliding, *Geophysics 47*(1), 1-15.

Wojtal, S., and G. Mitra (1986). Strain hardening and strain softening in fault zones from foreland thrusts, *Geological Society of America Bulletin 97*, 674-687.

Zhao, W.-L., D. M. Davis, F. A. Dahlen, and F. Suppe (1986). Origin of convex accretionary wedges: Evidence from Barbados, *Journal of Geophysical Research 91*(B 10), 10,246-10,258.

Zoback, M. D., M. L. Zoback, V. S. Mount, J. Suppe, J. P. Eaton, J. H. Healy, D. Oppenheimer, P. Reasenberg, L. Jones, C. B. Raleigh, I. G. Wong, O. Scotti, and C. Wentworth (1987). New evidence on the state of stress of the San Andreas Fault system, *Science*, 1105.

8

COCORP and Fluids in the Crust

JACK E. OLIVER
Cornell University

ABSTRACT

Using the seismic reflection profiling technique, COCORP has probed the continental basement along lines traversing a wide variety of major geological features. The results are correspondingly diverse, and they are informative on various aspects of crustal geology, including the geology of fluids, the particular aspect emphasized in this chapter. There are at least three ways in which COCORP data may relate to the understanding of fluids in the basement. First, some anomalously strong reflectors, typically near-horizontal and at depths of about 20 km, may correspond to pockets of magma or other fluids trapped within the basement. Second, as suggested by others, some or even all basement reflectors may correspond to zones of fluid-filled porous rocks. Third, new insights into structures suggest that tectonic models include inferences about the role of fluids in the crust. For example, fluids expelled tectonically from sediments buried in orogenic belts may play a role in migration of hydrocarbons, transport of minerals, diagenesis, crustal rheology, chemical remagnetization, and a variety of other phenomena.

INTRODUCTION

During a little more than a decade, COCORP (Consortium for Continental Reflection Profiling) has collected deep seismic reflection data in the United States for profiles totaling over 8000 km in length (Figure 8.1). These data, though trivial in quantity compared to the huge volume of industrial seismic data for the sedimentary basins, constitute the largest and most comprehensive set of land-based seismic reflection data on the deep crust of the continents anywhere. BIRPS (British Institutions Reflection Profiling Syndicate) has collected a comparable quantity of marine data on the deep crust beneath the continental shelf surrounding Great Britain, and at the time of this writing some 20 countries, including the United States and the United Kingdom, have surveyed a total of 20,000 to

25,000 km of seismic reflection profiles of the deep continental crust throughout the world. This total is exclusive of any proprietary industrial data on the deep crust.

Figure 8.1
Map showing deep seismic profiles surveyed by COCORP from the inception of the project through December 1986.

This quantity of data corresponds to exploration of only a small fraction of the total volume of the continents. Nevertheless, these data represent early exploration of a new frontier, and as such they have already provided surprising and important information on many aspects of continental crustal geology. This chapter, however, passes over most of these aspects and, because of the nature of this volume, focuses solely on just one aspect of crustal geology—fluids in the crust. Furthermore, the chapter is based almost exclusively on COCORP data, all taken in the 48 contiguous states. For a review of COCORP studies in general, see Brown et al. (1986).

COCORP observations relate to the study of fluids in the crust in at least three different ways and even more if various indirect lines of reasoning not emphasized here are pursued. First, the observations occasionally reveal, at mid-crustal depths, exceptionally strong, near-horizontal reflectors that are probably associated with magmas and/or other fluids. This information is some of the very best evidence in existence on spatial locations of fluids in deep basement rocks at the present time. Second, the COCORP observations reveal a variety of weaker reflectors throughout the continental basement, some, or even all according to one hypothesis, of which may be related to fluids in some manner. Third, COCORP data provide novel information on the structure and tectonics of the crust and thereby stimulate thinking about crustal tectonic models that have implications about fluid behavior in the crust. This chapter discusses these topics in order and with some emphasis on the third point because of the nature of this volume.

STRONG REFLECTORS IN THE MID-CRUST

In 1975 an early COCORP survey sensed a strong reflector at a depth of about 20 km in the crust beneath the Rio Grande Rift north of Socorro, New Mexico. This location is just where Allan Sanford (Sanford et al., 1973, 1977), basing his ideas on earthquake and other geophysical and geological data, had earlier postulated that a magma body exists. The COCORP reflection survey supported Sanford's hypothesis, delineated certain features of the body, and, if Sanford's well-supported hypothesis is indeed correct, provided a type example of reflection data for a magma body in the mid-crust.

Later, while profiling in Death Valley, COCORP detected a strong reflection that in many ways resembles that type example from beneath the Rio Grande Rift. The Death Valley reflector is at about the same depth, is nearly horizontal, and is in a similar extensional tectonic environment. An obvious possible, though not unrefutable, interpretation is that another magma body exists beneath that part of Death Valley (de Voogd et al., 1986). In fact, the data also reveal a dipping reflector that rises from the vicinity of the proposed magma body and extends to near the surface at a place where recent volcanic activity has occurred. It has thus been proposed (de Voogd et al., 1986; Serpa et al., 1987) that this dipping reflector corresponds to a fault zone that acted as a conduit for rising magma. Whether the fault zone continues to hold magma, in its lower if not upper reaches, is not resolved.

Since the Death Valley survey, COCORP has found at least two other comparable reflectors, one in Nevada and one in Arizona. Both locations are in the Basin and Range Province. These two reflectors are somewhat less prominent features, but they are of the same general character and at about the same depth as the Rio Grande Rift and Death Valley reflectors. They may also correspond to magma at some stage of cooling.

These strong mid-crustal reflectors, coupled with additional observations by COCORP and by others on various aspects of the buried continental crest, are prompting new speculation about the formation and evolution of the deep crest and its role in the evolution of modem surface geology. The subject of the structure, composition, evolution, and tectonics of the deep continental crest and uppermost mantle is a particularly important and exciting one at present. The speculation takes many forms and produces diverse hypothetical models. Figure 8.2 shows one example that corresponds to an old idea that magmas generated in the mantle intrude the mid-crest near the top of a zone of earlier intrusions. Subsidence and phase change continually convert the rocks at the base of the crest to denser and higher-velocity phases that become uppermost mantle or "anomalous" mantle. In a similar model (not shown) intrusion occurs at the base of the crest and successive intrusions build-up from below, adding to anomalous mantle. Other ideas to explain the nature of the deep crest without major intrusion are also under consideration; the two above are mentioned here because of their beating on fluid penetration of the crest and their partial base in COCORP data.

The examples cited above of strong mid-crustal reflectors all occur in the Basin and Range Province, a region characterized by extension in the most recent tectonic episode. COCORP has acquired data elsewhere on still another strong mid-crustal reflector, this one of somewhat different seismic character and somewhat different modem tectonic character (Brown, 1987). The reflector lies beneath the coastal plain of Georgia (i.e., a province distinctly different tectonically from the Basin and Range Province). This reflector is called the Surrency bright spot—"Surrency" because that is the name of a small town nearby and "bright spot" because the reflection resembles in many ways a bright spot of the type familiar to the petroleum industry. Oil industry bright spots typically occur in shallow sedimentary rocks. They are commonly drilled and commonly produce hydrocarbons. The typical bright spot is a consequence of occurrence of fluids that affect the acoustic impedance of porous rocks and of gravity which tends to make boundaries between fluids flat and horizontal. In spite of the striking similarity, the Surrency bright spot differs dramatically from a typical oil industry bright spot in that it occurs at a reflection time of about 6 s (i.e., a depth of about 15 to 18 km) and hence must be well within the basement that lies beneath the sediments of the coastal plain of Georgia.

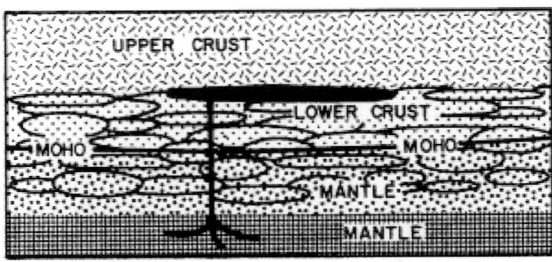

Figure 8.2

An example of one hypothesis proposed to explain the structure of the lower crust in some areas. Magma from mantle intrudes crust near top of zone of previous intrusion. Deeper portion of zone of intrusions is converted to mantle-like material (anomalous mantle) at base of crust. This hypothesis is but one example of many designed to explain structure and dynamics of lower crust and uppermost mantle.

The Surrency bright spot, which is described in more detail elsewhere (Brown, 1987), is located in the Paleozoic suture zone between former African and North American crest. How to interpret this unusual and very robust observation is not completely clear. Because of lack of any evidence for young volcanism or tectonic movements within the coastal plain, it seems unlikely that there is magma in the crest at this location. A lithologic boundary may be considered as an explanation, but a strikingly flat and horizontal boundary between two lithologies of such great impedance contrast as would be necessary to produce the observed reflection also seems unlikely in this highly deformed zone. Perhaps the most reasonable suggestion is that fluids caught up and trapped in the collision process, or trapped later in the deformed rocks of the collision zone, account for the observation. The fluids might be water, He, CO_2, hydrocarbons, or something else—just what is uncertain at present. But it seems obvious that information on such prominent, almost certainly fluid-related, features of the crest must be important to anyone attempting to understand the geology of fluids in the basement. Unfortunately, with only a small fraction of the total crustal volume so far explored, it is not possible to say whether this occurrence is unique or one of tens or hundreds or thousands. Nor is it known whether, if there are many such features, all are in suture zones or some are distributed through other parts of the crust as well. Detection of such features is so easy and straightforward technically, however, that a great opportunity for further study is readily apparent from this first striking observation.

THE REFLECTIVE CRUST

In addition to the rare, near-horizontal, strong reflectors of the mid-crust just discussed, in most areas the entire continental crust is generally reflective to some degree. Flat and curved reflectors, dipping reflectors, laminated reflectors and diffractors are common in the basement, although generally they are weaker than those in sedimentary rocks. Reflections within Phanerozoic, and in some cases Proterozoic, sedimentary basins are studied intensively by the petroleum industry. Such reflectors may correspond to lithologic changes, time-stratigraphic boundaries, faults, fluids, etc., and, because of the great industrial effort, understanding of them is relatively advanced. For reflectors within the crystalline basement, most not drilled and most so deep as to be undrillable, identification and interpretation are normally less certain. Sometimes a particular reflector can be identified by tracing it to the surface. Sometimes the spatial configuration of a reflector gives it a place in an interpretation of geologic structure that in turn gives confidence to speculation on the nature of the reflector. Commonly, however, it is not known for certain just what physical properties produce a particular reflector at a particular place. Furthermore, it is likely that a combination of factors is involved.

One possible explanation is that some, or in the extreme even all, such basement reflectors correspond to fluid-filled fractures in the crust. From drilling (Koslovski, 1984; Clarke *et al.*, 1986) it is known that, at least for one location, fluid-filled fractures prevail to depths of at least 11 km. That such fractures are reflective was demonstrated by Mair and Green (1981) and Green and Mair (1983) for a case of very shallow reflectors in a granitic intrusion. Matthews and Cheadle (1986) suggested that all reflectors throughout the basement are associated with fluids, an extreme but nevertheless provocative suggestion. Among other things, this suggestion implies that the reflection Moho, which is commonly taken as the base of the reflective zone, marks the base of fluid-filled fractures, and hence fluids, in the crust. An explanation in terms of a mobile deep fluid boundary is thereby provided for the vertical mobility of the Moho that is now well demonstrated by seismic reflection studies. This hypothesis is not without its difficulties. For example, mere draining of fluids from a part of the lower crust would likely not elevate velocities to sub-Moho levels, and hence the hypothesis would not account for the observed coincidence in reflection and refraction Mohos at most places.

Clearly the reflective crust is not yet fully understood and remains in a speculative stage. It would be risky, for example, to base hydrologic or geochemical studies of the crust on the concept that all basement reflectors necessarily indicate fluid-filled fractures of the crust, but it would also be risky to ignore the possibility that many of them do.

A SYNTHESIS OF INFORMATION ON CRUSTAL STRUCTURE, TECTONICS, AND FLUIDS IN THE CRUST

The COCORP data provide information on the structure and evolution of the continental crust. Many other sources contribute important, different, and highly diverse information on this subject. All such information must ultimately be fit into a common story. What follows is an attempt to synthesize observations from a wide variety of disciplines in earth science, all related in some way to the story of fluids in the crust. The role of COCORP data in this synthesis arises primarily from its contribution to the tectonic model and not so much from direct detection of fluids in the crust as described earlier.

The reader should be aware that (a) much of what follows is speculative. The hypothesis to be presented may be incorrect; some parts are surely controversial at present. Although a substantial amount of evidence in support of the hypothesis is presented, there remains much opportunity for further testing. Should the hypothesis be more or less correct, however, it will be of considerable importance because it links many geological observations of the continents to the great global geodynamical processes through the mechanism of plate tectonics. (b) Parts of what follow will be familiar to some, for the hypothesis and synthesis incorporate some concepts that have been proposed and discussed, though not necessarily agreed upon, previously. This chapter should not be mistaken as a claim for originality with regard to those concepts. Instead the chapter is an attempt at synthesis on an unusually broad scale. Appropriate references to concepts proposed earlier are provided insofar as possible. (c) The author is not a specialist in most of the wide variety of disciplines from which observations are drawn and would appreciate notification from specialists of any case in which data from a specialty have been misinterpreted, particularly if that misinterpretation affects the testing of the hypothesis. The synthesis of this chapter uses as a framework a hypothesis that may be described in general terms as follows (Oliver, 1986).

At the surface of the Earth, continents (and smaller land masses) drift about within the large sea that occupies most of the surface. The continents have limited freeboard, and their margins are generally awash in the sea. The continents are porous and permeable, at least in their upper parts, and are for the most part saturated with fluids. When two landmasses converge and collide, the process is an asymmetric one. One landmass is partially subducted beneath the other. Put another way, the margin of one landmass is commonly buried beneath the accretionary wedge on the leading edge of the other. According to the hypothesis, fluids from the sediments of the buried continental margin are expelled and travel toward the interior of the partially subducted continent carrying heat, minerals, and organic compounds (Figure 8.3). These fluids may leach or deposit materials as they travel; hence, they redistribute materials and leave a lasting record of their passage. As an illustration of the hypothesis, the Appalachian orogen can serve as the type example.

The Appalachian orogen is a consequence of convergence and closing of the proto-Atlantic Ocean during Paleozoic time. The process included several orogenies and culminated in the collision of the North American and African continents. Subduction during this closing was generally to the east, so a large accretionary wedge or thrust sheet was forced onto and over the sediments of the North American margin. Some sediments were buried, some were carried along with the thrust sheet, and some were reworked to become younger strata. The load of the thrust sheet caused a large depression near the sheet and a more distant forebulge, both changing dynamically as the process evolved. The depression, or foreland basin, filled with sediments from the thrust sheet. Some of these sediments were also buried by the advancing sheet. The sediments involved were generally porous and full of pore fluids, and they included hydrous minerals. Perhaps one-third to one-half of the total volume was made up of water in the pores and in the minerals. As the sediments were buried, water was expelled, some from shallow depths at modest temperatures and some from greater depths at higher temperatures. According to this hypothesis, some of these fluids, or brines, were expelled into rocks of adjacent parts of the continent.

Some of the fluids may be an important part of phenomena of the metamorphic core of the orogen, perhaps producing veins, dikes, and intrusions and facilitating thrusting, but they are not the phenomena of interest here. This chapter is about fluids that are expelled into sediments of the platform and foreland basin, perhaps eventually reaching the more distant continental interior. These fluids carry heat, minerals, and organic compounds, some from the marginal sediments and perhaps some leached from sediments along the way. Expelled by the tectonic load, and perhaps by heating that produces pressure of volatiles, perhaps by other volume changes, they leave the orogen and travel long distances in the style of long-distance hydrologic flow (Figure 8.3). Hydrologic flow for which meteoric waters are the source is thought to carry fluids for many hundreds of kilometers through parts of the continent, although the details of that flow are not yet well known (Sun, 1986). Fluids ejected tectonically into the continent may propagate regionally in similar manner. How much mixing occurs between water from meteoric sources and tectonic fluids is not specified here, but it seems that such mixing will be dependent on various local parameters and so may vary greatly from one region to another. Nor does this hypothesis specify such matters as the chemistry of the migrating brines or the particular place and time when organics mature into hydrocarbons. Such maturation presumably may occur at many places— in the sedimentary trough of the margin prior to subduction, during the subduction process, or in the foreland basin sediments almost independently of the tectonics. It remains to be seen whether data on the details of migration for a particular hydrocarbon province can be fit into this hypothesis, but, as shown later, more general information on hydrocarbon migration seems compatible with it.

Fluids injected into the hydrologic system must perturb the thermal regime. Precisely what form such perturbations take cannot easily be specified in the absence of information on parameters such as channeling, depths, rates of subduction, etc. Nevertheless, as the temperature of tectonic fluids from depth is likely to be high, the vertical temperature gradient is likely to be increased, thereby enhancing certain kinds of metamorphism.

Not all features or processes that might be associated with this model are closely specified here. At this stage of development it seems prudent to maintain flexibility in the hypothesis, or model, so as to preserve the opportunity to

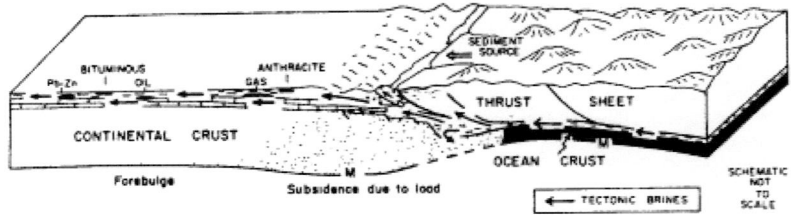

Figure 8.3
Block diagram of orogen as accretionary wedge or thrust sheet overrides preexisting continental margin. Subduction is to the east. Heavy arrows schematically illustrate flow of tectonic brines expelled from buried sediments. Gas and anthracite deposits are closer to orogen than oil and bituminous coal, respectively. Continental crust is -35 km thick; horizontal dimension of diagram is ~500 km (from Oliver, 1986).

refine the model as observations dictate. Let us now consider some of the data that provide support for the model. All observations reported here are taken from the literature. The data are many; only a summary is presented here. For a more complete discussion of the observations, see Oliver (1990).

Coal

It is consistently observed that in foreland basins coal is generally metamorphosed to higher grade near the orogenic belt; the grade decreases more or less gradually toward the continent. Thus, in Pennsylvania anthracite is found to the east and successively lower grades to the west. This point has been demonstrated not only for the Appalachian orogen but also for the Ouachita and Cordilleran orogens by Thom (1934). This pattern is conventionally explained by deeper burial, and hence the higher temperatures needed to metamorphose the coal, near the orogen. In such explanations the deep burial is followed by sometimes rapid uncovering to bring the coal near the surface where it is mined (Levine, 1986). Some coal geologists are not in accord with the burial hypothesis, however, and have suggested lesser depths of burial and anomalous sources of heat such as intrusions or hydrothermal fluids (Damberger, 1974; Haquebard and Donaldson, 1974). An alternative explanation is that a pulse of heat carried by tectonic fluids caused the coal to be metamorphosed at depths shallower than those usually called upon under the burial depth hypothesis.

Brines

There is a great deal of evidence that suggests or is compatible with the idea that a pulse of hot mineral-laden brines propagated through permeable sediments of the continent at the time of orogeny in a nearby orogenic belt. The evidence comes from study of Mississippi Valley-type (MVT) lead-zinc ores, fluid inclusions, certain occurrences of dolomite, paleoremagnetization, and some other sources.

After some years of controversy and discussion, economic geologists seem to be settling on the origin of the strata-bound sphalerite and galena deposits commonly called MVT Pb-Zn ores as, quoting Guilbert and Park (1986), "deposition from connate basinal water that moved updip in response to compaction or other loading pressure." Some geologists call upon expulsion of fluids from compacting basins to provide such brines. The view of this chapter is that basins subjected to tectonic deformation and burial are more likely to expel fluids in greater volumes than are simple compacting ones, although there is no reason to conclude that compaction might not be a source in some cases, and of course some basins may be affected by both self-compaction and tectonic activity.

The map pattern of MVT lead-zinc occurrence, shown in Figure 8.4, illustrates that such occurrences bear a spatial relation to orogenic belts similar to that for hydrocarbons (see later section) except that, for reasons unknown, Pb-Zn deposits tend to occur between basins rather than within them. MVT deposits tend to occur in carbonates with a strong bias toward dolomites (Anderson and MacQueen, 1982). This observation suggests a common origin for MVT Pb-Zn and some dolomites.

The subject is not free of controversy, but in recent years at least sedimentologists have come to believe that some dolomites are formed at depth as a consequence of passage of basinal brines through limestones (see Zenger and Dunham, 1980, for a review). This view is in contrast to the hypothesis that all dolomites are formed at or near the surface. Gregg (1985) proposed that circulation of basinal brines through an underlying sandstone caused a 6-m layer of dolomite at the base of a limestone and shale horizon in southeastern Missouri. He associated this dolomitization with the emplacement of nearby MVT Pb-Zn ores and also with the mineralizing waters that Leach *et al.* (1984), on the basis of fluid inclusion data, had postulated as flow from the Arkoma Basin. In other words, dolomitization is apparently related to MVT ores in some cases because of related origin of both as a consequence of mineralizing migrating brines that may have originated in orogenic belts. Thus, brines producing the MVT deposits of Missouri likely originated during the Ouachita orogeny to the south, although the possibility of some effects from the Appalachian orogen to the east must also be kept in mind.

The dating of emplacement of MVT Pb-Zn deposits is not easy, but some information is available. Sphalerite (ZnS) twins in deformation. Taylor *et al.* (1983) interpreted sphalerite twinning in intermediate layers of multi-layered sphalerite in eastern Tennessee to indicate deposition during the Alleghenian orogeny and attributed the mineralization to brines expelled from the shale basin to the east. Beales *et al.* (1980) used paleomagnetic data to determine an Upper Pennsylvanian age for MVT ores and speculated on the role of tectonic processes in ore formation and hydrocarbon migration. Such dates, as available, for MVT ore deposition support the tectonic fluids hypothesis.

Fluid inclusions are an important source of information on the history of fluids in the crust. Roedder (1984) prepared an excellent summary of this subject, and his book includes many points relevant to the subject matter of this chapter. For example, Roedder noted that evidence from fluid inclusions in Ohio, New York, Iowa, Missouri, and Tennessee are similar and hence suggest that scattered occurrences of sphalerite, etc., formed from the same kinds of fluids as MVT ores. This point implies a possible pulse, or pulses, of mineralizing brines that covered large regions. Roedder also noted that organic material is common in the inclusions of both ores and scattered minerals. Leach (1973) attributed fluids from sphalerite in a coal mine and MVT PB-Zn to a single episode of fluid flow. Leach et al. (1984), Rowan et al. (1984), and Leach and Rowan (1986) studied fluid inclusions from Missouri, Kansas, Arkansas, and Oklahoma and found lateral gradients in temperature increasing to the south. They suggest that fluids expelled tectonically from the Ouachita Arkoma basin at 200° to 300°C migrated northerly to deposit MVT Pb-Zn. Much evidence from fluid inclusion studies supports the concept of tectonic expulsion of fluids from orogenic belts.

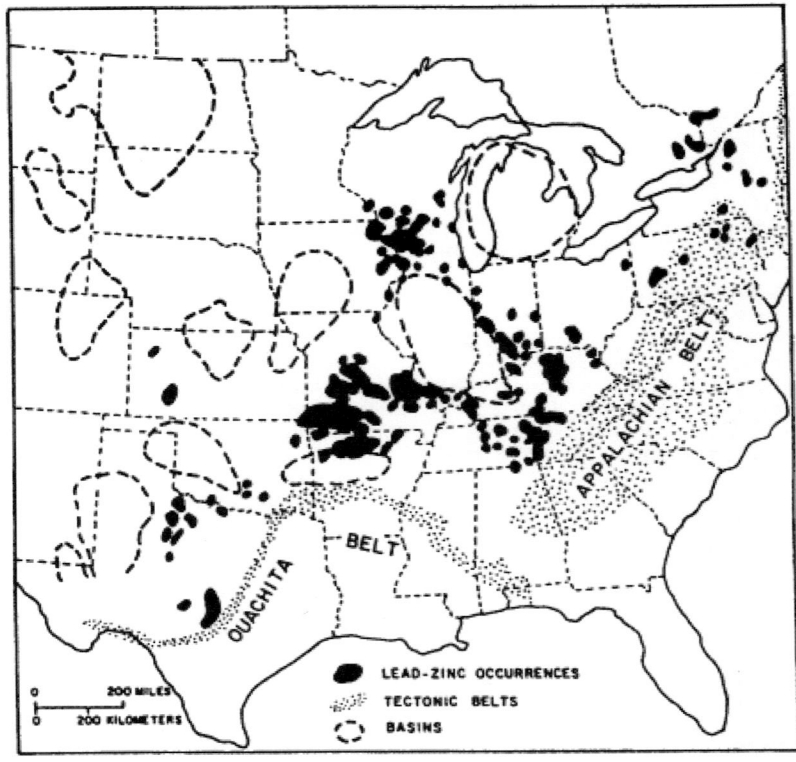

Figure 8.4
Map, modified from Cathles and Smith (1983), showing Mississippi Valley-type lead-zinc occurrences. This map is an approximation. Occurrences in small quantities may be more widespread.

Certain aspects of diagenesis, a term used here in its general sense to include epigenesis, of sediments may be a consequence of migrating tectonic fluids. Morton (1985) dated, by the Rb-Sr method, diagenetic illite from the Upper Devonian black shales of Texas and found a date equivalent to the time of the Ouachita orogeny. He suggested that brines from the Ouachita tectonic zone were the cause and, following Dickinson (1974), that hydrocarbons may also have migrated to West Texas at that time.

Authigenesis, the enlargement by overgrowth, of potassium feldspars in Cambro-Ordovician rocks of western Maryland was dated by Hearn and Sutter (1985) using the $^{40}Ar/^{39}Ar$ method. They found ages corresponding to the Alleghenian orogeny and suggested that brines from the orogenic zone produced the authigenesis and affected hydrocarbon migration and ore deposition. Hearn et al. (1985) obtained results that were consistent with and that extended these conclusions for feldspars in Pennsylvania, Virginia, and Tennessee.

Authigenesis by migrating brines is playing a role of growing importance in the subject of paleoremagnetization, a major new dimension in the field of paleomagnetism. In recent years paleomagneticists have demonstrated widespread remagnetization of sedimentary rocks for large parts of North America (Van der Voo, 1986). One possible explanation is based on thermally activated viscous magnetization. The other, and apparently the preferred one, is chemical remagnetization through authigenesis. The remagnetization is dated as the time of the Alleghenian orogeny in some areas. Van der Voo and French (1977) found Alleghenian dates in Virginia, West Virginia, and Pennsylvania. Scotese *et al.* (1982) found similar dates from a study of Silurian and Devonian rocks in New York. McCabe *et al.* (1983) studied remagnetization of folded Silurian-Devonian rocks in New York using the fold test and found that remagnetization occurred during folding (i.e., during the Alleghenian orogeny). McCabe *et al.* (1984) found similar dates in Quebec, Ontario, and New York.

Kent and Opdyke (1985) revised an earlier study contrasting paleolatitudes of the New England-Canadian Maritime region and the North American craton and showed an error as a result of paleoremagnetization that negated earlier and enigmatic conclusions about major relative tectonic movement between these two provinces. To summarize, there is widespread and growing evidence from studies of paleomagnetism that supports the concept of authigenesis of magnetic minerals in sediments of the continental interior as a consequence of brines migrating at the time of orogeny.

Hydrocarbons

In the preceding portion of this chapter evidence is presented in support of the concept of long-distance migration of brines expelled from orogenic belts. This section considers the possible effects on hydrocarbons, particularly the spatial distribution of hydrocarbons as a consequence of that brine migration.

Petroleum geologists are divided over the question of long-distance migration of hydrocarbons. Some believe little or no such migration occurs, others call for substantial migration, and still others take intermediate positions.

This chapter takes the position that for some hydrocarbon provinces—those unaffected by tectonic fluid flow— lateral migration is typically modest at most. For those provinces affected by tectonic fluids, however, lateral migration up to distances of many hundreds of kilometers may occur. The spatial pattern of hydrocarbons throughout the world may be examined in this framework.

A second factor influencing the occurrence and abundance of hydrocarbons in the case of tectonically influenced deposits is the nature of material on the down-going slab of the subduction zone. If the sediments are voluminous and organic-rich, as in the case of a large delta, the development of hydrocarbons is enhanced. Of course, many other factors that are part of the rich literature of petroleum geology but not discussed here, such as seals, traps, and the nature of organic materials in source rocks, influence the spatial pattern of hydrocarbon deposits; however, the emphasis here is on active tectonism and the factors discussed above.

As an example consider the Gulf Coast province, which holds one of the Earth's largest hydrocarbon deposits. Drainage from a large part of the North American continent has resulted in a large and complex delta with organics of terrestrial and marine origin. Organics are buried to maturation conditions and converted to hydrocarbons. The hydrocarbons migrate vertically and, at most, to modest distances (perhaps tens of kilometers) horizontally and are found in traps relatively near to where the hydrocarbons were formed.

Now imagine what would happen if the Gulf of Mexico were to close, so that the North American continent was partially subducted to the south beneath an accretionary wedge on the leading edge of a landmass impinging from the south. Fluids from the Gulf Coast sediments would be expelled and, according to the hypothesis, travel toward the interior of the continent (i.e., to the north). The fluids would be brines carrying minerals and both organics and hydrocarbons. Some organics may have already matured; some might encounter conditions for maturation along the way. The foreland basin resulting from the tectonic loading would fill with sediments. These sediments might produce some of their own hydrocarbons that might or might not be affected by the flow of tectonically generated fluids. Nevertheless, the general pattern of the hydrocarbon occurrences that resulted would reflect the effect of the migrating fluids.

This experiment need not be done solely in the mind. It may represent about what happened at the time of the Ouachita orogeny. Thus, the spatial pattern of the hydrocarbon province extending from West Texas through Oklahoma and Kansas (Figure 8.5) may be at least partly a consequence of the Ouachita orogeny and the expulsion of fluids from the Ouachita orogenic belt. The pattern is sharply defined by a smooth boundary along the south and east sides. Along the north and west it has the feather edge that might be expected of the squeegeeing effect of Ouachita overthrusting. This view resembles that of Salisbury (1968), who suggested years ago that such a phenomenon occurred in the case of the Marathon orogenic belt and associated hydrocarbons to the north of that belt. Gas in the Ouachita hydrocarbon province tends to occur near the orogen, oil farther away. The hypothesis suggests that some hydrocarbons of this province have not migrated much, but others have traveled as organics or hydrocarbons long distances from where the former continental margin lies buried beneath the Ouachita orogen.

To the east and north the similarity of the pattern of the boundary of hydrocarbon occurrences in Kentucky and Illinois with that of the Ouachita orogen raises the speculation that some flushing of these hydrocarbons to the north may have occurred.

In like manner one can interpret other hydrocarbon provinces of the map of Figure 8.5. Williston Basin, Michi gan Basin, and California hydrocarbons are found essentially where they were formed, with only modest migration within the particular basin. Appalachian oil has migrated to the west as described by Woodward (1958) as a consequence of burial of an organic-rich continental margin by convergence and collision during Paleozoic time. The oil of the Findlay arch in Ohio and Indiana was part of the westward migration in the early Paleozoic and concentrated in the arch when it was formed as a forebulge resulting from the loading of the crust by later Appalachian tectonics.

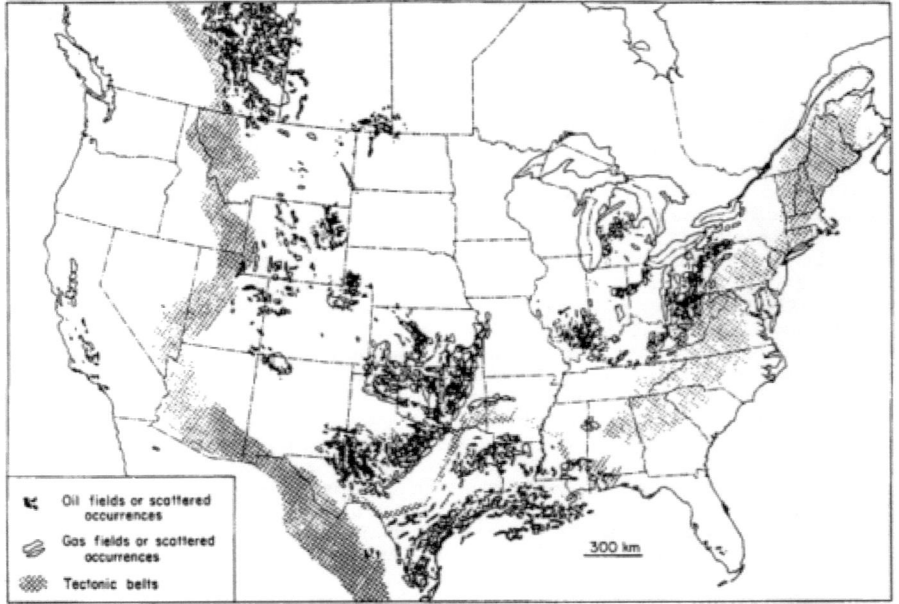

Figure 8.5
Map, generalized from PennWell map (Wilkerson, 1982), showing regions of oil and gas fields in the United States and adjoining parts of Canada and Mexico. Also shown are orogenic belts. The spatial distribution of hydrocarbons in the Appalachians, Alberta, the Cordilleran province, the West Texas-Oklahoma-Kansas province, and possibly the Illinois basin are in part a consequence of fluid migration from nearby orogenic belts according to the hypothesis discussed here.

Albertan hydrocarbons migrated to the east, the consequence of burial of an organic-rich margin beneath over-thrust slices from the west. Demaison (1977) proposed such an explanation for the Alberta tar sands (and also for large tar sands elsewhere). He called upon burial of a foreland basin delta, with source to the east, beneath thrusting from the west that drove the hydrocarbons updip through the permeable beds of the delta. The pattern of Cordillera hydrocarbons, though grossly similar to that of Alberta, is not so simple. This region has been disrupted by more recent tectonics in the foreland basin that must have affected the spatial distribution of hydrocarbons. Furthermore, the Cordillera hydrocarbons may not be associated with a subducted continental margin as rich in organics as that of Alberta. Nevertheless, the pattern is not in disagreement with the hypothesis.

Information on directions of migration of hydrocarbons in the 48 states is summarized in a map by Momper (1978) reproduced in Figure 8.6. Note that all of the directions on Momper's map are compatible with those predicted by the tectonics-based hypothesis discussed here, even to the extent of asymmetrical migration from the Illinois basin as opposed to symmetrical migration from other interior basins (i.e., Williston and Michigan). The surprisingly good fit of an independently collected set of data with a hypothesis based initially on entirely different considerations must be considered strong support for the hypothesis.

Similar reasoning about the spatial distribution of hydrocarbons may be applied to other parts of the world. For example, the petroliferous deltas of Africa and the rifts of China are places unaffected by major compressional orogeny and also where hydrocarbons have not migrated far from the place of maturation.

The hydrocarbons of the Middle East, as was pointed out by Dickinson (1974) in one of the first and most important early attempts to relate hydrocarbon accumulations to plate tectonics, likely migrated updip to the west from the Zagros orogenic belt as a consequence of subduction of the leading edge of the African continent, which then included the Arabian peninsula, to the east beneath the overriding Asian plate. Dickinson focused on hydrocarbon migration and not the associated brines. However, in the context of this chapter, one might speculate that, as a consequence of the paleodrainage of the African continent, a large organic-rich delta existed on the leading edge of the African plate that was subducted in the Zagros, with the consequence that fluids from the delta were expelled into parts of Iran and the Arabian peninsula. The Red Sea, a later feature, was not part of the proposed paleodrainage pattern. An interesting point here concerns the Persian Gulf, which has been in existence through much of the period of subduction and hence through the period of proposed hydrocarbon migration. Consequently, in this area there was not topographic relief with elevations decreasing from the orogenic belt into the foreland basin (i.e., towards Arabia). Thus, conditions were not favorable for hydrologic flow of meteoric waters from the orogenic belt toward the foreland basin in a manner comparable to that described by Garven and Freeze (1984) for Alberta. Consequently, such hydrologic flow is not likely the cause of hydrocarbon migration in the Middle East, and the migration is more likely a consequence of tectonic expulsion and updip transport of already formed hydrocarbon fluids and perhaps immature organic materials.

The locations of the hydrocarbons of the North Slope of Alaska may be at least partly a consequence of expulsion of fluids from the overthrust zone or subduction zone of the Brooks Range to the south. Difficulties of associating source rock and oil (Magoon and Claypool, 1985) in this region may be a consequence of failure to make full use of the tectonic framework as a basis for sample selection and interpretation of data.

Within the context of this hypothesis one might ask why greater accumulations of hydrocarbons are not found in India in the foredeep of the Himalayas, for example. A possible reason is that a large delta never existed on the leading edge of India, the edge that was subducted beneath the Himalayas. As India drifted across the Tethys, the paleodrainage was limited to the Indian subcontinent. The area of this subcontinent is not very large, and the major drainage may have terminated at other parts of the margin, a hypothesis that could be tested by offshore seismic studies. Prior to the time of drift the portion of India that made up the margin of Gondwanaland may not have been the site of a major delta.

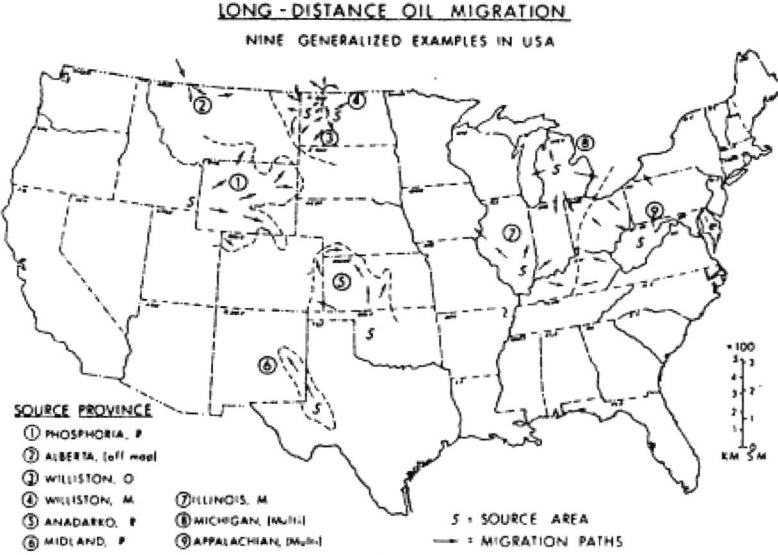

Figure 8.6
Map of Momper (1978) showing summary of information on migration direction of hydrocarbons in the United States. Note general agreement between the migration directions of this map and those suggested by the hypothesis on fluid expulsion from orogenic belts as described in the text.

The eastern margin of the United States seaward of the Appalachian hydrocarbon province is not very productive of hydrocarbons, whereas the Gulf Coast that occupies a similar position relative to the Ouachita hydrocarbon province is. According to the hypothesis, this situation occurs because the Gulf Coast is and has been the deposition area for drainage of a large part of the North American continent and the East Coast has not. Of course, other factors such as climate, conditions of deposition, and marine sources must be involved.

To summarize this section, it appears that a large amount of evidence from a wide diversity of specialties favors the hypothesis that fluids expelled from orogenic belts play an important role in many geological phenomena such as the spatial distribution of certain hydrocarbon and mineral deposits, authigenesis, diagenesis, coal metamorphism, dolomitization, and paleoremagnetism. The case is not closed, however, for there remains opportunity for further testing, and the attempt at such broad-based synthesis is so new that many aspects require further attention. Should the hypothesis be more or less correct, however, a means for relating many types of geological observations of the continent to plate tectonics, and hence the great geodynamical processes of the Earth, may be in sight.

ACKNOWLEDGMENTS

This chapter, as are all papers based on COCORP data, is highly dependent upon the efforts of many scientists and others who are part of the COCORP project. The COCORP project is funded by National Science Foundation grant EAR-8418157 and by the Cornell Program for the Study of the Continents. The story of fluids expelled from orogenic belts is not strictly a part of the COCORP project but is a speculative synthesis stimulated by COCORP and other data. This is contribution no. 79, INSTOC, Cornell University.

References

Anderson, G. M., and R. W. Macqueen (1982). Ore deposit models. 6. Mississippi Valley-type lead-zinc deposits , *Geoscience Canada 9*, 108-117.

Beales, F. W., K. C. Jackson, E. C. Jowett, G. W. Pearce, and Y. Wu (1980). Paleomagnetism applied to the study of timing in stratigraphy with special emphasis to ore and petroleum problems, in The Continental Crust and its Mineral Deposits, D. W. Strangway, ed., *Geological Association of Canada Special Paper 20*, 789-804.

Brown, L. D. (1987). Lower continental crust: Variations mapped by COCORP deep seismic profiling, *Annales Geophysicae 5B*(4), 325-330.

Brown, L. D., M. Barazangi, S. Kaufman, and J. Oliver (1986). The first decade of COCORP, 1974-1984, in *Reflection Profiling: A Global Perspective*, M. Barazangi and L. Brown, eds., Geodynamics Series, vol. 13, American Geophysical Union, Washington, D.C., pp. 107-120.

Cathles, L. M., and A. T. Smith (1983). Thermal constraints on the formation of Mississippi Valley-type lead-zinc deposits and their implications for episodic basin dewatering and deposit genesis. *Economic Geology 78*, 983-1002.

Clarke, J. W., R. C. McDowell, J. R. Matzko, P. P. Hearn, D. J. Milton, D. J. Percious, D. B. Vitaliano, and G. Ulmishek (1986). The Kola superdeep drill hole [detail summary of *Kol'skaya sverkhglubokaya* by Ye. A. Kozlovskiy (1984)], *U.S. Geological Survey Open File Report 86-517*, 249 pp.

Damberger, H. H. (1974). qualification patterns of Pennsylvanian coal basins of the eastern United States, in *Carbonaceous Materials as Indicators of Metamorphism*, R. E. Dutcher, ed., Special Paper 153, Geological Society of America, Boulder, Colo., pp. 53-74.

Demaison, G. J. (1977). Tar sands and supergiant oil fields, *American Association of Petroleum Geologists Bulletin 61*, 1950-1961.

de Voogd, V., L. Serpa, L. Brown, E. Hauser, S. Kaufman, J. Oliver, B. W. Troxel, J. Willemin, and L. A. Wright (1986). Death Valley bright spot: A mid-crustal magma body in the southern Great Basin, *Geology 14*, 64-67.

Dickinson, W. R. (1974). Subduction and oil migration, *Geology 2*, 421-424.

Garven, G., and R. A. Freeze (1984). Theoretical analysis of the role of groundwater flow in the genesis of stratabound ore deposits. 1. Mathematical and numerical 'model, *American Journal of Science 284*, 1085-1124.

Green, A. G., and J. A. Mair (1983). Subhorizontal fractures in a granitic pluton: Their detection and implications for radioactive waste disposal, *Geophysics 48*, 1428-1449.

Gregg, J. M. (1985). Regional epigenetic dolomitization in the Benneterre dolomite (Cambrian), southeastern Missouri, *Geology 13*, 503-506.

Guilbert, J. M., and C. F. Park (1986). *The Geology of Ore Deposits* , W. H. Freeman, New York, 985 pp.

Haquebard, P. A., and J. R. Donaldson (1974). Rank studies of coal in the Rocky Mountains and inner foothills belt, Canada, in *Carbonaceous Materials as Indicators of Metamorphism*, R. E. Dutcher, ed., Special Paper 153, Geological Society of America, Boulder, Colo., pp. 75-93.

Hearn, P. P., and J. F. Sutter (1985). Authigenic potassium feldspar in Cambrian carbonates: Evidence of Alleghenian brine migration, *Science 228*, 1529-1531.

Hearn, P. P., J. F. Sutter, M. J. Munk, and H. D. Belkin (1985). Evidence for Alleghenian brine migration in the central and southern Appalachians: Implications for Mississippi Valley-type sulfide mineralization, *Geological Society of America Abstracts with Programs 17*, 606.

Kent, D. V., and N. D. Opdyke (1985). Multicomponent magnetizations from the Mississippian Mauch Church formation of the central Appalachians and their tectonic implications, *Journal of Geophysical Research 90*, 5371-5383.

Koslovski, Ye. A. (1984). *Kol'skaya sverkhglubokaya*, Moscow, Nedra, 490 pp.

Leach, D. L. (1973). Possible relationship of Pb-Zn mineralization in the Ozarks to the Ouachita orogeny, *Geological Society of America Abstracts with Programs 5*, 269.

Leach, D. L., and E. L. Rowan (1986). Genetic link between Ouachita foldbelt tectonism and the Mississippi Valley-type lead-zinc deposits of the Ozarks, *Geology 14*, 931-935.

Leach, D. L., J. G. Viets, and L. Rowan (1984). Appalachian-Ouachita orogeny and Mississippi Valley-type lead-zinc deposits, *Geological Society of America Abstracts with Programs 16*, 572.

Levine, J. R. (1986). Deep burial of coal-bearing strata, anthracite region, Pennsylvania: Sedimentation or tectonics? *Geology 14*, 577-580.

Magoon, L. B., and G. E. Claypool, eds. (1985). *Alaska North Slope Oil/Rock Correlation Study*, Studies in Geology 20, American Association of Petroleum Geologists, Tulsa, Okla., 682 pp.

Mair, J. A., and A. G. Green (1981). High-resolution seismic reflection profiles reveal fracture zones within a homogeneous granite batholith, *Nature 294*, 439-442.

Matthews, D. H., and M. J. Cheadle (1986). Deep reflections from the Caledonides and Variscides west of Britain and Comparison with the Himalayas, in *Reflection Seisinology: A Global Perspective*, M. Barazangi and L. Brown, eds., Geodynamics Series, vol. 13, American Geophysical Union, Washington, D.C., pp. 5-19.

McCabe, C., R. Van der Voo, D. Pracor, C. R. Scotese, and R. Freeman (1983). Diagenetic magnetite carries ancient yet secondary remanence in some Paleozoic sedimentary carbonates, *Geology 11*, 221-223.

McCabe, C., R. Van der Voo, and M. M. Ballard (1984). Late Paleozoic remagnetization of the Trenton limestone, *Geophysical Research Letters 11*, 979-982.

Momper, J. A. (1978). Oil migration limitations suggested by geological and geochemical considerations, in *AAPG Continuing Education Course Note Series 8*, American Association of Petroleum Geologists, Tulsa, Okla., B1-B60.

Morton, J. P. (1985). Rb-Sr dating of diagenesis and source age of clays in Upper Devonian black shales of Texas, *Geological Society of America Bulletin 96*, 1043-1049.

Oliver, J. (1986). Fluids expelled tectonically from orogenic belts: Their role in hydrocarbon migration and other geologic phenomena, *Geology 14*, 99-102.

Oliver, J. (1990). The spots and stains of plate tectonics, in preparation.

Roedder, E. (1984). *Fluid Inclusions*, Reviews in Mineralogy, Vol. 12, Mineralogical Society of America, Washington, D.C., 644 pp.

Rowan, L., D. L. Leach, and J. G. Viets (1984). Evidence for a late Pennsylvanian-Early Permian regional thermal event in Missouri, Kansas, Arkansas, and Oklahoma, *Geological Society of America Abstracts with Programs 16*, 640.

Salisbury, G. (1968). Natural gas in Devonian and Silurian rocks of Permian basin, in *Natural Gases of North America; A Symposium*, B. W. Beebe, ed., Memoir 9 (2), American Association of Petroleum Geologists, Tulsa, Okla., pp. 1433-1445.

Sanford, A. R., O. S. Alptekin, and T. R. Toppozada (1973). Use of reflection phases on microearthquake seismographs to map an unusual discontinuity beneath the Rio Grande rift, *Seismological Society of America Bulletin 63*, 2021-2034.

Sanford, A. R., R. P. Mott, Jr., P. J. Shuleski, E. J. Rinehart, F. J. Caravella, R. M. Ward, and T. C. Wallace (1977). Geophysical evidence for a magma body in the crust in the vicinity of Socorro, New Mexico, in *The Earth's Crust*, J. G. Heacock, ed., Monograph 20, American Geophysical Union, Washington, D.C., pp. 385-403.

Scotese, C. R., R. Van der Voo, and C. McCabe (1982). Paleomagnetism of the Upper Silurian and Lower Devonian carbonates of New York State: Evidence for secondary magnetizations residing in magnetite, *Physics of Earth and Planetary Interiors 30*, 385-395.

Serpa, L., B. de Voogd, L. Wright, J. Willemin, J. Oliver, E. Hauser, and B. Troxel (1987). Structure of the central Death Valley pull-apart basin from COCORP profiles in the southern Great Basin, *Geological Society of America Bulletin 100*, 1437-1450.

Sun, R. J., ed. (1986). Regional aquifer-system analysis program of the U.S. Geological Survey, Summary of projects, 1974-84, *U.S. Geological Survey Circular 1002*, 264 pp.

Taylor, M., W. C. Kelly, S. E. Kesler, J. E. McCormick, F. D. Resnick, and W. V. Mellon (1983). Relationship of zinc mineralization in east Tennessee to Appalachian orogenic events, in *International Conference on Mississippi Valley Type Lead-Zinc Deposits, Proceedings Volume* , G. Kisvarsanyi, S. Grant, W. Pratt, and J. Koenig, eds., University of Missouri-Rolla, pp. 271-288.

Thom, W. T., Jr. (1934). Present status of the carbon-ratio theory, in *Problems of Petroleum Geology*, W. E. Wrather and F. H. Lahee, eds., American Association of Petroleum Geologists, Tulsa, Okla., pp. 69-95.

Van der Voo, R. (1986). Late Paleozoic remagnetization in the Appalachians and their foreland: Is it related to fluid migration? Oral presentation at 14th Meeting, COPSTOC, Cornell University, April 15-16, 1986.

Van Der Voo, R., and R. B. French (1977). Paleomagnetization of the Late Ordovician Juniata formation and the remagnetization hypothesis, *Journal of Geophysical Research 82*, 5796-5802.

Wilkerson, R. M., ed. (1982). *Oil and Gas Fields of the U.S.*, PennWell Publishing Co., Tulsa, Okla., 1-p. map.

Woodward, H. P. (1958). Emplacement of oil and gas in Appalachian basin, in *Habitat of Oil, A Symposium*, L. G. Weeks, ed., American Association of Petroleum Geologists, Tulsa, Okla., pp. 494-510.

Zenger, D. H., and J. B. Dunham (1980). Concepts and models of dolomitization: An introduction, in *Special Publication No. 28*, Society of Economic Paleontologists and Mineralogists, Tulsa, Okla., pp. 1-9.

9

Smoluchowski's Dilemma Revisited: an Note On the Fluid Pressure History of the Central Appalachian Fold-Thrust Belt

TERRY ENGELDER
The Pennsylvania State University

ABSTRACT

Cross-fold joints in the Central Appalachian fold-thrust belt propagated during periods of abnormally high fluid pressure prior to tectonic compaction and the development of first-order Alleghanian structures in the valley and ridge. These early joints, found in both the valley and ridge province and the plateau province, are organized in sets forming patterns that correlate across the Allegheny Front. Examples of early joints are found in the Devonian Brallier and Trimmers Rock Formations of the Pennsylvania Valley and Ridge and in the Genesee Group of the Appalachian Plateau. One interpretation is that high fluid pressures were generated by topographically driven flow across the Appalachian Basin as a consequence of uplift of the core of the Appalachians early in the Alleghanian Orogeny. The high fluid pressures accompanying this topographically-driven flow system later facilitated the development of first-order structures in the valley and ridge. Later joint sets that do not correlate across the Allegheny Front are more likely to be a consequence of fluid pressure pulses developed during local tectonic compaction and the development of first-order Alleghanian structures. These later joint sets vary in number and orientation from location to location.

INTRODUCTION

Smoluchowski's (1909) famous dilemma is that thrust sheets are too wide for emplacement by "dry" frictional sliding. Theoretically, the back end of wide thrust sheets should collapse under the large tectonic stress necessary to push the entire thrust sheet against frictional resistance. The most popular solution to Smoluchowski's dilemma was presented by Hubbert and Rubey (1959) and Rubey and Hubbert (1959), who pointed out that the theoretical width of thrust sheets is greatly increased by an increase in fluid pressure and concomitant reduction in effective normal stress across the basal décollement. Because frictional resistance is directly proportional to effective nor mal stress, a reduction in effective normal stress has the net effect of reducing the push (i.e., tectonic stress) necessary for emplacement of wide thrust sheets. Lower tectonic stress reduces the tendency for fracture and thickening at the back end of the thrust sheet (Davis *et al.*, 1983).

A reduction in effective normal stress occurs if the base of the thrust sheet cuts through a stratigraphic section containing abnormally high fluid pressures. How this high fluid pressure evolves is still subject to debate. Two mechanisms for generating high fluid pressures as mentioned by Hubbert and Rubey (1959) are artesian flow and mechanical compaction of water-filled pores. In their companion paper Rubey and Hubbert (1959) focus on the generation of abnormally high fluid pressures by three mechanisms: (1) the uplift of sealed sand lenses, (2) tectonic compaction, and (3) compaction by overburden weight. They gave no further consideration to artesian flow. There are, of course, other mechanisms such as aquathermal pressuring (Barker, 1972), diagenetic dewatering of clays (Schmidt, 1973), and generation of CO_2 and CH_4 during the breakdown of hydrocarbons (Spencer, 1987).

Rubey and Hubbert (1959) do not examine artesian flow as a mechanism for generation of abnormal fluid pressures in foreland fold-thrust belts. By implication they consider it less important than tectonic compaction as a source for abnormal fluid pressures in overthrust terrain. The problem is that tectonic compaction occurs well after the initiation of thrusting. The development of overthrust terrain would be greatly facilitated if high fluid pressures developed before the onset of thrusting. Artesian flow may permit such a buildup in fluid pressure. Furthermore, experience in the Alberta Basin suggests that it should be taken seriously as a model for generating high fluid pressures in foreland basins (Tôth, 1980).

Artesian flow is commonly understood to be groundwater flow from a topographically high recharge area to a topographically low discharge area. This type of flow, also called topographically driven flow, is modeled by Tôth (1962, 1980) using a flow net first illustrated by Hubbert (1940). One consequence of topographically driven flow is that the discharge area is subject to pore water pressures in excess of hydrostatic developed because the mechanical energy per unit volume of pore fluid is highest in the recharge area and lowest in the discharge area. A fortuitous combination of aquitards and topography can lead to near-lithostatic fluid pressures in the discharge area (Engelder and Bethke, 1985). A topographically driven flow system is steady state; leakage is balanced by recharge. This is in direct contrast to compaction-driven flow, where fluid pressure gradually returns to hydrostatic once compaction stops.

Are there geological structures that enable the geologist to distinguish topographically driven flow from other mechanisms including tectonic compaction that might have been the source of high fluid pressures in a foreland fold-thrust belt? In principle, regional joint sets could serve as such structures. This paper presents further evidence supporting the regional correlation of cross-fold joint sets (joints with normals subparallel to regional fold axes) in the Appalachian foreland fold-thrust belt and then deals with the geological consequences of regionally developed cross-fold joint sets in terms of a mechanism for generating the necessary pore fluid pressure.

CORRELATION OF CROSS-FOLD JOINTING ACROSS THE CENTRAL APPALACHIAN FORELAND FOLD-THRUST BELT

Cross-fold joints are very prominent in Upper Devonian outcrops along the edges of the Finger Lakes of New York State (Figure 9.1). By the first decade of the twentieth century geologists recognized that these cross-fold joints were organized into more than one set (Sheldon, 1912). While tracing these cross-fold joints along strike of the New York Plateau for more than 200 km, Parker (1942) recognized that they maintained an orientation normal to fold axes despite a 30° change in strike of the fold axes. Parker made no judgment about whether cross-fold joints on either end of the map area are part of the same joint set. Nickelsen and Hough (1967) were the first to map joints as systematic sets in the Central Appalachians. They identified five cross-fold joint sets in sandstones of the Appalachian Plateau in Pennsylvania. By extrapolating to New York State, they identified three joint sets in Parker's map area. On mapping in the Appalachian Valley and Ridge, Nickelsen (1979) and Orkan and Voight (1985) attempted to correlate joint sets between the valley and ridge and plateau of Pennsylvania. Orkan and Voight (1985) identified six cross-fold joint sets in the valley and ridge (Figure 9.2).

Nickelsen and Hough's (1967) and Orkan and Voight's (1985) technique for correlation of joints along strike depends largely on the orientation of joints. Their assumption is that joints of one set have similar orientations over large regions. If a suite of joints at an outcrop is misoriented by, say, 15° from an established joint set, this suite belongs to another joint set regardless of its orientation with respect to local structures. As is illustrated in Figure 9.2, the consequence of this assumption is that members of a joint set do not change orientation even as fold axes swing through the Central Appalachians. On a regional basis the change in strike of fold axes is accommodated by the overlap of joint sets of different orientations. The notion for overlapping joint sets is supported by outcrops containing more than one joint set. Nickelsen and Hough's (1967) correlation strategy was developed in response to the observation that first-order folds in the Pennsylvania Valley and Ridge are kink folds (e.g., Faill, 1973) with straight axes (Nickelsen, 1987, personal communication). Nickelsen's idea is that the valley and ridge developed as overlapping thrust sheets cored with duplexes moving toward the craton with straight-axis kink folds delimiting the thrust duplexes. The curvature of the Central Appalachian Valley and Ridge is accommodated by abrupt changes in the orientation of the first-order folds. Strictly parallel joint sets reflect the kinematics of thrust sheets associated with straight-axis kink folds. Presumably the motion of various sheets is independent, so various joint sets are unrelated in time and space. If these assumptions hold, the use of joint sets to draw stress trajectories over the whole mountain belt is invalid.

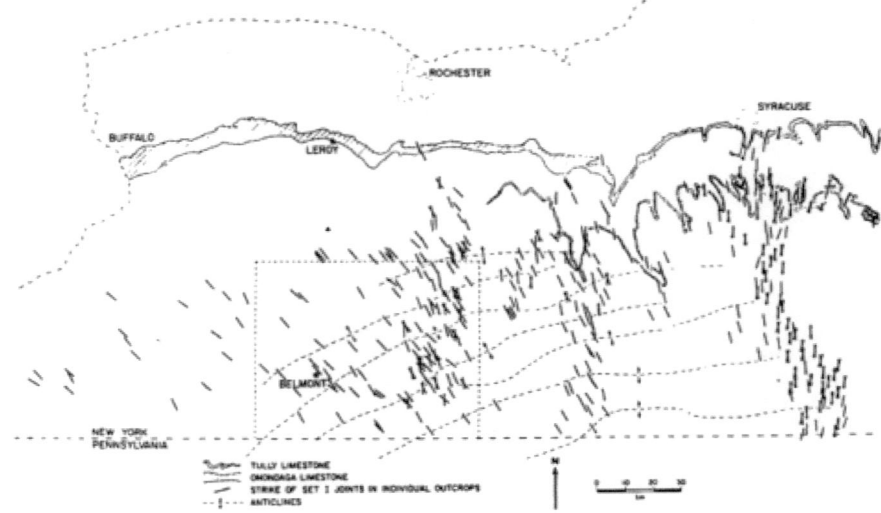

Figure 9.1
A map of cross-fold joints in the Middle and Upper Devonian rocks of the Appalachian Plateau of New York State (after Engelder and Geiser, 1980).

While remapping cross-fold joints on the Appalachian Plateau, Engelder and Geiser (1980) took a different approach to the correlation of joints along strike. They assumed that joint sets change orientation gradually to remain roughly perpendicular to local fold axes. Implicit in Engelder and Geiser's (1980) assumption is that stress trajectories associated with an orogenic pulse are curved and regional in extent. In the vicinity of Ithaca and Watkins Glen, New York, the apparent abrupt change in orientation as interpreted by Nickelsen and Hough (1967) is a manifestation of joint sets restricted to particular lithologies where joints in siltstones have a different orientation than joints in shales. An outcrop with two lithologies commonly exhibits two joint sets forming at different angles to the same fold axis. This pattern is not a manifestation of one joint set giving way to another set while moving around an oroclinal bend. Tracing both joint sets in their particular lithologies supports the notion that a single joint set can change orientation along with local fold axes (Engelder, 1985). In an area of the New York Plateau where Nickelsen and Hough (1967) and Orkan and Voight (1985) identified three joint sets, Engelder and Geiser (1980) argue that there are two with joint sets A and D in the western portion of the plateau being equivalent to joint sets D and E, respectively, in the eastern portion. All of this is said to make the point that the regional correlation of joint sets is not trivial.

Although correlation of joints across the Allegheny Front

has many of the same difficulties as correlation along strike of the Appalachians, both Nickelsen and Hough (1967) and Orkan and Voight (1985) feel that joints correlate across the Allegheny Front. [The Allegheny Front, which is the boundary between the Appalachian Valley and Ridge and Plateau, is largely controlled by the southeastern edge of the Silurian salt basin where décollement faulting climbed up from the Cambrian shales into the salt beds. Low strength of the salt changed the character of the Appalachian foreland tectonics from duplex structures of the valley and ridge to layer parallel shortening of the Appalachian Plateau (Davis and Engelder, 1987).] A correlation may be based on the common occurrence of a clockwise rotation of joint propagation in both the valley and ridge and the Appalachian Plateau. [Although the clockwise rotation of joint propagation is common throughout the region, Helgeson and Aydin (1989) report that a counter-clockwise rotation is well developed in some outcrops.] At Bear Valley Strip Mine Nickelsen (1979) identified eight stages of deformation with the first three being two phases of jointing followed by layer parallel shortening. These prefolding events witness a prefolding compression that rotates clockwise (Geiser and Engelder, 1983; Engelder, 1985). All along the Allegheny Front from Williamsport to State College, Pennsylvania, early cross-fold jointing shows a sequence indicating a clockwise rotation of compression (Lacazette, The Pennsylvania State University, personal communication). This same sequence is well displayed in the Devonian Brallier Formation at Huntingdon, Pennsylvania.

The Devonian Brallier Formation of the central Appalachian Valley and Ridge is equivalent in age and lithologic composition to the Genesee Group of the Appalachian Plateau. Of all the lithologies in the valley and ridge from Cambrian carbonates up through Carboniferous flu-vial deposits, none carry joints that more closely resemble those seen on the Devonian section of the Appalachian Plateau. At an outcrop just south of Huntingdon, Pennsylvania, the Brallier dips to the southeast at about 15°. Like joints in the sandstone-shale beds of the Genesee Group on the Appalachian Plateau, two sets of cross-fold joints cut the Brallier, with the finer-grained beds carrying joints striking about 140° and the coarser beds carrying joints striking about 158°. The relative time of propagation of the joint sets may be determined using a joint spacing criterion developed by DeGraff et al. (1987) in the Genesee Group at Taughannock Falls, New York. Toward the north end of the Brallier outcrop joints in siltstone beds can be seen propagating upward from joints in shale beds. Based on the spacing criterion, joints in the silty shale

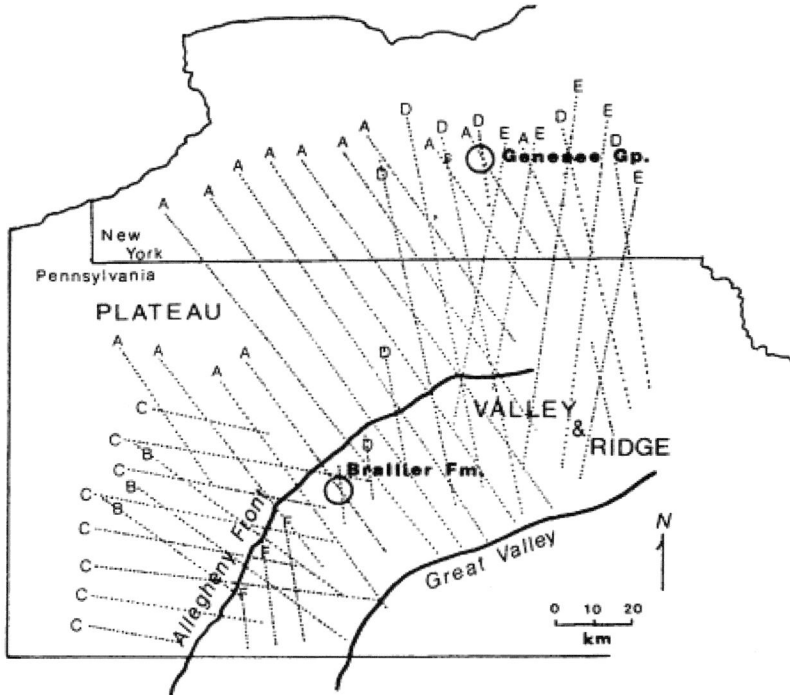

Figure 9.2 Orkan and Voight's (1985) map of regional joint sets within the Central Appalachian fold-thrust belt. Sets A through E are those of Nickelsen and Hough (1967). Set F was identified by Orkan and Voight (1985). Regional joint sets are based on the data of Nickelsen and Hough (1967) and Engelder and Geiser (1980).

(140°) propagated prior to those in the sandstone (158°) and, hence, show the same clockwise rotation as seen throughout the Appalachian foreland.

Aside from the fact that these joints look like those found in the flat-lying Devonian rocks of the Appalachian Plateau, two pieces of evidence suggest that the cross-fold joints in the Brallier preceded folding. First, the joints have been rotated to dip between 84° and 87° to the southwest. If the present dip of the Brallier is removed, these joints are vertical, presumably the orientation at which they propagated. Second, some joints in the silty shale beds are decorated with slickensides and fibrous calcite, indicating a left-lateral shear. This is the type of slip expected for the compression responsible for later folding. Furthermore, the orientation of the calcite fibers indicates that slip direction has a shallower plunge than bedding dip. These are some of the same arguments used by Nickelsen (1979) to demonstrate early jointing at Bear Valley.

It is likely that early joint sets propagated prior to the formation of the Allegheny Front. The Allegheny Front became significant only with the development of first-order structures of the valley and ridge, an event that took place long after early joint propagation, as shown by Nickelsen (1979) at Bear Valley. Joints correlate across the Allegheny Front largely because that structural front did not exist at the time the joints formed. The mechanism for early joint propagation must precede and be independent of the development of first-order structures of the Appalachian Valley and Ridge. In summary, the Upper Devonian shales and siltstones of the entire Central Appalachian foreland contain Alleghanian cross-fold joints that predate both major tectonic compaction and the development of first-order folds.

FLUID PRESSURE AND JOINTING

The regionally developed cross-fold joints of the Appalachian foreland formed at depth in the crust of the Earth, where the propagation of such joints requires the development of effective tensile stresses within the rock (e.g., Nickelsen, 1979; Narr and Currie, 1982). Effective tensile stresses are possible under conditions of cooling (Voight and St. Pierre, 1974), curvature above the neutral fiber of a fold (Price, 1974), or significantly high fluid pressures (Secor, 1965). In the foreland portion of mountain belts where folding would favor the propagation of strike joints (joints striking parallel to fold axes), curvature can be ruled out as a likely driving mechanism for cross-fold joints. At full depth of burial rocks have not cooled appreciably, so thermal cracking can also be ruled out as a driving mechanism. In contrast, a growing body of evidence suggests that high fluid pressure serves as the driving mechanism for joints at depth. Joints filled with such minerals as quartz, calcite, and chlorite are often cited as a manifestation of fluid pressure-driven joint propagation (i.e., hydraulic fracturing; e.g., Beach, 1977). The multiple fracture of crack-seal veins (e.g., Ramsay, 1980) and the repeated arrest of joints during propagation (e.g., Engelder, 1985) are fracture-related structures associated with the cracking of rock under the influence of high fluid pressure.

An understanding of the extent to which joints correlate in both time and space is critical to identifying the mechanisms for generation of high fluid pressures in a mountain belt. Although rapid joint propagation occurs on the scale of outcrops, the timing of joint propagation at different locations across a foreland is less certain. Presently it is not clear whether fluid pressure increases simultaneously everywhere across a foreland or whether high fluid pressure occurs as local pulses affecting only small parts of the mountain range at one time. This, of course, leads to uncertainty about whether joint sets of the same orientation should correlate across distances of tens to hundreds of kilometers. Certainly, based on data discussed above, current dogma for the Appalachians is that early joint sets do correlate over large distances (e.g., Nickelsen and Hough, 1967; Engelder and Geiser, 1980; and Orkan and Voight, 1985).

Figure 9.2 suggests that joint sets, such as "set A," extend from the Great Valley of Pennsylvania to the far reaches of the Appalachian Plateau near Buffalo, New York. This is the Orkan and Voight (1985) interpretation of regional joint sets where joint development cuts across tectonic boundaries such as the Allegheny Front. If a correlation across the Allegheny Front is valid, the evolution of a high pore pressure must have been a foreland-wide event. Not all mechanisms for generation of high pore pressure are regional in extent. Although depositional and diagenetic mechanisms for the generation of high fluid pressure may be regional, they are considered unlikely mechanisms for a regional pore pressure event in the Appalachian fold-thrust belt because the earliest cross-fold joint sets are Alleghanian and, hence, developed long after deposition and diagenesis of the foreland sediments containing the joints. Although tectonic compaction affects an entire foreland, it was not uniform, as indicated by a variation is strain. The upper crust does not have the strength to simultaneously compact across the foreland until the core of the mountain belt has built into a sizable wedge (Davis et al., 1983). Furthermore, strain in forelands is much too low to have been continuously active at plate tectonic rates for the duration of an orogenic event such as the Alleghanian orogeny in the Appalachians. For these reasons tectonic compaction seems unlikely to have contributed to a foreland-wide pore pressure event of the type required for early jointing. The only mechanism for generating a regional pore pressure event that cannot be rejected out of hand is the topographically driven flow system. Therefore, it is assumed to be the most likely source for high fluid pressures causing the simultaneous development of a joint set across the foreland, particularly during early stages of foreland development.

DISCUSSION: OROGENIC PULSES AND THE GENERATION OF ABNORMALLY HIGH FLUID PRESSURE

Early foreland-wide joint sets may be reconciled with topographically driven flow systems. In this case a regional flow system may have formed in response to uplift of mountains to the southeast of the Great Valley. To generate the high pressures for joint propagation, such a topographically driven flow system is, of course, going to require regional aquitards and a significant topographic gradient across the foreland. Because the upper Paleozoic section of the Central Appalachians developed very few through-going thrust faults, it may have served as a regional aquitard. Furthermore, evidence is accumulating that suggests that during the Alleghanian orogeny the Central Appalachians southeast of the Allegheny Front was quite thick (Levine, 1983; Paxton, 1983; Orkan and Voight, 1985). Vitrinite reflectance and fission track data suggest that the Devonian and Carboniferous of New York and Pennsylvania may have been buried to a depth of 6 km (Friedman and Sanders, 1982). Current studies of crustal flexure suggest that external forces were necessary for the magnitude of crustal depression necessary for the depth of burial found in Pennsylvania (Beaumont, 1981). Such crustal loading can be accomplished during continent-continent collisions. The Alleghanian Orogeny was a period during which the continent of Africa collided with North America, producing continental edges having a topography similar to the India-Asia collision. This interpretation of regional joint sets requires that uplift at the core of the mountain belt preceded the development of first-order structures in the foreland. The early development of a regional flow system with elevated pore pressures facilitates later thrusting and the development of first-order structures, particularly in the discharge area of the foreland.

Regardless of their correlation, everyone agrees that some cross-fold joints in the central Appalachian foreland fold-thrust belt propagated early and are organized into discrete sets rather than being distributed randomly or uniformly. The existence of multiple joint sets indicates that the syntectonic stress field changed in orientation during the evolution of the foreland fold-thrust belt. This regional organization of joints leads to the inference that joint propagation took place during punctuated events. Not only did the orientation of the stress field change with time but the magnitude of the effective stress varied. Fluid pressures were not continuously at a level necessary for joint propagation, but rather some poorly understood events caused fluid pressure to fluctuate up and down throughout a region. Such events took place a finite number of times during the development of the foreland portion of the Central Appalachians.

Mountain belts include a complex combination of diachronous structures superimposed over periods as long as 1 billion years ago. During the evolution of foreland fold-thrust belts, deformation is punctuated rather than continuous. Punctuated events called orogenic pulses are identified on the basis of the appearance of an arbitrarily chosen set of structures within the mountain belt. For example, a regionally developed disjunctive cleavage may be attributed to one orogenic pulse, whereas a second cross-cutting cleavage may be attributed to a later orogenic pulse. With few exceptions the duration of an orogenic pulse is extremely difficult to measure.

The intensity of an orogenic pulse is often correlated with the finite strain within rocks or the regional shortening associated with folding and faulting. Orogenic pulses become increasingly hard to discriminate as the finite strain or regional shortening decreases. Although the case may be argued that major structures such as folds are the signature of a single orogenic pulse, multiple joint sets within folds are themselves witness for multiple orogenic pulses prior to the folding event. Regional joints, particularly sensitive indicators of individual orogenic pulses, are commonly found in the unmetamorphosed foreland where more than one set may cross-cut. Cross-fold joints may propagate even during very mild orogenic pulses and in many instances before significant bed rotation. These mild orogenic pulses in the foreland may reflect uplift events in the core of the mountain belt or periods of rapid tectonic compaction. Unlike faults, folds, or finite strain markers of any sort, the propagation of joints is so close to instantaneous that one moment in the history of mountain building is recorded. The convenience of joint sets is that stress trajectories associated with an orogenic pulse can be mapped with reasonable confidence (Ode, 1957).

The development of several joint sets suggests that fluid pressures were not continuously lithostatic throughout the Alleghanian orogeny. If fluid pressures were continuously at lithostatic during realignment of the stress field, then joints should have a uniform distribution of orientations rather than appear as isolated joint sets. Multiple joint sets suggest that fluid pressures rise to lithostatic levels during short-lived events before pore fluids leak off to drop the pressure well below that needed for joint propagation. Fluid pressures rise again once the stress field is realigned. In order for topographically driven flow to account for multiple joint sets, the topography may have bounced up and down several times during the development of the foreland fold-thrust belt. An example of such vertical movements of crust is found in the multiple development of black shale basins in the Appalachian Catskill delta of the Acadian Orogeny (Ettensohn, 1985).

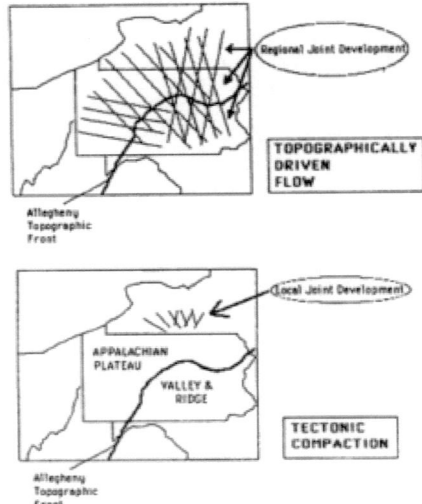

Figure 9.3
A schematic drawing illustrating the differences between local and regional joint development.

Although rapid uplift and erosion of a Himalayan-like mountain chain may be reasonable, admissible strain rates suggest that tectonic compaction as envisioned by Rubey and Hubbert (1959) may also be a likely mechanism for punctuated events in the fluid pressure history later during the Alleghanian orogeny (Evans et al., 1989). After initial jointing events the Appalachian Plateau was shortened by about 10 percent during the Alleghanian Orogeny, an event that may have lasted more than 50 m.y. (Engelder and Engelder, 1977). Evidence for more than one orogenic pulse suggests that shortening was discontinuous throughout the 50-m.y. period, in which case the strain rate would have exceeded 10^{-16} s^{-1} during individual events. If the porosity reduction rate is on this order, then fluid-pressures would build toward lithostatic pressures in shales and well-cemented siltstones within periods less than 1 m.y. (Walder and Nur, 1984). If this is the case, fluid pressure events were not foreland wide, and so there may be little reason to correlate cross-fold joints in the valley and ridge with those of the Appalachian Plateau (Figure 9.3).

Although a judgment is highly subjective, I would attribute set A in Figure 9.2 to a topographically driven flow system. Set A is equivalent to joints cutting siltstones in the Genesee Group (Engelder, 1985). Fluid pressures associated with joint set D, which propagated in the direction of Alleghanian layer-parallel shortening (Engelder and Geiser, 1980), are more likely to have been generated by tectonic compaction. Because I am not familiar with joint sets B, C, and F, I cannot make a judgment concerning them.

CONCLUSIONS

The mechanism responsible for high fluid pressures in the Appalachian foreland may be identified on the basis of the regional correlation of joint sets. The correlation of early regional joint sets across the boundary of structural provinces suggests that topographically driven flow was active prior to the development of first-order structures. This is then the solution to Smoluchowski's dilemma for the Appalachian fold-thrust belt where the high fluid pressures from a topographically driven flow system facilitated the development of first-order structures where such development is highly dependent on the reduction of effective stress. In contrast, some later joint sets developed as a consequence of fluid-pressure pulses during local tectonic compaction.

ACKNOWLEDGMENTS

Craig Bethke introduced me to the significance of topographically driven flow in foreland basins. Gerald Friedman, Alfred Lacazette, and Don Wise are thanked for reviewing an early version of the manuscript. This work was supported by a contract from the Electric Power Research Institute (No. RP-2556-24).

References

Barker, C. (1972). Aquathermal pressuring: Role of temperature in development of abnormal-pressure zones, *American Association of Petroleum Geologists Bulletin 56*, 2068-2071.
Beach, A. (1977). Vein arrays, hydraulic fractures, and pressure solution structures in a deformed flysch sequence, S.W., England, *Tectonophysics 40*, 201-225.
Beaumont, C. (1981). Foreland basins, *Geophysical Journal 65*, 291-329.
Davis, D. M., and T. Engelder (1987). Thin-skinned deformation over salt, in *Dynamical Geology of Salt and Related*

Structures, I. Lerche and J. J. O'Brien, eds., Academic Press, Orlando, Fla., pp. 301-338.

Davis, D., J. Suppe, and F. A. Dahlen (1983). Mechanics of fold-and-thrust belts and accretionary wedges, *Journal of Geophysical Research 88*, 1153-1172.

DeGraff, J. M., D. E. Helgeson, and A. Aydin (1987). Transient and stabilized spacing of thermal and tectonic joints, *Geological Society of America Abstracts with Programs 19*, 638.

Engelder, T. (1985). Loading paths to joint propagation during. a tectonic cycle: An example from the Appalachian Plateau, U.S.A., *Journal of Structural Geology 7*, 459-476.

Engelder, T., and C. Bethke (1986). Reexamination of the Gulf Coast Model used by the Rubey-Hubbert Hypothesis for thrust belt tectonics, *Geological Society of America Abstracts with Programs 10*, 595.

Engelder, T., and R. Engelder (1977). Fossil distortion and décollement tectonics of the Appalachian Plateau, *Geology 5*, 457-460.

Engelder, T., and P. A. Geiser (1980). On the use of regional joint sets as trajectories of paleostress fields during the development of the Appalachian Plateau, New York, *Journal of Geophysical Research 85*, 6319-6341.

Ettensohn, F. R. (1985). The Catskill Delta complex and the Acadian Orogeny: A model, in *The Catskill Delta*, D. L. Woodrow and W. D. Sevon, eds., Special Paper 201, Geological Society of America, Boulder, Colo., pp. 39-49.

Evans, K., G. Oertel, and T. Engelder (1989). Correlated anomalies of in situ stress and compaction in Devonian shales of the Appalachian Plateau: An insight into the nature of Alleghanian deformation, *Journal of Geophysical Research 94*, 7155-7170.

Faill, R. T. (1973). Kink band folding, Valley and Ridge Province, Pennsylvania, *Geological Society of America Bulletin 84*, 1289-1314.

Friedman, G. M., and J. E. Sanders (1982). Time-temperature burial significance of Devonian anthracite implies former great (6.5 km) depth of burial of Catskill Mountains, *Geology 10*, 93-96.

Geiser, P., and T. Engelder (1983). The distribution of layer parallel shortening fabrics in the Appalachian foreland of New York and Pennsylvania: Evidence for two noncoaxial phases of the Alleghanian orogeny, in *Contributions to the Tectonics and Geophysics of Mountain Chains*, R. D. Hatcher et al., eds., Memoir 158, Geological Society of America, Boulder, Colo., pp. 161-175.

Helgeson, D., and A. Aydin (1989). Use of surface features for interpretation of the propagation, interactiona, and intersection of joints, *Geological Society of America Abstracts with Programs 21*, A64.

Hubbert, M. K. (1940). The theory of groundwater motion, *Journal of Geology 48*, 785-944.

Hubbert, M. K., and W. W. Rubey (1959). Role of fluid pressures in mechanics of overthrust faulting: I. Mechanics of fluid-filled porous solids and its application to overthrust faulting, *Geological Society of America Bulletin 70*, 115-166.

Levine, J. R. (1983). Tectonic history of coal-bearing sediments in eastern Pennsylvania using coal reflectance anisotropy, Ph.D. thesis, Pennsylvania State University, 315 pp.

Narr, W., and J. B. Currie (1982). Origin of fracture porosity: Example from Altamont field, Utah, *American Association of Petroleum Geologists Bulletin 66*, 1231-1247.

Nickelsen R. P. (1979). Sequence of structural stages of the Allegheny orogeny at the Bear Valley Strip Mine, Shamokin, Pennsylvania, *American Journal of Science 279*, 225-271.

Nickelsen, R. P., and V. D. Hough (1967). Jointing in the Appalachian Plateau of Pennsylvania, *Geological Society of America Bulletin 78*, 609-630.

Ode, H. (1957). Mechanical analysis of the dyke pattern of the Spanish Peaks area, Colorado, *Geological Society of America Bulletin 68*, 567-576.

Orkan, N., and B. Voight (1985). Regional joint evolution in the Valley and Ridge Province of Pennsylvania in relation to the Allegheny Orogeny, *Guidebook, 50th Annual Field Conference on Pennsylvanian Geology*, Bureau of Topographic and Geological Survey, Harrisburg, Pa., 144-164.

Parker, J. M. (1942). Regional systematic jointing in slightly deformed sedimentary rocks, *Geological Society of America Bulletin 53*, 381-408.

Paxton, S. T. (1983). Relationships between Pennsylvania-age lithic sandstone and mudrock diagenesis and coal rank in the Central Appalachians, Ph.D. thesis, Pennsylvania State University, 503 pp.

Price, N. J. (1974). The development of stress systems and fracture patterns in undeformed sediments, *Advances in Rock Mechanics, Proceedings 3rd Congress ISRM*, 487-496.

Ramsay, J. G. (1980). The crack-seal mechanism of rock deformation, *Nature 284*, 135-139.

Rubey, W. W., and M. K. Hubbert (1959). Role of fluid pressures in mechanics of overthrust faulting: II. Overthrust belt in geosynclinal area of western Wyoming in light of fluid pressure hypothesis, *Geological Society of America Bulletin 70*, 167-205.

Schmidt, G. W. (1973). Interstitial water composition and geochemistry of deep Gulf Coast shales and sandstones, *American Association of Petroleum Geologists Bulletin 57*, 321-337.

Secor, D. T. (1965). Role of fluid pressure in jointing, *American Journal of Science 263*, 633-646.

Sheldon, P. (1912). Some observations and experiments on joint planes, *Journal of Geology 20*, 53-70.

Smoluchowski, M. (1909). Some remarks on the mechanics of overthrusts, *Geological Magazine 6*.

Spencer, C. W. (1987). Hydrocarbon generation as a mechanism for overpressuring in Rocky Mountain Region, *American Association of Petroleum Geologists Bulletin 71*, 368-388.

Tôth, J. (1962). A theory of groundwater motion in small drainage basins in central Alberta, *Journal of Geophysical Research 67*, 4375-4387.

Tôth, J. (1980). Cross-formational gravity-flow of groundwater: A mechanism of the transport and accumulation of petroleum (the generalized hydraulic theory of petroleum migration), *American Association of Petroleum Geologists Bulletin 64*, 121-167.

Voight, B., and B. H. P. St. Pierre (1974). Stress history and rock stress, *Advances in Rock Mechanics, Proceedings 3rd Congress ISRM*, 580-582.

Walder, J., and A. Nur (19.84). Porosity reduction and crustal pore pressure development, *Journal of Geophysical Research 89*, 11,539-11,548.

10

Fluid Pressure History in Subduction Zones: Evidence from Fluid Inclusions in the Kodiak Accretionary Complex, Alaska

PETER VROLIJK[1]
GEORGIANNA MYERS[2]
University of California, Santa Cruz

ABSTRACT

Fluid inclusions offer a simple and elegant means to examine the history of fluid pressure, temperature, and composition at intermediate levels in subduction zones. At deeper levels ductile deformation mechanisms tend to disrupt fluid inclusion relationships in veins. Fluid inclusions provide a direct record of the fluid phase and apparently preserve relatively short-term changes in fluid pressure. Moreover, fluid inclusions are found in veins that can be directly tied to the development of structural fabrics in the rock. In syntectonic veins of the Kodiak accretionary complex, Alaska, fluid inclusions record fluctuating fluid pressures. These fluctuations reflect a history of fracture opening, continued fracture growth, and creation of an interconnected fracture network, with inferred fluid flow toward shallower levels within the subduction zone.

Fluid inclusions also record fluid temperature at the time of inclusion entrapment and thereby indicate the temperature of fluids involved in the deformation of rocks in the décollement zone during active subduction. Data from the Kodiak accretionary complex indicate that fluid temperatures at 10 to 15 km in the ancient Kodiak subduction zone were two to three times higher than predicted by conductive heat flow models. The cause of this temperature anomaly appears to be the migration of warm fluids along the décollement zone from deeper structural levels.

These data, in conjunction with emerging data from both modem and ancient subduction complexes, suggest that high fluid pressures are common in subduc

---

[1] Current Address: Exxon Production Research Company, Houston, Texas
[2] Current Address: Minnesota Pollution Control Agency

tion zones and that the presence of high fluid pressures has a profound influence on the geology at convergent margins. The manifestations of high fluid pressures include the style of structural deformation, heat flow, modifications of diagenesis and metamorphism by tectonically driven fluid flow, and effects on the global water and geochemical budgets.

INTRODUCTION

High fluid pressures are probably more common in subduction zones than in any other tectonic environment. In subduction zones water-rich sediments are rapidly dragged with the subducting oceanic crust to great depth. The sediments directly overlying the oceanic crust, which have the greatest potential for being subducted, are typically fine grained and pelagic with low intergranular permeability (Moore, 1975; Shepherd and Bryant, 1983). The rapid sediment burial afforded by plate convergence and subduction, in combination with the common occurrence of fine-grained sediments, should theoretically lead to the pervasive development of high fluid pressures (Walder and Nur, 1984).

Understanding the fluid pressure history in subduction zones has regional and global significance. Many aspects of the geology, including the shape of accretionary wedges (Davis et al., 1983), the development of thrust faulting (Hubbert and Rubey, 1959), fold vergence (Seely, 1977), heat transport (Reck, 1987; Vrolijk et al., 1988), the diagenetic and metamorphic history (Etheridge et al., 1983; Ritger et al., 1987), and the support of benthic organisms (Suess et al., 1985), are affected by high fluid pressures and resulting fluid flow. The global water and geochemical budgets also depend on the fluid pressure distribution in subduction zones because mass balance calculations suggest that water must return to the ocean through subduction complexes to prevent the world's oceans from being subducted into the mantle (e.g., Fyfe et al., 1978; Ito et al., 1983).

Fluid pressures appear to play a direct role in mineral diagenesis (e.g., Bird, 1984; Koster van Groos and Guggenheim, 1984, 1986, 1987) and therefore may in part determine the depth at which water is expelled from hydrous minerals. In the presence of high fluid pressure, montmorillonite retains interlayer water to much higher confining pressures than under hydrostatic conditions (Hall et al., 1986; Colten-Bradley, 1987; Koster van Groos and Guggenheim, 1987). Smectites in ocean crust and sediments subducted beneath accretionary prisms may lose water according to the fluid pressure distribution. Determining where water-rich minerals dehydrate in a subduction zone also has important implications for the genesis of arc magmatism (e.g., Gill, 1981).

EVIDENCE FOR HIGH FLUID PRESSURES

Investigations in modem subduction zone environments have led to the discovery of several tantalizing results. On DSDP Leg 78A in the Barbados Ridge Complex, Moore and Biju-Duval (1984) inferred high, near-lithostatic fluid pressures along the décollement zone, or detachment horizon, between the subducting Atlantic and overriding Caribbean plates. In the Japan trench, von Huene et al. (1980) attributed lower sediment density in highly fractured cores to fracture porosity and overpressuring. Similarly, wells in subduction zone forearc settings record fluid pressures far above hydrostatic (Shouldice, 1971; Hottman et al., 1979). The common occurrence of mud volcanoes and mud diapirs in accretionary complexes also suggests the presence of high fluid pressures (Shouldice, 1971; von Huene, 1972; Brown and Westbrook, 1987; Langseth et al., 1988). Whereas none of these studies have determined the distribution of fluid pressures in subduction zones, they strongly suggest that high fluid pressures may be common.

Indirect evidence for high fluid pressures in modem environments comes from pore water chemical and thermal data (e.g., Yamano et al., 1982; Davis and Hussong, 1984; Kulm et al., 1986; Moore et al., 1987; Ritger et al., 1987; Gieskes et al., 1989). Both types of data indicate anomalies along subduction zones, and further consideration of these data suggest that anomalies can be sustained only by relatively rapid fluid flow; which is likely aided by high fluid pressures.

Mechanical models of subduction zones also suggest that high fluid pressures are present (e.g., Davis et al., 1983; Shi and Wang, 1988). In addition, a conceptual model of subducting sediments suggests that fluids cannot easily escape along sedimentary layers toward the seafloor (Figure 10.1). In the model presented in Figure 10.1 a tectonically undisturbed sedimentary layer extends from in front of the deformation front far beneath the accretionary prism; this bed remains within the subducting plate throughout the model. The geometry of the subduction zone is modeled after the Eastern Aleutian trench (von Huene et al., 1985). During subduction a fluid pressure gradient will develop along the sedimentary bed because of the increased load of the accretionary prism, which will encourage fluid flow from beneath it. However, in a fixed reference frame (e.g., relative to the sea surface), fluid flow in the sedimentary bed must exceed the subduction rate to result in net migration toward the surface. For the parameters described in Figure 10.1, the sedimentary bed must be silt sized or coarser to permit Darcian flow. In settings where subduction is more rapid than 5 cm/yr (e.g., the northeast Pacific during the late Mesozoic; Engebretson et al., 1984), the permeability of subducted sediments must be correspondingly higher to allow fluids to effectively migrate toward the seafloor.

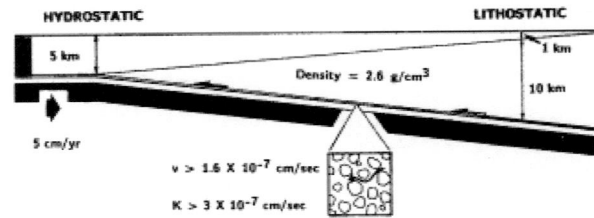

Figure 10.1
Simple model of hydrologic conditions in sediments subducted beneath an accretionary wedge. Half arrows indicate displacement along the décollement zone. Fluid pressure in the subducting sedimentary bed is hydrostatic at the deformation front and lithostatic beneath the décollement zone and accretionary prism. Assuming a water depth of 5 km at the deformation front and 1 km above the accretionary prism (given the modeled geometry), a 10-km-thick accretionary prism, and bulk density of 2.6 g/cm^3 within the prism (Bray and Karig, 1985), the hydraulic gradient along the sedimentary bed is 0.306. If the plate convergence rate is 5 cm/yr (i.e., 1.6×10^{-7} cm/s), sediments within the subducting plate must have hydraulic conductivities (K) $> 3 \times 10^{-7}$ cm/s for fluids to migrate via Darcian flow up-dip along sedimentary beds in a sea-level-based reference frame. Silt-sized sediments have K in this range (Freeze and Cherry, 1979). Because fine-grained sediments are common in subduction zone settings (Shepherd and Bryant, 1983), up-dip fluid escape along sedimentary strata may often be untenable. In these cases if fluid is to escape from the subducting plate, it may rise vertically, in which case it must ultimately intersect the décollement zone, indicating the fundamental importance of the décollement zone in the hydrogeology of subduction zones.

Because subducted sediments are mostly clay sized to silt sized with low intrinsic permeability, fluids will either be subducted into the mantle or must find an alternate path of escape. Any resistance to fluid escape will cause fluid pressures to rise further. Fluids may rise vertically until they reach the décollement zone and associated faults, which then may act as a fracture network for fluid flow. Alternatively, low-angle, bedding-parallel natural hydrofractures may develop, as hypothesized along the Barbados accretionary complex by Westbrook and Smith (1983).

Moore et al. (1987) calculated from measured hydrologic parameters that fluid could only flow up-dip along sandy sedimentary horizons of the subducting Atlantic plate along the ODP Leg 110 transect in the Barbados accretionary complex, an hypothesis that was supported by geochemical and geothermal observations. These same geochemical and thermal anomalies were also detected in the décollement zone, suggesting that the décollement zone and the sandstone beds have similar hydrologic roles but with fluid flow in the décollement zone controlled by the history of faulting.

The recognition of extensive networks of veins in ancient rocks also provides evidence for high fluid pressures (Cloos, 1984; Vrolijk, 1986, 1987). In ancient rocks syntectonic veins are interpreted as natural hydrofractures that formed under conditions of high fluid pressure and low effective stress (Secor, 1965).

EVIDENCE OF FLUID PRESSURE HISTORY FROM FLUID INCLUSION STUDIES

Examining ancient rocks is imperative in understanding small-scale processes at levels deeper than about 1 km, or the maximum depth of drilling in active margins. In contrast, remote sensing techniques, such as seismic reflection and electrical conductivity profiles, are useful for deciphering relatively large-scale variations in physical character. Fluid inclusion analyses in syntectonic veins have proven useful in examining the fluid pressure history in a variety of tectonic environments.

One reason fluid inclusions are valuable is that fluids reach thermodynamic equilibrium with each other far faster than solid phases. If fluid pressures in a deforming rock mass fluctuate, fluid inclusions trapped in continuously growing crystals may preserve some record of the fluid pressure history and thermochemical evolution. The kinetics of the growth of quartz crystals, a common vein-filling mineral, suggest that variations on the order of days may be preserved (Rimstidt and Barnes, 1980). In contrast, preservation of evidence of the fluid pressure history within solid phases would require creating compositional zoning patterns during mineral growth. Zoning must exist on a scale large enough to escape chemical diffusion and also be analytically identifiable, difficult criteria to fulfill when conditions change rapidly. Another useful feature of fluid inclusions is their common occurrence in veins that appear at all scales. Because veins can be mapped at the outcrop and microscopic scales, it is relatively easy to tie veins to the development of structural fabrics in the rock.

Complex Fluid Pressure History

An example of the fluid pressure history recorded by fluid inclusions in a subduction zone is presented by Vrolijk

(1987). In this study syntectonic veins formed during deformation in active fault zones were analyzed in rocks from the Kodiak accretionary complex, Alaska. A transect across half of each vein revealed that the density of methane in fluid inclusions varied dramatically during the growth of the veins (Figure 10.2). Vrolijk (1987) suggested that during the growth of each vein fluid pressures dropped from the near-lithostatic values present during initial crack opening to pressures as much as 45 percent lower (Figure 10.3). The fluid pressure history was interpreted by using the pressure-volume-temperature (P-V-T) characteristics of methane in conjunction with independent determinations of the quartz crystallization temperature, resolved by analysis of coeval water-rich fluid inclusions. Moreover, considering each fluid inclusion analysis as a single paleofluid pressure measurement, justified on the basis that fluid temperatures remained constant during crystal growth, suggests that during the growth of veins fluid pressures fluctuated widely, and fluid pressures close to initial, fracture-forming, near-lithostatic values appeared repeatedly. These successive high pressure pulses are interpreted to reflect further widening of the fracture (Figure 10.4).

One important observation drawn from this work is the inextricable link between the formation and growth of fractures and the fluid pressure history. This point is illustrated in Figure 10.4, in which an inferred record of fluid pressure fluctuations is plotted. The important parts of this diagram include the following: (1) the initial fluid pressure builds up to some value near lithostatic; although ancient rocks contain no record of this stage, it is inferred from theory (e.g., Walder and Nur, 1984). (2) The fluid pressure exceeds the least principal stress and the tensile strength of the rock (e.g., Secor, 1965; Etheridge et al., 1984), creating a fracture. When the fracture forms, new voids are formed, causing the fluid pressure to drop. Vrolijk (1987) suggested that this fluid pressure drop caused silica to become oversaturated in the fluid, leading to quartz precipitation. Once the fracture exists, it is reutilized as a fluid pathway and continues to accommodate local extensional strain within the rock. (3) Fluid pressure must repeatedly rise to widen the fracture, but with each increment of growth the fluid pressure drops, creating a cyclic fluid pressure history. (4) The fracture seals with a final pressure decrease. Vrolijk (1987) hypothesized that the gradual drop at the lowest fluid pressures represented growth of an interconnected fracture network along which fluids escaped to shallower levels of the subduction zone.

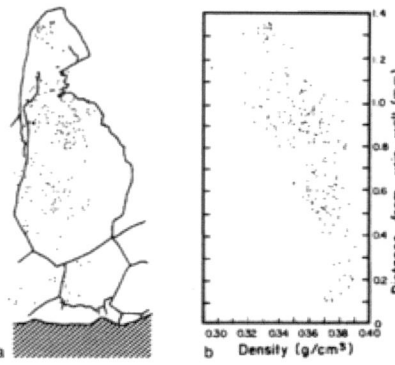

Figure 10.2
(a) Drawing of quartz crystal from syntectonic vein from a boudin formed during mélange deformation. Dots indicate distribution of fluid inclusions within quartz crystal. Cross-hatched area at bottom indicates sandstone of vein wall. (b) Methane densities of fluid inclusions plotted versus distance from vein wall, indicating drop in density with crystal growth (from Vrolijk, 1987).

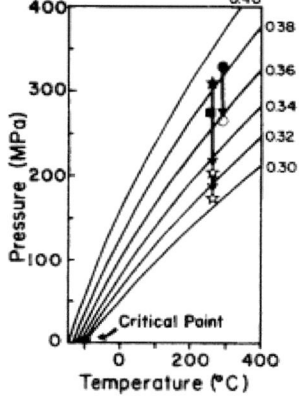

Figure 10.3
Methane densities (g/cm^3) in P-T space, plotted from Angus et al. (1976). Also plotted are high and low fluid pressures from two samples of the Ghost Rocks Formation (stars) and one sample of the Uyak Complex (circles), Kodiak accretionary complex, Alaska; note that high-pressure points for both Ghost Rocks samples fall on same P-T point. Solid symbols are interpreted as probable near-lithostatic fluid pressures, open symbols as lowest fluid pressure in each vein. Square point represents a later vein deposited in strike-slip fault of Ghost Rocks Formation, indicating that fluid pressure drop in boudin veins is not related to uplift (from Vrolijk, 1987).

The presence of repeated high fluid pressures in these rocks probably played an important role in determining the style of deformation, following the ideas of Hubbert and Rubey (1959). The ubiquitous presence of faults, fractures, and shear zones in all of the units on the Kodiak Islands (e.g., Moore and Wheeler, 1978; Moore and Allwardt, 1980; Byrne, 1984; Sample and Moore, 1987), in contrast to the relative rarity of folds larger than a single outcrop, suggests that high fluid pressures may have contributed to strain being accommodated most easily along fractures.

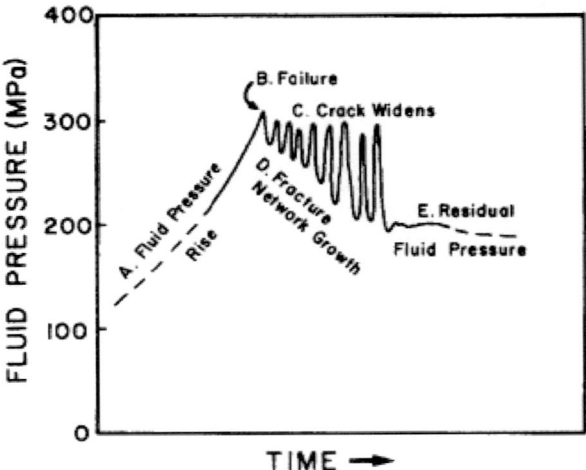

Figure 10.4
Interpretive fluid pressure evolution in extensional fracture. Fluid pressure values modeled after data presented in Figures 10.2 and 10.3. A: Deformation first creates fluid reservoir by dilating rock mass and increasing porosity, then causes fluid pressure to rise. B: Rock failure occurs along thrust faults in mélange matrix; where faults intersect boudins, extensional fractures develop. Fluid pressure at this point equals least principal stress plus tensile strength of rock. C: Fluid pressure drops to some value near least principal stress (interpreted as lithostatic pressure) as extensional fracture widens. D: Lowest fluid pressure decreases during each increment of crack growth as interconnected fracture network grows toward increasingly shallower levels. E: Fluid reservoir built up during initial dilatant deformation becomes exhausted, and local deformation wanes. Fluid pressure equilibrates along fracture network as last voids are sealed (from Vrolijk, 1987).

Consequences of High Fluid Pressure

Understanding the fluid pressure history of individual fractures and how the fluid inclusion record within veins reflects that history makes fluid inclusions useful for paleobarometry and paleothermometry. Studying fluid inclusions proves a useful analytical method because the inclusions provide a direct record of the fluid phase, and migrating fluids may often be the best medium for transporting heat and dissolved chemical constituents through the rock.

Vrolijk *et al.* (1988) and Myers (1987) used methane-rich and water-rich fluid inclusions in syntectonic veins from three units of the Kodiak accretionary complex, Alaska, to investigate the fluid temperature history of vein-forming fluids. The veins chosen for this study were interpreted to have formed during deformation of sediments within the décollement zone between the subducting oceanic and overriding North American plate. The principal conclusion of this study is that fluid temperatures within the décollement zone (Figure 10.5) were substantially higher than temperatures predicted by conductive heat flow models (e.g., Oxburgh and Turcotte, 1971; Ernst, 1974; Wang and Shi, 1984).

Warm fluids in fault zones of the Kodiak accretionary complex were suggested by Vrolijk *et al.* (1988) to arise from the migration of fluids along faults faster than heat dissipated from the fluid, although the presence of young ocean crust during the formation of these units could not be completely ruled out as a significant heat source. Other potential heat sources, such as intruding magmas, enhanced radioactive decay, and frictional heating, can be ruled out by field observations. Plate reconstructions (Wells *et al.*, 1984; Engebretson *et al.*, 1984) suggest that low thermal gradients (e.g., 5° to 10°C/km) should have been produced along the Kodiak margin because plate convergence was fast. The hypothesis put forth by Vrolijk *et al.* (1988) suggests that deformation within the décollement zone allowed faults and fractures to open and permitted fluids to migrate structurally up-dip. Fluid migration was sufficiently rapid that heat advection outpaced conduction into the surrounding rock (Figure 10.6).

The décollement zone in this example appears to have focused fluid flow along its surface. Because of the décollement zone's apparent hydrologic importance, the term *tectonic aquifer* is introduced in Figure 10.6. *Aquifer* is used to highlight the observations that fluid flow is enhanced along the décollement zone, and the modifier *tectonic* signifies the role that deformation along the décollement zone plays in increasing permeability, thereby allowing the décollement zone to support enhanced fluid flow.

The implications of the hypothesis of warm fluid flow along fault zones touch on the metamorphic history of subduction zones. If fluid flow is short lived and episodic, the isotherms in subduction zones may have complicated, temporally variable shapes controlled by the growth of faults and fractures (Figure 10.7). On the other hand, fluid flow may be pervasive and persistent, generating warmer temperatures throughout the accretionary complex (Figure 10.7). Fully resolving this problem will require information from (1) further studies of the metamorphic history of subducted and accreted materials; (2) studies and models of the amount of water present in subducting plates, where that water is released during subduction, and how the fluid migrates within the subduction zone; and (3) coupled hydrologic/thermal models that more realistically incorporate fluid migration mechanisms than has so far been attempted.

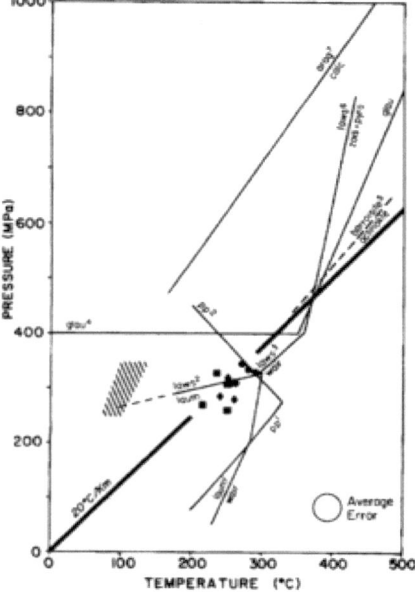

Figure 10.5
Rock P-fluid T measurements from veins formed during mélange deformation, Ghost Rocks Formation (Diamonds), Kodiak Formation (Squares), and Uyak Complex (Circles), Kodiak accretionary complex, Alaska. Plotted for comparison are a line describing a thermal gradient of 20°C/km (as compared to gradients $\geq$ 7 C/km from model calculations, e.g., Wang and Shi, 1984), assuming a sediment bulk density of 2500 kg/m^3 and a corresponding pressure-depth ratio of 1 kbar/4 km (Bray and Karig, 1985) and the lower stability limits of typical blueschistfacies minerals. Notice that if the Kodiak samples had been subducted more deeply, they may not have developed the expected blueschist mineralogy, even though they formed in the décollement zone of a subduction zone. Reactions (1) laumontite $\rightleftarrows$ wairakite + fluid, (2) laumontite $\rightleftarrows$ lawsonite + quartz + fluid, and (3) wairakite $\rightleftarrows$ lawsonite + quartz are from Liou (1971). The glaucophane stability boundary (4) is plotted from Maresch (1977). The boundary between barroisitic and actinolitic amphiboles (5) is drawn from Ernst (1979). Nitsch (1972) defined reaction (6), lawsonite + quartz $\rightleftarrows$ zoisite + pyrophyllite + water, and the calcite $\rightleftarrows$ aragonite inversion (7) follows Johannes and Puhan (1971). Boundary pp^1 marks the transition from prehnite-pumpellyite facies (low temperature) to prehnite-actinolite facies (pumpellyite + quartz $\rightleftarrows$ zoisite + prehnite + chlorite + fluid), and pp^2 limits the prehnite-pumpellyite facies to the low temperature side and pumpellyite-actinolite facies to the high temperature side (prehnite + chlorite + quartz $\rightleftarrows$ pumpellyite + tremolite + fluid); both reactions are for the model metabasite system of Liou et al. (1985) (from Vrolijk et al., 1988).

In addition to thermal anomalies, chemical anomalies also appear to be associated with fault zones. In the Barbados Ridge complex, Moore et al. (1987) described the importance of the décollement zone in focusing fluid flow and in separating chemically defined hydrogeologic regimes. In the ancient Kodiak accretionary complex, Vrolijk (1986, 1987) distinguished veins formed in mélanges from veins formed in structurally coherent units by comparing oxygen isotope ratios of vein-forming fluids. Both examples suggest that fluids in fault zones migrated from deeper structural levels in the subduction zone along active faults, principally the décollement zone. Further geochemical analyses of pore waters and vein-filling minerals will serve to trace the origin and paths of fluids, thereby more closely constraining the fluid migration history.

Similar Fluid Inclusion Studies

Fluid inclusion studies have proven useful in unraveling tectonic and fluid histories in a number of regions and in various tectonic settings. In the western Alps, Mullis (1976, 1979) pioneered the use of methane plus water fluid inclusions in tectonic studies. Large quartz crystals growing into cavities during Alpine tectonism trapped fluid inclusions; these inclusions, like the Kodiak samples, similarly record fluctuating fluid pressures. Mullis (1976) interpreted these changes to have occurred on a longer time scale than the Kodiak veins, and he correlated quartz crystal growth stages and changing P-T conditions with successive nappe stacking. Recently, however, Mullis (1988) interpreted fluid pressure fluctuations recorded in samples from the Apennines, Italy, as a record of the expansion of crystal-filled cavities.

In another collisional tectonic setting, Orkan and Voight (1985) used methane-rich and water-rich fluid inclusions to determine P-T conditions of joint formation during the Alleghany Orogeny in the Valley and Ridge Province of Pennsylvania. This study systematically combined detailed structural and kinematic data from joints and other structural features with precise P-T measurements. Such data will help clarify the sorts of problems regarding the evolution of joints referred to by Engelder (Chapter 9, this volume).

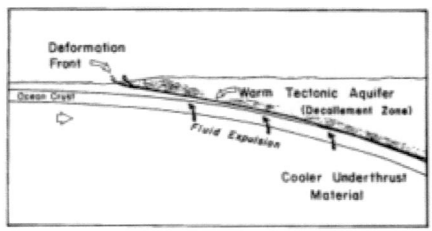

Figure 10.6
Heat redistribution by warm fluid migration in a subduction zone. Fluid is liberated from subducted oceanic crust and sediments and escapes laterally back toward the surface along the décollement zone. Because subduction acts as a con-veyor belt to drag water to depth, and because the décollement zone is an areally limited structural zone, fluid flow through the décollement zone may be relatively rapid. Fluid escape thereby serves to warm active fault zones at intermediate to shallow levels. However, because no in situ physical or chemical process is generating significant quantities of heat, the warming effect at shallow levels must be balanced by a corresponding loss of heat and minor cooling at deeper levels where fluids originate.

Fluid inclusions have also been profitably used in extensional tectonic environments by Parry and Bruhn (1986). By examining CO_2 + H_2O fluid inclusions in rocks exposed in the footwall of the Wasatch normal fault in north-em Utah, Parry and Bruhn (1986) suggested that fluid pressures along the fault shifted from near-lithostatic to near-hydrostatic as the footwall rocks were uplifted through the brittle/ductile transition. In a related study, Parry and Bruhn (1987) used the same CO_2-H_2O-NaCl fluid inclusion system to determine that rocks now at the surface were once 11 km deep, thereby constraining the amount of offset along the fault.

Future Prospects for Fluid Inclusion Research

Methane and water are increasingly being recognized as the most important fluid constituents in subduction zones (Ernst, 1972; Magaritz and Taylor, 1976; Cloos, 1984; Moore et al., 1987; Vrolijk, 1987). In reconnaissance examinations of syntectonic veins from accretionary complexes in Japan, Washington, and Papua New Guinea, methane-rich and water-rich fluid inclusions have been recognized, suggesting that analysis of fluid inclusions may prove useful in accretionary complexes around the world.

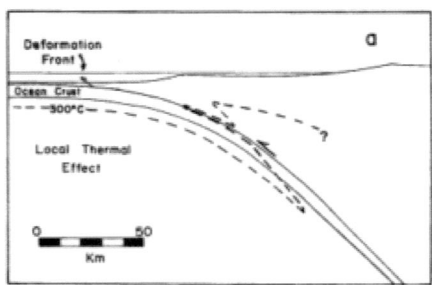

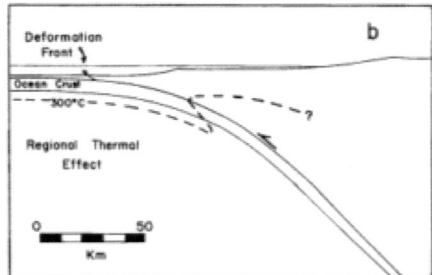

Figure 10.7
Extent of heating due to fluid escape in a subduction zone. Half arrow indicates displacement along the décollement zone. A: Local thermal effect. The gross temperature structure of subduction zones can be described by conductive heat flow models (the 300°C isotherm drawn here is taken from Ernst, 1970), but fluid escape along the décollement zone heats only rock immediately adjacent the fault zone (i.e., heat conduction out of the décollement zone is minimal). This thermal configuration only occurs periodically during the life of a subduction zone and results from episodic fluid flow. B: Extensive regional thermal effect. Fluids migrating along the décollement zone have a profound effect on temperatures in the subduction zone, and isotherms are only mildly depressed as conduction of heat from the décollement keeps pace with heat conducted into the subducting plate. In this case fluid flow is more continuous than in (A), and there is greater net heat flux upward by fluid flow. The true thermal structure of subduction zones where fluid escape is an important component probably lies somewhere between these two end-members. Temperatures may be somewhat higher along the décollement zone, but heat conduction probably smoothes out this anomaly (from Vrolijk et al., 1988).

CONCLUSIONS

Subduction zones may represent the tectonic environment most strongly influenced by high fluid pressures. Deformation, diagenesis, and metamorphism of subducted water-rich sediments and rocks leads to increasing fluid pressures because subduction effectively "buries" materials faster than fluid can escape through intergranular pore spaces. Evidence from modem and ancient subduction zone settings is beginning to uncover details of the fluid pressure history and the chemical and thermal evolution of fluids. However, this research remains in its infancy. More information regarding the physical properties of sediments is required to better understand fluid flow and its effect on how sediments deform. Research into the source, path, and migration history of fluids is required to better understand the gross hydrogeology of subduction zones and to determine how the fluid pressure history is intertwined with hydrogeology.

Ancient rocks preserve a record of phenomena occurring at depths too great to be directly sampled from the surface. Fluid inclusions in syntectonic veins offer a means to examine processes occurring on a relatively short time scale, much shorter than those that can be studied by most other techniques. Moreover, several studies suggest that the fluid inclusion method may be more accurate than other techniques for determining P-T at incipient metamorphic conditions. Fluid inclusion research may be broadly applied to a host of different tectonic problems around the world.

ACKNOWLEDGMENTS

The research reported here was supported by National Science Foundation grants EAR 84-07720 and EAR 86-08337 to J. C. Moore; by ARCO, Union, Sohio, and Mobil oil companies; and by the U.S. Geological Survey. Many ideas discussed here evolved from years of discussions with J. C. Moore and J. C. Sample, although the authors bear full responsibility for the paper's contents. W. G. Ernst, H. Gibbons, and M. Reid kindly provided careful and helpful reviews and comments.

References

Angus, S., B. Armstrong, and K. M. de Reuck (1976). *International Thermodynamic Tables of the Fluid State—5 : Methane*, International Union of Pure and Applied Chemistry, Chemical Data Series, No. 16, Pergamon Press, New York, 247 pp.

Bird, P. (1984). Hydration-phase diagrams and friction of montmorillonite under laboratory and geologic conditions, with implications for shale compaction, slope stability, and strength of fault gouge, *Tectonophysics 107*, 235-260.

Bray, C. J., and D. E. Karig (1985). Porosity of sediments in accretionary prisms and some implications for dewatering processes, *Journal of Geophysical Research 90*, 768-778. Brown, K. M., and G. K. Westbrook (1987). The tectonic fabric of the Barbados Ridge accretionary complex, *Marine Petroleum Geology 4*, 71-81.

Byrne, T. (1984). Early deformation in mélange terranes of the Ghost Rocks Formation, Kodiak Islands, Alaska, in *Mélanges: Their Nature, Origin, and Significance*, L. A. Raymond, ed., Special Paper 198, Geological Society of America, Boulder, Colo., pp. 21-52.

Cloos, M. (1984). Landward-dipping reflectors in accretionary wedges: Active dewatering conduits? *Geology 12*, 519-522.

Colten-Bradley, V. A. (1987). Role of pressure in smectite dehydration: Effects on geopressure and smectite-to-illite transformation, *American Association of Petroleum Geologists Bulletin 71*, 1414-1427.

Davis, D. M., and D. M. Hussong (1984). Geothermal observations during DSDP Leg 78A, in *Initial Reports, Deep-Sea Drilling Project 78A*, B. Biju-Duval and J. C. Moore et al., eds., U.S. Government Printing Office, Washington, D.C., pp. 593-598.

Davis, D., J. Suppe, and F. Z. Dahlen (1983). The mechanics of fold-and-thrust belts, *Journal of Geophysical Research 88*, 1153-1172.

Engebretson, D. C., A. Cox, and R. G. Gordon (1984). Relative motions between oceanic plates of the Pacific Basin , *Journal of Geophysical Research 89*, 10,291-10,310.

Ernst, W. G. (1970). Tectonic contact between the Franciscan mélange and the Great Valley sequence, crustal expression of a Late Mesozoic Benioff Zone, *Journal of Geophysical Research 75*, 886-902.

Ernst, W. G. (1972). CO_2-poor composition of the fluid attending Franciscan and Sanbagawa low-grade metamorphism, *Geochimica et Cosmochimica Acta 36*, 497-504.

Ernst, W. G. (1974). Metamorphism and ancient continental margins, in *The Geology of Continental Margins*, C. A. Burk and C. L. Drake, eds., Springer-Verlag, New York, pp. 907-919.

Ernst, W. G. (1979). Coexisting sodic and calcic amphiboles from high-pressure metamorphic belts and the stability of barroisitic amphibole, *Mineralogical Magazine 43*, 269-278.

Etheridge, M. A., V. J. Wall, and R. H. Vernon (1983). The role of the fluid phase during regional metamorphism and deformation, *Journal of Metamorphic Geology 1*, 205-226.

Etheridge, M. A., V. J. Wall, and S. F. Cox (1984). High fluid pressures during regional metamorphism and deformation: Implications for mass transport and deformation mechanisms, *Journal of Geophysical Research 89*, 4344-4358.

Freeze, R. A., and J. A. Cherry (1979). *Groundwater*, Prentice-Hall Inc., Englewood Cliffs, N.J., 604 pp.

Fyfe, W. S., N. J. Price, and A. B. Thompson (1978). *Fluids in the Earth's Crust*, Developments in Geochemistry 1, Elsevier Scientific Publishing Co., New York, 383 pp.

Gieskes, J., G. Blanc, P. Vrolijk, J. C. Moore, A. Mascle, E. Taylor, P. Andreiff, F. Alvarez, R. Barnes, C. Beck, J. Behrmann, K. Brown, M. Clark, J. Dolan, A. Fisher, M.

Hounslow, P. McLellan, K. Moran, Y. Ogawa, T. Sakai, J. Schoonmaker, R. Wilkens, C. Williams, 1989. Hydrogeochemistry in the Barbados accretionary complex: Leg 110 ODP, *Palaeogeography, Palaeoclimatology, Palaeoecology 71*, 83-96.

Gill, J. (1981). *Orogenic Andesites and Plate Tectonics*, Springer-Verlag, New York, 400 pp.

Hall, P. L., D. M. Astill, and J. D.C. McConnell (1986). Thermodynamics and structural aspects of the dehydration of smectites in sedimentary rocks, *Clay Minerals 21*, 633-648.

Hottman, C. E., J. H. Smith, and W. R. Purcell (1979). Relationships among Earth stresses, pore pressure, and drilling problems, offshore Gulf of Alaska, *Journal of Petroleum Technology 31*, 1477-1484

Hubbert, M. K., and W. W. Rubey (1959). Role of fluid pressure in the mechanics of overthrust faulting, I: Mechanics of fluid-filled porous solids and its application to overthrust faulting, *Geological Society of America Bulletin 70*, 115-166.

Ito, E., D. M. Harris, and A. T. Anderson (1983). Alteration of oceanic crust and geologic cycling of chlorine and water, *Geochimica et Cosmochimica Acta 47*, 1613-1624.

Johannes, W., and D. Puhan (1971). The aragonite-calcite transition, reinvestigated, *Contributions to Mineralogy and Petrology 31*, 28-38.

Koster van Groos, A. F., and S. Guggenheim (1984). The effect of pressure on the dehydration reaction of interlayered water in Na-montmorillonite, *American Mineralogist 69*, 872-879.

Koster van Groos, A. F, and S. Guggenheim (1986). Dehydration of K-exchanged montmorillonite at elevated temperature and pressures, *Clays and Clay Minerals 34*, 281-286.

Koster van Groos, A. F., and S. Guggenheim (1987). Dehydration of a Ca-and a Mg-exchanged montmorillonite (SWy-1) at elevated pressures, *American Mineralogist 72*, 292-298.

Kulm, L. D., E. Suess, J. C. Moore, B. Carson, B. T. Lewis, S. D. Ritger, D. C. Kadko, T. M. Thornberg, R. W. Embley, W. D. Rugh, G. J. Mossoth, M. G. Langseth, G. R. Cochran, and R. L. Scamman (1986). Oregon subduction zone: Venting, fauna, and carbonates, *Science 231*, 561-566.

Langseth, M. G., G. K. Westbrook, and M. A. Hobart (1988). Geophysical survey of a mud volcano seaward of the Barbados Ridge accretionary complex, *Journal of Geophysical Research 93*, 1049-1061

Liou, J. G. (1971). P-T stabilities of laumontite, waikarite, lawsonite, and related minerals in the system $CaAl_2Si_2O_8$-SiO_2-H_2O, *Journal of Petrology 12*, 379-411.

Liou, J. G., S. Maruyama, and M. Cho (1985). Phase equilibria and mineral parageneses of metabasites in low-grade metamorphism, *Mineralogical Magazine 49*, 321-333.

Magaritz, M., and H. P. Taylor, Jr. (1976). Oxygen, hydrogen, and carbon isotope studies of the Franciscan formation, Coast Ranges, California, *Geochimica et Cosmochimica Acta 40*, 215-234.

Maresch, W. V. (1977). Experimental studies on glaucophane: An analysis of present knowledge, *Tectonophysics 43*, 109-125.

Moore, J. C. (1975). Selective subduction, *Geology 3*, 530-532.

Moore, J. C., and A. Allwardt (1980). Progressive deformation of a Tertiary trench slope, Kodiak Islands, Alaska, *Journal of Geophysical Research 85*, 5741-4756.

Moore, J. C., and B. Biju-Duval (1984). Tectonic synthesis Deep Sea Drilling Project Leg 78A: Structural evolution of offscraped and underthrust sediment, northern Barbados Ridge complex , in *Initial Reports, Dea-Sea Drilling Project 78A*, B. Biju-Duval, J. C. Moore et al., eds., U.S. Government Printing Office, Washington, D.C., pp. 601-621.

Moore, J. C., and R. W. Wheeler (1978). Structural fabric of a mélange, Kodiak Islands, Alaska, *American Journal of Science 278*, 739-765.

Moore, J. C., A. Mascle, E. Taylor, P. Andreiff, F. Alvarez, R. Barnes, C. Beck, J. Behrmann, G. Blanc, K. Brown, M. Clark, J. Dolan, A. Fisher, J. Gieskes, M. Hounslow, P. McLellan, K. Moran, Y. Ogawa, T. Sakai, J. Schoonmaker, P. Vrolijk, R. Wilkens, and C. Williams (1987). Expulsion of fluids from depth along a subduction-zone décollement horizon, *Nature 326*, 785-788.

Mullis, J. (1976). Das Wachstumsmilieu der Quarzkristalle im Val d'Illiez (Wallis, Schweiz), *Schweiz. Mineral. Petrogr. Mitt. 56*, 219-268.

Mullis, J. (1979). The system methane-water as a geologic thermometer and barometer from the external part of the Central Alps, *Societe Francaise de Mineralogle et de Cristallographie, Bulletin 102*, 526-536.

Mullis, J. (1988). Rapid subsidence and upthrusting in the North-em Appenines deduced by fluid inclusions studies in quartz crystals from Poretta Terme, Schweiz, Mineral. Petrogr. Mitt. 68, 157-170.

Myers, G. (1987). Fluid expulsion during the underplating of the Kodiak Formation: A fluid inclusion study, M.S. thesis, University of California, Santa Cruz, 41 pp.

Nitsch, K.-H. (1972). Das P-T-XCO_2-stabilitaetsfeld von lawsonit, *Contributions to Mineralogy and Petrology 34*, 116-134.

Orkan, N., and B. Voight (1985). Regional joint evolution in the Valley and Ridge Province of Pennsylvania in relation to the Alleghany Orogeny, in *Guidebook for the 50th Annual Field Conference of Pennsylvania Geologists*, Bureau of Topographic and Geological Survey, Harrisburg, Pa., pp. 144-164.

Oxburgh, E. R., and D. L. Turcotte (1971). Origin of paired metamorphic belts and crustal dilation in island arc regions, *Journal of Geophysical Research 76*, 1315-1327.

Parry, W. T., and R. L. Bruhn (1986). Pore fluid and seismogenic characteristics of fault rock at depth on the Wasatch Fault, Utah, *Journal of Geophysical Research 91*, 730-744.

Reck, B. H. (1987). Implications of measured thermal gradients for water movement through the northeast Japan accretionary prism, *Journal of Geophysical Research 92*, 3683-3690.

Rimstidt, J. D., and H. L. Barnes (1980). The kinetics of silica-water reactions, *Geochimica et Cosmochimica Acta 44*, 1683-1699.

Ritger, S., B. Carson, and E. Suess (1987). Methane-derived authigenic carbonates formed by subduction-induced pore-water expulsion along the Oregon/Washington margin, *Geological Society of America Bulletin 98*, 147-156.

Sample, J. C., and J. C. Moore (1987). Structural style and kinematics of an underplated slate belt, Kodiak Islands, Alaska, *Geological Society of America Bulletin 99*, 7-20.

Secor, D. T. (1965). Role of fluid pressure in jointing, *American Journal of Science 263*, 633-646.

Seely, D. R. (1977). The significance of landward vergence and oblique structural trends on trench inner slopes, in *Island Arcs, Deep Sea Trenches, and Back-Arc Basins*, M. Talwani and S. C. Pitmann, eds., Maurice Ewing Series 1, American Geophysical Union, Washington, D.C., pp. 187-198.

Shepherd, L. E., and W. R. Bryant (1983). Geotechnical properties of lower trench inner-slope sediments, *Tectonophysics 99*, 279-312.

Shi, Y., and C.-Y. Wang (1988). Generation of high pore pressures in accretionary prisms: Inferences from the Barbados subduction complex, *Journal of Geophysical Research 93*, 8893-8910.

Shouldice, D. H. (1971). Geology of the western Canadian continental shelf, *Canadian Petroleum Geology Bulletin 19*, 405-436.

Suess, E., B. Carson, S. Ritger, J. C. Moore, M. Jones, L. D. Kulm, and G. Cochran (1985). Biological communities at vent sites along the subduction zones off Oregon, *Bulletin of the Biological Society of Washington 6* (special issue, The Hydrothermal Vents of the Eastern Pacific: An Overview, M. L. Jones, ed.), 475-484.

von Huene, R. (1972). Structure of the continental margin and tectonism at the Eastern Aleutian Trench, *Geological Society of America Bulletin 83*, 3613-3626.

von Huene, R., M. Langseth, N. Nasu, and H. Okada (1980). Summary, Japan Trench Transect, in *Initial Reports, Deep-Sea Drilling Project 56 and 57* (part 1), M. Lee and L. Stout, eds., U.S. Government Printing Office, Washington, D.C., pp. 473-488.

von Huene, R., S. Box, R. Detterman, M. Fisher, J. C. Moore, and H. Pulpan (1985). *A-2 Kodiak to Kuskokwin, Alaska*, Continent/Ocean Transect, vol. 6, Geological Society of America, Boulder, Colo., 1 sheet, scale 1:500,000.

Vrolijk, P. J. (1986). Channelized fluid flow along mélanges of the Ghost Rocks Fm., Kodiak accretionary complex, Alaska (abs.), *EOS 67*, 1205.

Vrolijk, P. J. (1987). Paleohydrogeology and fluid evolution of the Kodiak accretionary complex, Alaska, Ph.D. thesis, University of California, Santa Cruz, 232 pp.

Vrolijk, P., G. Myers, and J. C. Moore (1988). Warm fluid migration along tectonic mélanges in the Kodiak accretionary complex, Alaska, *Journal of Geophysical Research 93*, 10,313-10,324.

Walder, J., and A. Nur (1984). Porosity reduction and crustal pore pressure development, *Journal of Geophysical Research 89*, 11539-11548.

Wang, C.-Y., and Y.-L. Shi (1984). On the thermal structure of subduction complexes: A preliminary study, *Journal of Geophysical Research 89*, 7709-7718.

Wells, R. E., D. C. Engebretson, P. D. Snavely, Jr., and R. S. Coe (1984). Cenozoic plate motions and the volcano-tectonic evolution of western Oregon and Washington, *Tectonics 3*, 275-294.

Westbrook, G. K., and M. J. Smith (1983). Long décollements and mud volcanoes: Evidence from the Barbados Ridge Complex for the role of high pore fluid pressure in the development of an accretionary complex, *Geology 11*, 279-283.

Yamano, M., S. Uyeda, Y. Aoki, and T. H. Shipley (1982). Estimates of heat flow derived from gas hydrates, *Geology 10*, 339-343.

11

Degassing of Carbon Dioxide As an Possible Source of High Pore Pressures in the Crust

JOHN D. BREDEHOEFT and STEVEN E. INGEBRITSEN
U.S. Geological Survey, Menlo Park

INTRODUCTION

Increased pore pressures, especially pore pressures approaching lithostatic loads, change the state of effective stress and greatly reduce the work necessary for tectonic deformation (Hubbert and Rubey, 1959; Rubey and Hubbert, 1959). A number of mechanisms have been proposed that increase pore pressures (Hanshaw and Zen, 1965). One of the more interesting of the proposed mechanisms is the movement of carbon dioxide (CO_2) through the crust, suggested by Irwin and Barnes (Irwin and Barnes, 1975; Barnes *et al.*, 1978, 1984). The purpose of this chapter is to investigate the possible role of CO_2 as a source of high pore pressure by examining the following question: Given the current best estimate of the rate of CO_2 degassing, how low would the permeability have to be in order to generate pore pressures approaching lithostatic values?

Barnes *et al.* (1978, 1984) compiled worldwide data on the distribution of CO_2 discharge from the crust. They showed that most of the discharges are concentrated in two areas of the world: (1) a narrow circum-Pacific belt and (2) a broad mountainous area that extends across central and southern Europe and Asia Minor. They went on to note that the CO_2 discharge areas coincide with areas that are seismically active at the present time. Their clearest statement of the role of CO_2 in tectonic processes was presented in Irwin and Barnes (1975).

There is a continuing debate about the nature of volatiles in the mantle. Gold (see, e.g., Gold and Soter, 1980) argues that the carbon in the mantle is present largely as methane. On the other hand, most of the geological community reports the observations of carbon emanating from the mantle as CO_2 (see, e.g., Leavitt, 1982; Gerlach, 1988). The current consensus of the geological community seems to be that CO_2 is the more likely form of volatile carbon in the mantle. For the purpose of our simple experiment we have assumed a source of free CO_2 either in the mantle or within the crust.

Carbon dioxide is thought to come from three different sources: (1) organic material, (2) metamorphism of marine carbonate rock, and (3) degassing of the mantle (Barnes *et al.*, 1984). Each source is thought to have a different ratio of carbon isotopes. The evidence for the isotopic composition of mantle-derived CO_2 comes from analyses of fluid inclusions from volcanic rocks erupted along oceanic spreading ridges. Moore *et al.* (1977) reported $\delta^{13}C$ values ranging from -4.7 to -5.8 for CO_2 inclusions from basalts in the Pacific. Pineau *et al.* (1976) found similar values ($\delta^{13}C$ of -7.6 ± 0.5 ‰) for fluid inclusions in tholeiitic rocks from the mid-Atlantic ridge. Carbon dioxide with $\delta^{13}C$ in the range of -4.7 to -8.0 is thought to indicate mantle-derived CO_2, although the carbon isotope ratios alone do not provide an unambiguous indication of a mantle derivation. $\delta^{13}C$ in the range of -4.7 to -8.0 can also be derived by mixing an organic CO_2 source, typically -20 ‰, with a marine carbonate CO_2 source, typically 0 ‰. However, in the case of the oceanic basalts a mantle source seems to be indicated as carbon isotope data is combined with other geologic information in order to interpret the source of the CO_2.

Magmatic degassing may be a major source of free CO_2 within the crust. Harris (1981) showed that the solubility of CO_2 in tholeiitic basalts was strongly pressure dependent; thus, CO_2 that is in solution in magma at great depth will exsolve as the magma migrates upward in the crust and the pressure is reduced. Gerlach (1986, 1988) suggested that most of the CO_2 is degassed from plutons at depths of 5 to 10 or 12 km. Basaltic magmas associated with hot spots, such as Kilauea, have a higher CO_2 content than mid-ocean ridge basalts (Gerlach, 1988), so that they will tend to degas at higher pressures and greater depths.

GLOBAL FLUX OF CO2

To estimate the permeability necessary to maintain near-lithostatic pore pressure, one must first estimate the flux of CO_2. A number of investigators have made estimates of the CO_2 flux from deep in the Earth's interior to the surface. Leavitt (1982) summarized earlier estimates; we have extended Leavitt's summary to include more recent data. The various estimates are summarized in Table 11.1. They range from 1.7×10^{10} to 2×10^{13} moles of CO_2 per year; most estimates are in the range of 10^{12} moles per year. The scatter in the estimates is not as wide as one might expect given the nature of the data.

If one assumes that the ultimate source of all carbon in the Earth is the mantle, one can place the current estimates of CO_2 flux in perspective by calculating a constant flux rate that would generate the known carbon reservoir. Current estimates of the global carbon reservoir are summarized by Sundquist (1985). One of these estimates (perhaps the best) is presented in Table 11.2. Assuming a constant rate of outgassing over 4.5 billion years yields a rate of outgassing of 2×10^{12} moles of CO_2 per year. These are grossly simple assumptions; however, the calculation suggests that the current estimated rate of CO_2 outgassing would approximately account for the global carbon reservoir. Other workers (e.g., Marty and Jambon, 1987; Gerlach, 1988) have also commented that the current rate is sufficient to generate the total carbon reservoir.

There are obvious complications, for example, differing rates of degassing in the geologic past and recycling of carbonate rocks in subduction zones. Des Marais (1985) noted that a rate of 10×10^{12} moles of CO_2 per year from the mantle would generate the global carbon inventory in approximately 700 million years and went on to point out the subduction tends to recycle carbon from the crust into the mantle. He estimates that perhaps half of the current flux is recycled carbon.

MODELING CO2 FLUX IN THE CRUST

We have investigated the effects of a CO_2 flux at mid-crustal depths utilizing a numerical simulation model. The model simulates the simultaneous transport of mass and

TABLE 11.1 Reported CO2 Inputs into the Atmosphere from the Earth's Interior

Reference	CO_2 Source	CO_2 Released (10^{12} moles/yr)
Borchert (1951)	Igneous and metamorphic	6.7
	Igneous only	1.3
Rubey (1951)	Total "excess CO_2" since Earth's origin, including that from hot springs	0.5
Plass (1956)	CO_2 "released from the interior of the Earth"	2
Li (1972)	Total CO_2 at Earth's surface	1.0
Libby and Libby (1972)	Volcanic CO_2	0.017
Buddemeier and Puccetti (1974)	Hawaiian estimate	2.0
Anderson (1975)	Outgassing of oceanic crust	0.23
Baes et al. (1976)	Volcanoes, fumaroles, and hot springs	1.7 to 8.3
Leavitt (1982)	Volcanic eruptions	0.15
Javoy et al. (1982)	Mid-ocean ridges	20
Des Marais (1985)	Mid-ocean ridges	1 to 8
Marty and Jambon (1987)	C/^{3}He, mid-ocean ridges	2.0
Gerlach (1988)	Mid-ocean ridges	0.3

heat by solving the appropriate set of coupled partial differential equations. It was originally developed to simulate geothermal reservoirs (Bodvarsson, 1982); we modified it to describe the properties of CO_2 rather than H_2O.

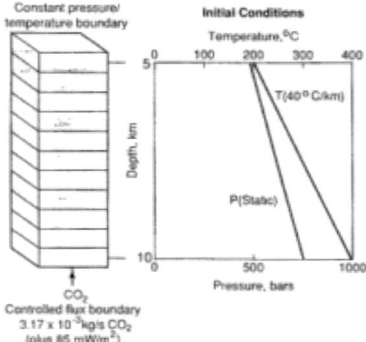

Figure 11.1
Schematic diagram of the one-dimensional model used to simulate CO_2 movement through the crust. Graph shows initial temperature and pressure profiles. The CO_2 flux at the lower boundary is derived by assuming that the global flux of CO_2 (about 3×10^{12} moles per year) is distributed over 1 percent of the Earth's surface area (1.3×10^6 km^2). The heat flow is typical of tectonically active areas.

In our particular case we assumed one-dimensional flow vertically upward through a prism of crust, and we specify at the bottom of the prism what we believe to be a reasonable flux of CO_2 and heat. We then adjusted the permeability in the course of a set of numerical experiments. A schematic diagram of the model is shown in Figure 11.1. For the purposes of calculation we assumed that the pore fluid is pure CO_2. The fluid properties of CO_2 were obtained from Kennedy and Holser (1966), Jacobs and Kerrick (1981), Vargaftik (1975), and Atkins (1978). The published density and viscosity data had to be extrapolated to higher pressures. Over the pressure and temperature ranges considered, CO_2 appears to have transport properties that are quite similar to water (Figure 11.2).

We assumed as the lower boundary condition a constant influx of CO_2 and heat. One can hypothesize other conditions at this lower boundary; however, for an initial calculation, constant flux provides insight. We have taken the flux as 3×10^{12} moles of CO_2 per year, which seems well within the estimates given in Table 11.2. This flux is distributed over the tectonically active area, which we have taken as 1 percent of the Earth's surface; it is then something of an average and may be substantially higher or lower locally.

Given this set of simple assumptions, we made a series of calculations to determine how low permeability would have to be in order to cause pore pressures approaching lithostatic conditions by simple permeation (i.e., without invoking enhanced transport or focusing effects). The results are summarized in Table 11.3. Since the model is run in a transient mode, the results are presented in terms of the time to reach lithostatic fluid pressure at 10 km depth, the lowermost cell of our simulated column of crust. At a permeability of 10^{-7} darcies, pore pressures do not approach lithostatic at steady state (infinite time). Given our assumed flux, the permeability must be on the order of 10^{-8} darcies or lower to generate pore pressures near lithostatic. Figure 11.3 shows pressure profiles at various times as the pressure builds to lithostatic for a case in which permeability is 10^{-9} darcies. In Figures 11.3 through 11.6, k is permeability, C is rock compressibility, and   is porosity. Typical ranges of rock permeability are given by Brace (1980) and are discussed below (see Figures 11.7 and 11.8).

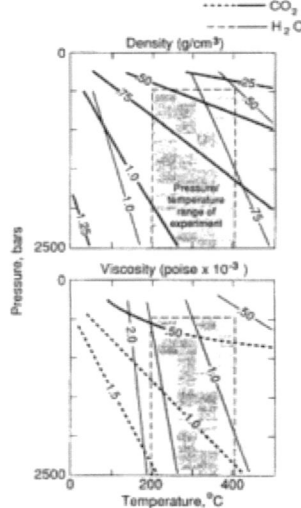

Figure 11.2
Density and viscosity of CO_2 and H_2O as functions of pressure and temperature. CO_2 properties (bold contours) are dashed where extrapolated from published data. H_2O properties are from Burnham et al. (1969), Keenan et al. (1969), and Haar et al. (1984). The pressure-temperature range of our experiment is shown by shading.

TABLE 11.2 Global Carbon Reservoir (from Sundquist, 1985)

Reservoir	Carbon (10^{15} tonnes)
Atmosphere	0.036
Oceans	6.4
Continents	7.3
Carbonate rocks	
Oceans	28
Continents	54
Metamorphic rocks	10
Total	106

TABLE 11.3 Time to Reach Lithostatic Pressure at a Depth of 10 km

Permeability (darcy)	Porosity	Time (10^6 yr)
10^{-9}	0.01	0.15
	0.02	0.30
10^{-8}	0.01	1.5
	0.02	3.0

HYDRAULIC FRACTURING

If the boundary condition at the base of a rock column of low permeability has a constant influx of CO_2, it is possible to calculate arbitrarily high pore pressures. At some point increased fluid pressure will generate either a hydraulic fracture or a shear failure, which would increase the local permeability. The nature of the failure depends on the local state of stress. In the case of a hydraulic fracture, a true tensional opening, the fracture will occur normal to the least principal stress. The fracture will occur when the pore pressure exceeds the sum of the least principal stress and the tensile stress of the rick. If the least principal stress is horizontal, the hydraulic fracture will tend to be a vertical opening. A variety of fracture orientations are possible depending on whether they represent shear or tensional openings and the local state of stress. When the vertical permeability is increased locally, pressure effects are distributed upward very quickly. The lithospheric load is a convenient upper bound for failure.

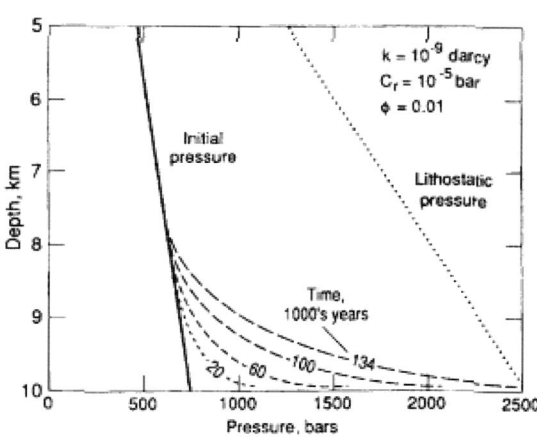

Figure 11.3
Pressure-depth profiles for various times after initiating flux of CO_2 into bottom of column (see Figure 11.4).

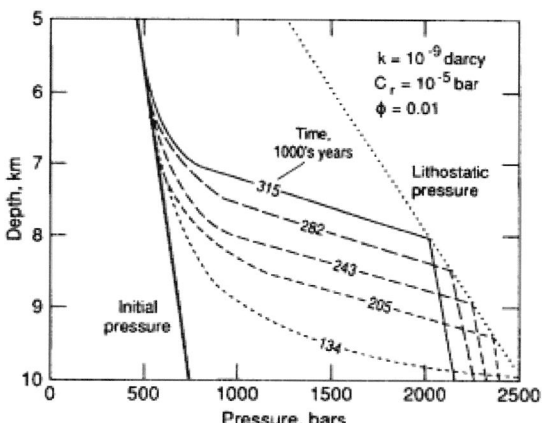

Figure 11.4
Pressure-depth profiles for various times. Once lithostatic pressure is reached, permeability is increased 1000 times, simulating the effect of fracturing.

We have attempted to simulate a system in which hydraulic fractures are created. It seems that two possible processes can occur following fracturing. Once the pressure falls following the break, the fissure can (1) remain open, thereby increasing the local permeability or (2) seal itself, and return to something approaching its initial permeability. We have attempted to simulate both occurrences.

Figure 11.4 illustrates the pressure history in the lower portion of the column in the case where the fracture permeability remains high following a break. In this case permeability is increased 1000 times once the pressure within the simulated rock block reaches lithostatic. Note that lithostatic pore pressure migrates upward with time and that once breaks occur in the lower rock units they do not reach lithostatic pressures again. Figures 11.5 and 11.6 show what happens when the breaks are resealed, that is, permeability is returned to its original value following a break. The pressure once again builds to a lithostatic level at the bottom of the column, a second break occurs, and this sequence continues to repeat itself. As in the case where the fissure remains open, the breaking will migrate upward (Figure 11.6). An interesting feature of this model is the pulsing nature of the pore pressure at a given depth (see Figures 11.5 and 11.6) Gold and Soter (1985) suggested a similar mechanism in considering the migration of fluids through the crust. Repeated episodes of fracturing are also suggested by studies of hydrothermal ore deposits (see Titley, Chapter 3, this volume).

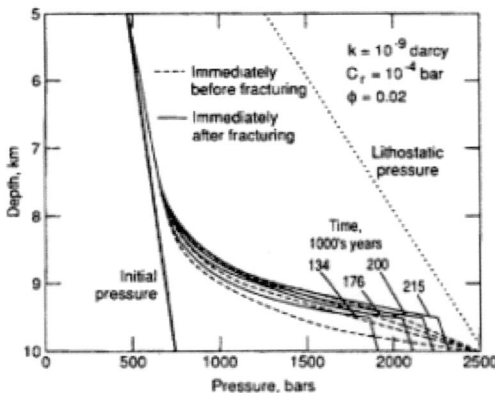

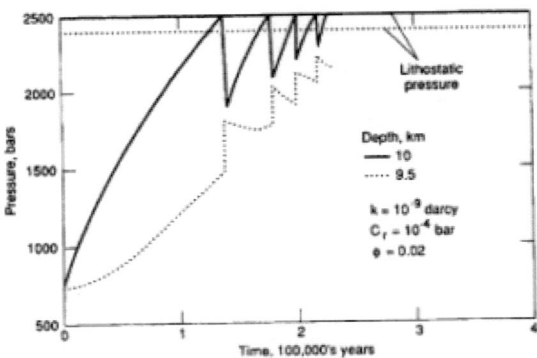

Figure 11.5
Pressure-depth profiles for various times for a system in which fracturing occurs once pressure reaches lithostatic and fractures quickly reseal themselves. In the simulation the following sequence is followed: (1) once pressure reaches lithostatic, permeability is temporarily increased 1000 times; (2) pressure drops quickly; and (3) once pressure drops, permeability is set back to its initial value, and the cycle repeats itself. There are two sets of profiles: the solid profiles are immediately before the fracturing and the dashed profiles immediately after. Pressure increases consistently with time within each set.

Figure 11.6
Pressure versus time for a system in which fracturing occurs once pressure reaches lithostatic and fractures quickly reseal themselves.

DISCUSSION

The results of our analysis suggest that degassing of CO_2 could be a source of high pore pressure, provided that the permeability of the rocks is sufficiently low. We must now consider whether such low permeabilities might reasonably be expected deep within the crust.

Brace (1980) has compiled both laboratory and field-measured permeability values for crystalline and argillaceous rocks. Figures 11.7 and 11.8 are adapted from Brace. Clearly, the values suggested by our simulation are in the lower range of what has been measured, both in the laboratory and in situ. If one examines only the in situ values, they are very near the low end of what has been measured. However, most of the in situ values have been measured near the Earth's surface, usually at depths above 500 m. It is our judgment that rock permeabilities in the range of 10^{-8} to 10^{-9} darcies are low but within the realm of expectation deep within the crust.

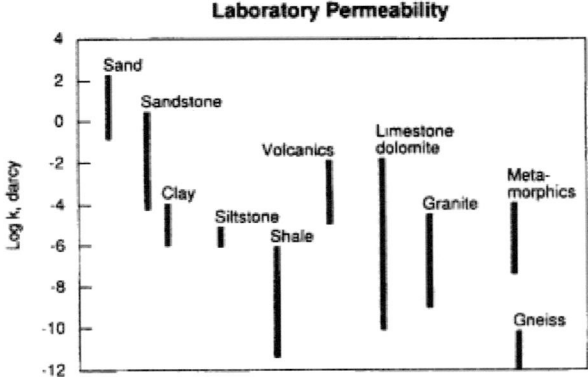

Figure 11.7
Range of laboratory permeabilities for different rock types (from Brace, 1980).

We have assumed in our calculations that the pore fluid is entirely CO_2. There is likely to be some H_2O present in the crust in the pressure-temperature range of the experiments. If significant amounts of both H_2O and CO_2 are present, the pore fluid would be a homogeneous single-phase mixture of H_2O and CO_2 at temperatures ≥ 300 and a heterogenous two-phase mixture of H_2O-rich liquid and CO_2-rich vapor at lower temperatures. The presence of NaCl or other electrolytes would extend the two-phase region to higher temperatures (Bowers and Helgeson, 1983).

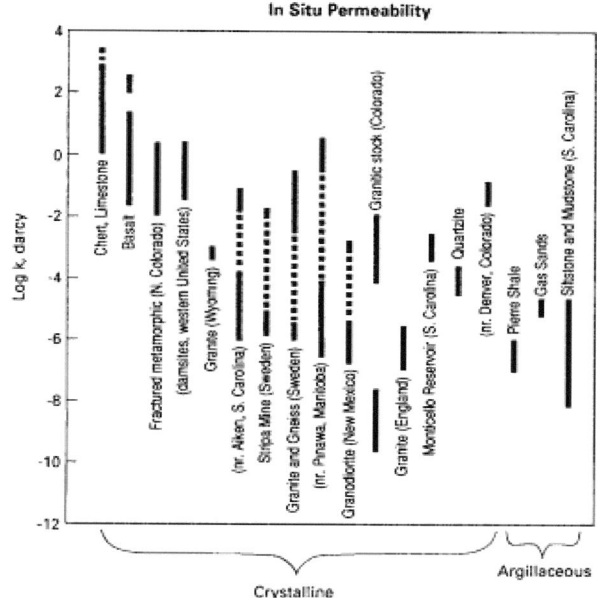

Figure 11.8
Range of in situ permeability measurements for crystalline and argillaceous rocks from various sites around the world (from Brace, 1980).

The limited solubility of CO_2 in H_2O in the pressure-temperature range considered (Takenouchi and Kennedy, 1964; Gehrig, 1980) implies that the volumetric flow rate required to transport a given flux of CO_2 in solution or in a two-phase mixture would generally be somewhat greater than the flow rate required to transport the same CO_2 as a separate phase. Since the density and viscosity of H_2O and CO_2 are comparable (Figure 11.2), the limiting permeabilities suggested here for anhydrous systems may be reasonably limiting values for hydrous systems as well.

Chemical controls on the buildup of CO_2 pressure may significantly constrain the applicability of our analyses. CO_2 pressures in natural systems at temperatures of 200° to 400°C will generally be limited by the reaction of Ca-Al-silicates, H_2O, and CO_2 to form calcite and mica and/or clay. At 300°C, for example, the partial pressure of CO_2 is likely to be on the order of tens of bars (Giggenbach, 1986). Where the available feldspar is converted to calcite and mica and/or clay, or the supply of H_2O is limited, the partial pressure of CO_2 is not fixed and may increase to greater values. This might occur where the flux of CO_2 is greatest (e.g., in fault zones and volcanic terranes). In general, CO_2 flux may be lower and/or permeabilities higher than those required to create high pore pressures; otherwise, we would expect most Ca-Al-silicates in the upper crust to be altered to calcite.

The question of fluid movement and pore pressure within the deep crust is complex. The intent of this chapter was a simple bounding calculation; we hope that it will provoke further thought, debate, and analysis. On balance it seems possible that fluid CO_2 migrating through the crust may yield high pore pressure where the rocks deep in the crust are sufficiently "tight" and the local geochemistry does not preclude high CO_2 fluid pressures.

ACKNOWLEDGMENTS

We thank R. O. Fournier and H. R. Shaw for helpful reviews of this manuscript. Our brief discussion of chemical controls on CO_2 pressure is based largely on Fournier's comments.

References

Anderson, A. T. (1975). Some basaltic and andesitic gases, *Reviews of Geophysics and Space Physics 13*, 37-55.
Atkins, P. W. (1978). *Physical Chemistry*, W. H. Freeman and Company, San Francisco.
Baes, C. F., H. E. Goeller, J. S. Olsen, and R. M. Rotty (1976). The global carbon dioxide problem, Oak Ridge National Laboratory Report ORNL-5194, 78 pp.
Barnes, I., W. P. Irwin, and D. E. White (1978). Global distribution of carbon dioxide discharges and major zones of seismicity, U.S. Geological Survey, *Water Resources Investigations 78-39*, 12 pp.
Barnes, I., W. P. Irwin, and D. E. White (1984). Global distribution of carbon dioxide discharges and major zones of seismicity, U.S. Geological Survey, *Miscellaneous Investigations Map 1-1528*, 10 pp.
Bodvarsson, G. S. (1982). Mathematical modeling of geothermal systems, Ph.D. thesis, University of California, Berkeley.
Borchert, H. (1951). Zür geochemie des kohlenstoffs, *Geochimica et Cosmochimica Acta 2*, 62-75.
Bowers, T. S., and H. C. Helgeson (1983). Calculation of the thermodynamic and geochemical consequences of nonideal mixing in the system H_2O-CO_2-NaCl on phase relations in geologic systems: Equation of state for H_2O-CO_2-NaCl fluids at high pressures and temperatures, *Geochimica et Cosmochimica Acta 47*, 1247-1275.
Brace, W. F. (1980). Permeability of crystalline and argillaceous rocks, *International Journal of Rock Mechanics: Mineral Sciences and Geomechanics 17*, 241-251.
Buddemeier, R. W., and C. Puccetti (1974). C-14 dilution estimates of volcanic CO_2 emission rates (abstract), *EOS 55*, 488.
Burnham, C. W., J. R. Holloway, and N. F. Davis (1969). *Thermodynamic Properties of Water to 1,000°C and 10,000 bars*, Special Paper 132, Geological Society of America, Boulder, Colo., 96 pp.
Des Marais, D. (1985). Carbon exchange between the mantle and crust and its effect upon the atmosphere: Today compared to Archean time, in *The Carbon Cycle and Atmospheric CO_2*:

Natural Variations Archean to Present, Geophysical Monograph 32, American Geophysical Union, Washington, D.C., pp. 602-611.

Gehrig, M. (1980). Phasenglichgewichte und PVT-daten ternarer mischungen aus wasser, kohlendioxid und natriumchlorid bis 3 kbar und 550°C, Ph.D. thesis, University of Karlsruhe, Freiburg, Germany.

Gerlach, T. M. (1986). Carbon and sulphur isotopic composition of Kilauea parental magma, *Nature 319*, 480-483.

Gerlach, T. M. (1988). Plutonic degassing of carbon dioxide at transitional and mid-oceanic spreading centers, *Earth and Planetary Science Letters*.

Giggenbach, W. T. (1986). Graphical techniques for the evaluation of water/rock equilibrium conditions by use of Na, K, Mg, and Ca-contents of discharge waters, in *Proceedings of the Eighth New Zealand Geothermal Workshop*, Aukland, pp. 37-43.

Gold, T., and S. Soter (1980). The deep-Earth-gas hypothesis, *Scientific American 242*, 154-161.

Gold, T., and S. Soter (1985). Fluid ascent through the solid lithosphere and its relation to earthquakes, *PAGEOPH 122*, 492-530.

Haar, L., J. S. Gallagher, and G. S. Kell (1984). *NBS/NRC Steam Tables*, Hemisphere Publishing Corp., New York.

Hanshaw, B. B., and E. Zen (1965). Osmotic equilibrium and overthrust faulting, *Geological Society of America Bulletin 76*, 1379-1386.

Harris, D. M. (1981). The concentration of CO_2 in tholeiitic basalts, *Journal of Geology 89*, 689-701.

Hubbert, M. K., and W. W. Rubey (1959). Role of fluid pressure in mechanics of overthrust faulting, *Geological Society of America Bulletin 70*, 115-206.

Irwin, W. P., and I. Barnes (1975). Effect of geologic structure and metamorphic fluids on seismic behavior of San Andreas fault system in central and northern California, *Geology 3*, 713-716.

Jacobs, G. K., and D. M. Kerrick (1981). APL and Fortran programs for a new equation of state for H_2O, CO_2, and their mixtures at super-critical conditions, *Computers in Geoscience 7*, 131-143.

Javoy, M., F. Pineau, and C. J. Allegre (1982). Carbon geodynamic cycle, *Nature 300*, 171-173.

Keenan, J. H., F. G. Keyes, P. G. Hill, and J. G. Moore (1969). *Steam Tables*, John Wiley & Sons, New York.

Kennedy, G. C., and W. T. Holser (1966). Pressure-volume-temperature and phase relations of water and carbon dioxide, in *Handbook of Physical Constants*, Revised Edition, Memoir 97, Geological Society of America, Boulder, Colo., pp. 371-383.

Leavitt, S. W. (1982). Annual volcanic carbon dioxide emission: An estimate from eruption chronologies, *Environmental Geology 4*, 15-21.

Li, Y-H. (1972). Geochemical mass balance among lithosphere, hydrosphere, and atmosphere, *American Journal of Science 272*, 119-137.

Libby, L. M., and W. F. Libby (1972). Vulcanism and radiocarbon dates, in *Proceedings of the 8th International Conference on Radiocarbon Dating*, Wellington, New Zealand, pp. A72-A75.

Marty, B., and A. Jambon (1987). $C/^3He$ in volatile fluxes from the solid Earth: Implications for carbon geodynamics, *Earth and Planetary Science Letters 83*, 17-26.

Moore, J. G., J. N. Bachelder, and C. G. Cunningham (1977). CO_2-filled vesicles in mid-ocean basalt, *Journal of Volcanology and Geothermal Research 2*, 309-327.

Pineau, P., M. Javoy, and Y. Bottinga (1976). $^{13}C/^{12}C$ ratios of rocks and inclusions in popping rocks in Mid-Atlantic Ridge and their bearing on the problem of isotopic composition of deep seated carbon, *Earth and Planetary Science Letters 29*, 413-421.

Plass, G. N. (1956). The carbon dioxide theory of climate change, *Tellus 8*, 140-154.

Rubey, W. W. (1951). Geologic history of the sea, *Geological Society of America Bulletin 61*, 1111-1148.

Rubey, W. W., and M. K. Hubbert (1959). Role of fluid pressure in mechanics of overthrust faulting: (II) Overthrust belt in geosynclinal area of western Wyoming in light of fluid-pressure hypothesis, *Geological Society of America Bulletin 70*, 167-206.

Sundquist, E. T. (1985). Geological perspectives on carbon dioxide and the carbon cycle, in *The Carbon Cycle and Atmospheric CO_2: Natural Variations Archean to Present*, Geophysical Monograph 32, American Geophysical Union, Washington, D.C., pp. 5-59.

Takenouchi, S., and G. C. Kennedy (1964). The binary system H_2O-CO_2 at high temperatures and pressures, *American Journal of Science 262*, 1055-1074.

Vargaftik, N. B. (1975). *Tables on the Thermophysical Properties of Liquids and Gases in Normal and Disassociated States*, John Wiley & Sons, New York.

INDEX

A
Actinolite, 82
Advection, 9, 29-30, 31, 32, 33, 47-48
Africa, 132, 145
Alaska, 81, 137, 148-155
Alberta, 136, 137
Alps, 10, 32, 69, 153
Aluminum
  andalusite, 98, 101, 109
  chlorites, 77-78, 82, 83, 98, 99
  mica, 97, 98, 99, 100, 106, 163
  see also Feldspar
Amphiboles, 82, 88, 97, 98-99, 101, 106, 109, 153
Anatexis, 108-111
Andalusite, 98, 101, 109
Appalachian Mountains, 12, 35, 132, 133, 134-135, 136, 138, 140-146, 153
Aquifers, see Groundwater tables
Arizona, 38, 51, 52, 54, 55, 58, 130
Arrow Lake, 78
Asia, 145, 158
  see also geographical subunits
Atlantic Ocean, 150

B
Barbados, 150, 153
Barium, 15
Basalts
  carbon dioxide solubility, 159
  fluid-rock ratios, 65
  magma, 84-90, 159
Basement complexes

Hercynian orogeny, 98, 99, 100, 103, 106-110
  seismic reflection, 128-138
Batholiths, 33, 54, 75-80, 81, 87-88, 115
Bentonite, 33
Biotites, 77, 82, 99, 101, 103, 104, 105
Boreholes, see Drilling
Boundary conditions, 15, 16, 23
Breccias, 53, 80
Brines, 20, 125, 133-138
British Columbia, 73, 81
  cordilleran batholiths, 75-78
British Institutions Reflection Profiling Syndicate, 128
Brittleness, 5, 54, 73, 90, 111, 116

C
Calderas, 79-80, 81, 84-85, 89, 90
California, 21, 22, 81, 136
  San Andreas Fault, 12-13, 35-36, 116
Canada, 135, 136, 137
  see also geographic subunits
Carbonates, 64-65, 68, 98-105, 106, 109, 158
Carbon dioxide, 5, 39, 64, 65-66, 96, 130, 141, 154, 159-163
Carboniferous period, 97, 98, 99, 105, 106, 109
Carbon isotopes, 96, 158-159
Cascade Mountains, 80
Caspian Basin, 14, 36, 37
Cenozoic era, 51, 76, 84
  see also Quaternary period;

Tertiary period
Channelized flows, 9, 29, 45, 47-48, 69-70, 132
Chemical processes, 4, 5, 8, 9, 19
  carbon dioxide, 5, 39, 64, 65-66, 96, 130, 141, 154, 159-163
  diagenesis, 12, 13-15, 27, 36-37, 134, 149
  equations, 4, 10, 31-32
  fault zones, 153
  free water, 5
  global geochemical budget, 149
  hydration/dehydration, 15-16, 37-38, 56-57, 68
  metamorphism, 64-65
  phase equilibria, 4, 64, 65, 66
  quartz precipitation, 69-70
  research recommendations, 21
  salt and systems, 20, 59, 65, 133-135
  transport, 3, 31-33, 48
  volatiles, 39, 47, 67-68, 69, 132, 158
  see also Isotope geochemistry; Minerals and mineralogy
Chlorites, 77-78, 82, 83, 98, 99
Clay, 11, 14, 37, 67, 82, 150
  bentonite, 33
  montmorillonite-illite transformation, 14, 17, 37, 39, 134, 149
  pelite, 65, 68-70, 98-99, 100, 101, 102, 107, 108, 111
Coal, 133
Colorado, 22, 33, 81, 90

INDEX

Computer applications, 4, 22
   fracture processes, 19, 20
   heat/mass transport, 15
Conservation equations, 6-9, 28-32
Consortium for Continental Reflection Profiling, 128-138
Contamination, 10, 21, 27, 31-32
Continental margins, 64, 75-76, 132, 138
Convection, 6, 15-16, 27, 30, 33, 37, 38, 42, 50, 73, 77, 90, 110, 111
Cordilleran phenomena
   batholiths, British Columbia, 75-78
   hydrocarbons, 136
Cores, *see* Drilling
Coupling of processes, 3, 10, 14, 17, 22-23, 32, 38, 39, 48, 153, 160
Cretaceous period
   meteoric hydrothermal activity, 76, 81
   shale, 8, 33, 96
Crystalline rock and crystallization, 8, 19, 38
   basement complexes, 98-100, 103, 106, 107, 108, 109, 110, 128-138
   batholiths, 33, 54, 75-80, 81, 87-88, 115
   chlorites, 77-78, 82, 83, 98, 99
   geometry, 84
   granitic, 97
   magmatic, 16, 38, 45, 46, 47, 53-54, 80, 83-84, 87, 90-91
   paragenesis, 55-59
   permeability/porosity, 8, 47, 83-84
   pressure relations, 117
   see also Quartz

D

Darcy's Law, 7, 10, 28-29, 30, 32, 34, 37, 150
Deep crustal processes, 4, 5, 16-17, 38, 64, 96, 128-138
Density, 10, 15, 29, 32, 37-38
   fracturing, 58, 59
   pressure and, 38, 43, 68, 82
   rock/mineral, 43, 68
Devonian period, 35, 97, 135, 141, 143-144
Diagenesis, 12, 13-15, 27, 36-37, 134, 149
Differential equations, *see* Equations
Diffusion, 15, 29-31, 32, 117
   see also Advection; Convection
Dispersion processes, 29-30, 32
Drilling, 5, 58, 86, 90, 96, 111
Ductility, 5, 89, 90, 111

E

Earthquakes, 3, 5, 10, 21-22, 27, 33, 115, 124
Elasticity, 30, 45, 124
   visco, 22, 32
   see also Stress and strain
Electrical conductivity/resistivity, 5, 11, 33, 115, 150
Electromagnetic techniques, 5, 115
Elkhorn Mountains, 81
Energy factors, 47

Energy resources
   coal, 133
   hydrothermal, commercial use, 16, 38
   natural gas, 21, 27
   oil, 10, 21, 27, 135, 136
Energy transport, 3, 6-7, 27-32, 30
   see also Heat transport; Momentum
Eocene epoch, 78
   magmatism, hydrothermal effects, 78-80
Epidote, 82, 83
Equations, 3
   boundary conditions, 15, 16, 23
   carbon dioxide flux, 159-160
   conservation, 4, 6-9, 28-32
   coupling of processes, 3, 10, 14, 17, 22-23, 32, 38, 39, 48, 153, 160
   Darcy's Law, 7, 10, 28-29, 30, 32, 34, 37, 150
   density, 37-38, 68
   dispersion-diffusion-advection, 29-30, 31, 32, 33, 48
   geothermal reservoirs, 16, 38
   hydration/dehydration, 38-39
   magma-associated flow, 38, 42
   permeability, 60, 120
   pressure effects, 4, 10, 29, 32, 37, 42-48, 68-69, 120, 121, 122
   research recommendations, 22-23
   stress and strain, 31, 36-37
   tectonic stress, 35, 121
   time factors, 121-122
   volume, failure, 46-47, 48, 122, 123
Europe, 106, 158
   see also geographical subunits

F

Faults, 149, 150, 151-154
   Appalachian Fold-Thrust Belt, 35, 140-146
   joint/fracture systems, 54, 144-145, 146
   magma conduits, 129
   rift zones, 72, 86, 87, 89, 90, 91, 109, 110, 129-130
   San Andreas, 12-13, 35-36, 116
   strain rates, 121, 124
   see also Earthquakes
Feldspar, 73-75, 78, 81, 82, 83, 88, 99, 101, 104, 134, 163
   plagioclase, 82, 83
   sericite, 83, 98
   see also Granite and granitoids
Fick's Law, 29
Field studies, 17-19, 33
   fluid inclusion, 150-155, 158
   joint sets and systems, 51-61, 144-145, 146
Finger Lakes, 141
Flows
   artesian, 141
   channelized flows, 9, 29, 45, 47-48, 69-70, 132
   flow equations, 29
   grain boundaries, 8, 65, 69

lithostatic fluid pressure, 5, 10, 14, 21, 33, 37, 66, 67-69, 90, 91, 110, 111, 117, 120-123, 141, 145, 146, 149-151, 158, 159
mid-to lower-crustal levels, 5, 16-17, 64, 66-68
velocity, 20, 28, 29, 68-69
viscosity, 10, 22, 37, 46, 47, 66, 68-69, 82, 120
see also Advection; Convection; Diffusion; Permeability and porosity
Flux, 9, 30, 31, 64-66, 68-69
   carbon dioxide, 159-160, 163
   Darcy's Law, 7, 10, 28-29, 30, 32, 34, 37, 150
   mass/energy, 6-7
   thermal, 44
Fourier's Law, 30
Fracture processes
   brittleness, 5, 54, 73, 90, 111, 116
   channelized flows, 9, 29, 45, 47-48, 69-70, 132
   deep crustal, 5
   diffusive, 29
   failure criteria, 15, 27, 38, 44, 45-48, 50, 54, 122, 123, 124
   gabbros, 84
   healing/sealing, 118, 120-121, 122, 124, 151
   hydraulic, 6, 8, 13-16, 27-28, 36, 38, 42, 45, 54, 68, 69, 117-120, 122, 123, 144, 150, 151-152, 161-162
   hydrothermal, 50-61, 69
   joint sets and systems, 50-61, 144-145, 146
   magma, 44-48, 50-61
   mineralization, 4, 15, 27-28, 47-48, 53, 55-59, 69-70, 81, 82, 118, 120-121, 122, 124, 153
   ocean lithosphere, 123
   permeability, 6, 20-21, 27, 50-61, 78, 84, 90
   research recommendations, 17, 19
France, Pyrenees, 33, 65, 73, 89, 91, 96-111

G

Gabbros, 73, 81-84
Geometry
   crystallization, 84
   fractures, 19
   joint/fracture systems, 51, 54-59
   pores, 42, 45
Gold, 81
Gneisses, 97, 98-99, 101-110 (passim), 117
Gondwanaland, 137
Grain boundaries, 8, 65, 69, 118, 149
Granite and granitoids, 73, 76, 77, 98, 99, 100, 102, 105, 108
   crystallization, 97
   gabbros *vs*, 73, 83-84
   gneisses, 101

magmatic, 75, 78-79, 80, 81, 83, 84, 87-90, 91, 97, 103, 104, 107
pressure/porosity, 120
Gravitational processes
pressure, 42-43, 68
sliding, 10, 32, 140
Great Britain, 128
Greenland, 81, 84
Groundwater tables, 6
contamination, 10, 21, 31-32
free water, 5, 14, 113-125
heat transport, 15-16, 27, 30
magmatic interactions, 15, 27, 38, 78-90
mass transport, 9, 27
topography and, 11-12, 137, 141, 145, 146
Gulf Coast Basin, 14, 17, 36, 37, 39, 135, 138

H

Heat, *see* Hydrothermal processes; Thermal processes
Heat transport, 3, 5, 6, 15-19, 30, 31, 33, 38, 64, 82, 114, 132, 149, 152
carbon dioxide flux, 159-160
Helium, 86, 130
Hercynian orogeny, 98, 99, 100, 103, 106-110
Himalayas, 137, 146
Hydraulic processes, *see* Pressure processes
Hydrocarbons, 5, 21, 130, 134, 135-138
coal, 133
methane, 39, 141, 151, 152, 153
natural gas, 21, 27
oil, 10, 21, 27, 135, 136
Hydrogen isotopes, 59, 72, 75-77, 79, 81, 89, 91, 96, 105-106, 107, 108, 109, 114
Hydrothermal processes, 5, 27, 132
commercial use, 16, 38
continental margins, 64
convection, 6, 15-16, 27, 30, 33, 37, 38, 42, 50, 73, 77, 90, 110, 111
field studies, 19
fracture/joint systems, 50-61, 59
gabbro bodies, 73, 81-84
magmatic, 6, 15, 27, 38, 42-48, 50-61, 70, 76, 78-90
metamorphism, 17, 64, 65, 66, 70, 72-92, 96-111
meteoric, 18, 38, 51, 59, 73, 75-81, 82, 96, 104
modeling, 19
ore deposits, 3, 29, 38, 59, 114-115, 124
pressure and, 5-6, 15, 27, 37, 38, 39, 42, 82, 152-155, 162-163
salinity, 59
see also Heat transport, Plutons

I

Iceland, 80, 85-87, 110
Idaho, 84, 96
batholith, 78-80
Igneous phenomena
anatexis, 108-111
basalts, 65, 84-90, 159
batholiths, 33, 54, 75-80, 81, 87-88, 115
biotites, 77, 82, 99, 101, 103, 104, 105
gabbro, 73, 81-84
hydrogen isotopes, 105
monzonite, 81
ophiolites, 82
porphyry, 52, 53-54, 58, 59, 60
surface water circulation and, 72
xenoliths, 87-88
see also Feldspar; Granite and granitoids; Magma and magmatic processes; Plutons
Illinois, 136
Illite, 14, 17, 37, 39, 134, 149
India, 137, 145
Indiana, 136
Indian Ocean, 82, 87-88
Iron chlorites, 77-78, 82, 83, 98, 99
Islands and island arcs, 52, 54, 75, 80, 81-82, 87-88, 148-155
Isotope geochemistry, 16, 18, 55, 59, 66, 111
carbon, 96, 158-159
hydrogen, 59, 72, 75-77, 79, 81, 89, 91, 96, 105-106, 107, 108, 109, 114
oxygen, 33, 59, 65, 72-92, 96-97, 98, 99-105, 106, 107, 108, 109-110, 114
strontium, 99, 100, 106-107, 108, 109

J

Joint sets and systems, 50-61
fluid pressure and, 144-145, 146
magma processes, 51, 53-54
Jurassic period, 75

K

Kodiak Islands, 148-155
Kola Peninsula, 5, 21

L

Lakes and lacustrine geology, 77-78, 141
Lead, 15, 133, 134

M

Magma and magmatic processes
anatexis, 108-111
arc, 149
basaltic and rhyolitic, 65, 84-90, 159
carbon dioxide, 159
crystalline, 16, 38, 45, 47, 53-54, 80, 83-84, 87, 90-91
fracture processes, 44-48, 50-61
granitic, 75, 78-79, 80, 81, 83, 84, 87-90, 91, 97, 103, 104, 107
hydrothermal, 6, 15, 27, 38, 42-48, 50-61, 70, 76, 78-90
joint formation, 51, 53-54
mineralization, 27-28, 117
permeability, 33
pressure effects, 38, 42-48, 159
seismic reflection, 129-130
see also Plutons
Magnesium chlorites, 77-78, 82, 83, 98, 99

Magnetism, 5, 115, 134-135
Magnetite, 82
Mantle, 16-17, 39, 70, 130, 150, 158-159
oxygen isotopes, 96-97, 103, 105
Maryland, 134
Massifs, 98, 99, 101-107, 108, 109
Mass transport, 3, 6-7, 15, 27-32
carbon dioxide flux, 159-160
chemical, 9, 31-33
deep crustal, 5, 8
in equations, 6, 29
field studies, 17-19
Measurement instruments and techniques
computer applications, 4, 15, 19, 20, 22
deep-crustal fluids, 5, 128-138
drilling, 5, 58, 86, 90, 96, 111
electrical conductivity/resistivity, 5, 11, 33, 115, 150
fluid inclusion, 150-155, 158
microscopy, 118
radiometric dating, 55
reflection seismology, 5, 124-125, 128-138, 145
wells, 66-67
see also Isotope geochemistry
Melts and melting, 3, 111
anatexis, 108-111
granites, 98
ocean crust/lithosphere, 122-123
pelitic, 98-99, 108, 111
silicates, 52, 53-54, 87, 89, 90-91, 98
surface water systems and, 72-92
water budgets, 108-109
see also Magma and magmatic processes
Mesozoic era, 51, 52, 76, 78, 150
see also Cretaceous period; Jurassic period; Triassic period
Metals
formation, 15, 38
transport in ores, 4
see also specific metals
Metamorphism, 18, 33, 42, 66
biotites, 77, 82, 99, 101, 103, 104, 105
ductile, 5, 89, 90, 111
gneisses, 97, 98-99, 101-110 (passim), 117
granite, 97, 99, 100, 103, 104, 105, 107
Hercynian, 98, 99, 100, 103, 106, 107, 108, 109, 110
hydrothermal, 17, 64, 65, 66, 70, 72-92, 96-111
regional, 33, 64-70, 96-111
see also Crystalline rock and crystallization
Metasomatism, 27
Meteoric water, 51, 73, 105, 115
hydrothermal systems, 18, 38, 51, 59, 73, 75-81, 82, 96, 104
precipitation, 18, 38
salinity, 59
topography and, 11-12, 137, 141

Methane, 39, 141, 151, 152, 153
Methodology, *see* Measurement instruments and techniques
Mexico, 53, 81
Mica, 97, 98, 99, 100, 105-106, 163
Michigan, 136-137
Microscopy, 118
Middle East, 137
Minerals and mineralogy, 3, 9, 5, 15, 16-17, 21, 31-32, 64-65
  brines, 133-138
  density, 43, 68
  diagenesis, 12, 13-15, 27, 36-37, 134, 149
  fractures, 4, 15, 27-28, 47-48, 53, 55-59, 69-70, 81, 82, 118, 120-121, 122, 124, 153
  gabbros, oxygen isotopes, 73, 82
  gneisses, 97, 98-99, 101-110 (passim), 117
  hydration/dehydration, 15-16, 38-39, 56-57
  joint sets and systems, 51, 53, 144-145
  magmatic, 27-28, 117
  paragenesis, 55-59
  see also Clay;
    Crystalline rock and crystallization;
    Isotope geochemistry;
    Metamorphism;
    Ores;
    Rock processes;
    specific minerals
Miocene epoch
  meteoric hydrothermal systems, 79, 80
  stress directions, 55
Models
  analytical, 23
  artesian flow, 141
  carbon dioxide flux, 159-160, 163
  computer applications, 4, 15, 19, 20, 22
  crustal dynamics, 4
  geothermal reservoirs, 16, 38
  hydraulic, 13-14, 117-118
  parameterization, 6-9, 23, 28
  sedimentation-consolidation, 14-15
  tectonic, 129, 131-133, 150, 152-153
  see also Equations
Momentum, 6
Montana, 81
Montmorillonite, 14, 17, 37, 39, 134, 149
Monzonite, 81
Mountains, *see* Orography;
  specific mountain ranges
Muscovite, 105-106
  sericite, 83, 98

N
Natural gas, 21, 27
Navier-Stokes equation, 28
Nevada, 51, 90, 91, 92, 130
New York, 141, 142, 143, 144, 145
North America, 90, 138, 145
  meteoric hydrothermal systems, 75-81
  reflection seismology, 132, 134, 135, 136
  see also geographic subdivisions
    Nuclear wastes, *see* Radioactive wastes

O
Ocean crust, 19, 122-123, 149-155, 158, 159
  ophiolites, 82
Oceans, *see* Seawater;
  specific oceans
Ohio, 136
Oil, 10, 21, 27, 135, 136
Okanagan Lake, 77-78
Oklahoma, 135
Ophiolites, 82
Ordovician period, 98, 99, 107, 134
Oregon, 15, 21, 80
Ores, 27, 32, 33, 59, 133-134
  diagenesis, 12, 13-15, 27, 36-37, 134, 149
  hydrothermal processes, 3, 29, 38, 59, 114-115, 124
  metal transport, 4
Organic material, 21, 135, 136, 158
  see also Hydrocarbons
Orography
  cordilleras, 75-78, 136
  hydrothermal systems, metamorphism and anatexis, 96-111
  massifs, 98, 99, 101-107, 108, 109
  mountain fold belts, 64
  topographic flow, 12, 137, 141, 145, 146
  see also specific mountains
Orogeny, 51-52
  brine expulsion, 133-138
  fluid pressure, 140-146, 153
  Hercynian, 98, 99, 100, 103, 106, 107, 108, 109, 110
  overthrusting, 10, 32-33, 135, 136, 137, 140-146
  uplift, 12, 97, 141, 145, 146
Overthrusting, 10, 32-33, 135, 136, 137, 140-146
Oxygen isotopes, 33, 59, 65, 72-92, 96-110, 114

P
Pacific region, 51-52, 122, 150, 158
  see also geographic subunits
Paleocene epoch, 81
Paleomagnetism, 134-135
Paleozoic era, 97, 98, 99, 100, 104, 106, 108, 109, 110, 130, 132
  see also Carboniferous period;
    Ordovician period;
    Silurian period
Papua New Guinea, 54, 55, 58, 154
Paragenesis, 55-59
Parameterization, 6-9, 23, 28
Pelite, 65, 68, 69, 70, 98-99, 100, 101, 102, 107, 108, 111
Pennsylvania, 134, 135, 142, 143, 144, 145
Permeability and porosity
  conductivity related to, 28-29
  crustal, general, 33-39
  crystalline rock, 8, 47, 83-84
  Darcy's Law, 7, 10, 28-29, 30, 32, 34, 37, 150
  in equations, 60, 120
  fracture-related, 6, 20-21, 50-61, 78, 84, 90
  geometry of, 42, 45
  grain boundaries, 8, 65, 69, 118, 149
  infiltration, 107-108, 109, 110
  pressure and, 3, 12-14, 15, 31, 35-36, 38, 114, 117-125, 141, 150-163
  resistivity measurements, 5, 11, 33, 115, 150
  rock, general, 4, 5, 6, 8, 11, 17, 19-21, 27, 33, 38, 47, 69, 110, 117-118
  secondary, intrusion-centered systems, 52-53, 59
  time dependence, 116, 117-122, 124
Peru, 53
Phase equilibria, 4, 64, 65, 66
Physical processes, 7
  advection, 9, 29-30, 31, 32, 33, 47-48
  conservation equations, parameters, 6-9
  diffusion, 15, 29-30, 32, 117
  dispersion, 29-30, 32
  infiltration, 107-108, 109, 110
  shear forces, 12, 15, 27, 35, 98, 115, 118, 121, 152
  see also Convection;
    Pressure processes;
    Stress and strain
Plagioclase, 82, 83
Plutons
  gabbro, 81-82
  granitic, 75, 78-79, 80, 81, 83, 84, 87-90, 91, 97, 103, 104
  hydrothermal systems, 15, 38, 42, 61, 75, 78-79, 80, 81, 84, 107
  joint/fracture systems, 51-55, 61
  meteoric water, 75, 78-79, 80, 81
Porosity, *see* Permeability and porosity
Porphyry, 52, 53-54, 58, 59, 60
Potassium, 86
Precambrian era, 97
Precipitation, meteorology, 18, 38
  see also Meteoric water
Prehnite, 82
Pressure processes, 3, 33, 66, 141
  advective, 47-48
  Appalachian Fold-Thrust Belt, 35, 140-146
  artesian, 141
  carbon dioxide, 158-163
  conductivity, 28-29, 67
  depth-related, 42-44, 114-125, 132
  in equations, 4, 10, 29, 32, 37, 38, 42-48, 68-69, 120, 121, 122
  fracturing, 6, 8, 13-14, 15, 16, 27-28, 36, 38, 42, 45, 54, 68, 69, 117-120, 122, 123, 144, 150, 151-152, 161-162
  free water depth, 114-125
  gneiss formation, 97, 98-99, 101-110 (passim), 117
  gravitational, 42-43, 68
  groundwater, 11, 12
  high, 5, 10-11, 12, 14, 21, 33, 34-39, 42-48, 61, 66-67, 140-146, 149-150, 152-153, 158-163
  hydration/dehydration, 15-16, 38-39
  hydrothermal systems, 5-6, 15, 27, 37, 38, 39, 42, 82, 152-155, 162-163

joint sets and systems, 144-145, 146
lithostatic, 5, 10, 14, 21, 33, 37, 66, 67-69, 90, 91, 110, 111, 117, 120-123, 141, 145, 146, 149-151, 158, 159
magmatic processes, 38, 42-48, 159
metamorphism, 66-69
mineral hydration, 16-17
orogenic pulses, 145-146
pore, 3, 12-15, 31, 35-36, 38, 114, 117-125, 141, 150-163
rock compression, 5, 38
sedimentary basins, 12
seismicity, 115
subduction, 3, 36, 52, 115, 122-123, 124, 132, 135, 136, 137, 148-155
tectonic, 10-13, 17, 21, 32-33, 35-36, 114, 115, 141, 144
temperature and, 5-6, 15, 27, 37, 38, 39, 42, 82, 152-155, 162-163
time-dependent, 13, 113-115
topographic, 12, 141, 145, 146
water properties, 38, 42
see also Stress and strain
Pyrenees, 33, 65, 73, 89, 91, 96-111
Pyroxene, 82, 83

Q

Quartz, 48, 52, 55-59, 60, 61, 64, 65-66, 69-70, 73-75, 78, 81-84, 88, 99, 101, 104, 105, 118, 120-121, 150
see also Granite and granitoids
Quaternary period, 79-80

R

Radioactive wastes, 10, 20
Radiometric dating, 55
Reflection seismology, 5, 124-125, 128-138, 145
Research, recommended, 17-23
  field studies, 17-19
  fluid inclusion, 154
  fractured rocks, 17, 19
  permeability, 19-21
  tectonics, 22
Rhyolite, 52, 84-90, 159
Rift zones, 72, 86, 87, 89, 90, 91, 109, 110, 129-130
Rock processes, 4, 18, 45
  basement complexes, 97, 98
  compression, 5, 38
  deformation, 36
  density, 43, 68
  in equations, 43
  joint systems, 50-61, 144-145, 146
  permeability, 4, 5, 6, 8, 11, 17, 19-21, 27, 33, 38, 53, 69
  volume, 6, 43, 51, 52-53, 58, 59, 123
  see also Crystalline rock and crystallization;
    Fracture processes;
    Magma and magmatic processes;
    Metamorphism;
    Permeability and porosity

S

Salt and salt systems, 20, 59, 65, 133-135, 143, 154
  brines, 125, 133-138
  see also Seawater
San Andreas Fault, 12-13, 35-36, 116
San Juan Mountains, 81, 90
Saudi Arabia, 81
Scotland, 81-82
Seawater, 73, 75, 81, 90, 96, 104, 105, 106, 108, 109, 122-123
Sedimentology, 12
  basement complexes and, 97
  continental margins, 64
  diagenesis, 12, 13-15, 27, 36-37, 134, 149
  oxygen isotopes, 96
  pelite, 65, 68, 69, 70, 98, 100, 101, 102, 107, 108
  pore pressure, 14-15, 35, 36-37
  rock fractures, 19
  topography, 12
  see also Isotope geochemistry;
    Minerals and Mineralogy
Seismology, 115-116, 124
  orogeny, 51
  reflection, 5, 124-125, 128-138, 145
  see also Earthquakes
Sensitivity analysis, 23
Sericite, 83, 98
Seychelles Islands, 80, 87-88
Shale, 8, 20, 33, 35, 96, 99, 100, 106, 107, 109, 117, 142, 143, 146
Shear forces, 12, 15, 27, 35, 98, 115, 118, 121, 152
Silicates, 48, 52, 56-57, 64, 85, 86, 89, 102, 151, 163
  amphiboles, 82, 88, 97, 98-99, 101, 106, 109, 153
  andalusite, 98, 101, 109
  chlorites, 77-78, 82, 83, 98, 99
  melts, 52, 53-54, 87, 89, 90-91, 98
  mica, 97, 98, 99, 100, 106, 163
  pyroxene, 82, 83
  rhyolite, 52, 84-90, 159
  see also Feldspar
Silurian period, 135, 143
Silver, 81
Skaergaard intrusion, 15, 80-82, 84
Slocan Lake, 78
South Dakota, 8, 12, 35
Soviet Union, 5, 14, 21, 36, 37
Spatial dimension, 111
  hydrocarbons, 135-138
  joint sets, 144
  modeling, computer power needs, 22
Stress and strain, 27, 30-31
  in equations, 31, 36-37, 43
  failure criteria, 15, 27, 38, 44, 45-48, 50, 54, 122, 123, 124
  faults, 121, 124, 141, 145
  joint/fracture systems, 54-55
  pore pressure, 3, 12-14, 15, 31, 35-36, 38, 114, 117-125
  regional, 42, 51, 55
  tensile, 15, 27, 46, 54, 68, 144, 151

see also Fracture processes;
  Pressure processes;
  Shear forces;
  Tectonics
Strontium, 99, 100, 106-107, 108, 109
Subduction, 3, 36, 51-52, 55, 70, 115, 122-123, 124, 132, 135, 136, 137, 148-155
Subsidence, 15, 37, 44
Supercritical fluids, 6, 27, 37, 54, 82
Surface water, *see* Meteoric water;
  Seawater
Sutter, M. J., 134
Switzerland, 69

T

Tectonics
  basalt/rhyolitic magmas, 89, 90
  brines, 125, 133-138
  carbon dioxide degassing, 39
  continental margins, 64
  diagenesis and, 134
  dilation and compression, 12-13, 35-36
  in equations, 29, 35, 39
  joint/fracture systems, 51-52, 54, 55, 61
  models, 129, 131-133
  overthrusting, 10, 32-33, 135, 136, 137, 140-146
  pressure, 10-13, 17, 21, 32-33, 35-36, 114, 115, 141, 144, 158
  reflection seismology, 129, 131-138
  rift zones, 72, 86, 87, 89, 90, 91, 109, 110
  subsidence, 15, 37, 44
  surface water circulation, 72
  uplift, 12, 97, 141, 145, 146
  see also Earthquakes;
    Faults;
    Orogeny;
    Subduction
Tennessee, 134
Tension and tensile strength, 15, 27, 46, 54, 68, 144, 151
Tertiary period, 52, 75, 76, 78, 80, 81, 91
  see also Paleocene epoch
Texas, 134, 135
Thermal processes, 15-16, 37-38
  commercial use, 16, 38
  crack healing, 120-121
  cracking, 144
  in equations, 4, 10, 29, 30, 32, 42-48
  field studies, 19
  modeling, 19
  see also Heat transport;
    Hydrothermal processes;
    Magma and magmatic processes;
    Melts and melting
Time factors, 111
  hydraulics, 13, 36, 113-125
  isotope changes, abrupt, 89
  joint sets, 61, 144
  paragenesis, 56
  permeability/porosity, 116, 117-122, 124
  see also Isotope geochemistry;
    Radiometric dating

INDEX

Topographic effects, 11-12, 34-35, 64, 76, 137, 141, 145, 146
Transport properties, 4, 7-8, 11, 17, 48, 53
  chemical, 3, 31-33, 48
  energy, 3, 6-7, 27-32, 30
  grain boundary migration, 8, 65, 69, 118, 149
  infiltration, 107-108, 109, 110
  metamorphism, 107-109
  reactive, 11
  see also Advection;
    Convection;
    Equations;
    Heat transport;
    Mass transport
Triassic period, 75

U

Uplift, 12, 97, 141, 145, 146

V

Vancouver Island, 75
Velocity, 20, 28, 29, 68-69, 116, 124
Venting, 82
Virginia, 134-135
Viscoelasticity, 22, 32
Viscosity, 10, 22, 37, 46, 47, 66, 68-69, 82, 120
Volatiles, 39, 47, 67-68, 69, 132, 158
Volcanoes, 3, 78, 80, 84, 86, 158
  bentonite, 33
  calderas, 79-80, 81, 84-85, 89, 90
  joint sets and systems, 51-52, 53-54, 61
  orogeny, 51
  rhyolite, 52, 84-90
  seismic reflection, 129, 130
  subduction and, 122-123
  see also Magma and magmatic processes
Volume and volumetrics
  dehydration, 16-17, 38-39
  in equations, 46-47, 48, 122, 123
  fracture area, 46-47, 52-53, 55, 58
  oxygen isotopes, 104-105
  pore strain/pressure, 3, 12-14, 15, 30, 35-36, 114, 117-125
  porosity and, 45, 55
  rock, 6, 43, 51, 52-53, 58, 59
  water, shallow vs deep crust, 4, 5, 16, 38

W

Waste management, 10
  radioactive, 10, 20
Water systems
  free, 5, 14, 113-125
  precipitation, meteorology, 18, 38
  sources, 16-17, 38-39, 59, 122-123
  subduction, 122-123, 135, 136
  transport equations and, 3, 39
  see also Groundwater tables;
    Hydrothermal processes; Meteoric water; Seawater
Weathering, 9, 52, 55
Wells, 66-67
West Germany, 88-89
West Virginia, 135
Wetness and wetting, 67
Wollastonite, 64
Wyoming, 84

X Y Z

Xenoliths, 87-88
Zinc, 15, 133, 134